Einstein's Theory

Øyvind Grøn • Arne Næss

Einstein's Theory

A Rigorous Introduction for the Mathematically Untrained

Springer

Øyvind Grøn
Oslo University College
Oslo
Norway
oyvind.gron@fys.uio.no

Arne Næss
Oslo University College
Oslo
Norway

ISBN 978-1-4899-9732-6 ISBN 978-1-4614-0706-5 (eBook)
DOI 10.1007/978-1-4614-0706-5
Springer New York Dordrecht Heidelberg London

Softcover re-print of the Hardcover 1st edition 2011

Printed on acid-free paper

Springer is part of Springer Science+Business Media (www.springer.com)

Preface by Arne Næss

This introduction to The General Theory of Relativity and its mathematics is written for all those, young and old, who lack or have forgotten the necessary mathematical knowledge to cope with already published introductions. Some of these introductions seem, at the start to require only moderately much mathematics. Very soon, however, there are frightful 'jumps' in the exposition, or suddenly new concepts or notations appear as if nearly self evident. The present text starts at a lower level than any other, and leads the reader slowly and faithfully all the way to the heart of relativity: Einstein's field equations.

Who are those who seriously desire to get acquainted with General Relativity, but have practically no mathematical knowledge? There are tens of thousands of them, thanks to the great general interest in relativity, quantum physics and cosmology of every profession, including those with education only in the humanities.

Slowly many of these interested persons understand the truth of what one of the last Century's most brilliant physicists and populariser, Sir Arthur Eddington, told us already in the 1920s: that strictly speaking, mathematical physics cannot be understood through popularisations. Mathematics plays a role that is not merely instrumental, like cobalt chemicals for the paintings of Rembrandt. Mathematical concepts enter in an essential way, and readers of popularisations are mislead. Their intelligence is insulted and bulled when they ask intelligent questions that their popular text cannot answer except by absurdities. The honest readers may end up in a quagmire of paradoxes, and may get the usually false idea that there is something wrong with their intelligence. What they read they consider beyond their intellectual grasp.

It is a widespread expectation that a mathematical understanding of general relativity involves difficult calculations. Actually, coping with the few somewhat lengthy strings of symbols in this text may be felt as a relief from abstract thinking. What takes time is the thorough understanding of the relations between a few basic concepts. They require close, repeated attention and patient work. This surprising feature to some degree justifies that we have not included exercises in the text. To be honest, some should have been included, but they would have been of a rather strange kind: exercises in articulating conceptual relations.

But what about the formidable calculations one may read about in popularisations? They affect applications, for instance particular solutions of Einstein's field equations. Examples of such calculations are found in appendices B and C. Even if the reader should not expect to be an operator of relativistic physics, he or she should be well acquainted with it having read this book. We venture to suggest that the understanding acquired by the reader may be deeper than what is necessary for completing graduate courses intended to make the student an expert in calculations involving general relativity, but requiring only crude discussions of the all-important conceptual framework.

The present text shows, we hope, that only patience is needed—no special talent for mathematics. Personally I have never shown any such talent, only a persistent wonder at strange mathematical phenomena, like the endless number 3.1415... with the very short name 'pi'. (Caution: the length of the circumference of a circle divided by the diameter may have any value. Only in the special case of a flat space will you get the number 3.1415.... More about that later!) Again and again I refused to comply with the long streams of strange mathematical symbols which Øyvind Grøn, my patient Guru of mathematics and physics, rapidly wrote on our gigantic blackboard. "Stop, stop! I don't want that equation! How did you jump from that one to the next?" To the astonishment of both of us it was possible to break down the long deductions into small and easily understandable steps.

A serious weakness of those courses, in my view as a humanist, is the implicit appeal to make the student accept what is going on without wondering. Along the road to Einstein's field equations, feats of artistic conceptual imagination abound. Also postulates and assumptions of seemingly arbitrary kinds are made. Some of them are seen to have a rational aspect when properly understood, but far from all. Einstein did not find all this wonderful. What deeply moved him in his wonder was that the concepts and equations he (and others, like Minkowski) *invented*, could be tested in the real world and in part *confirmed*. Somehow there must be, wondered Einstein, a kind of 'pre-established harmony' between inventive conceptual imagination and aspects of reality itself.

The present text tries to keep wonder alive, a wonder *not* due to misunderstandings.

Some people, as part of their religion, creep on their knees all around the holy mountain Kailas in Tibet. The present text would not have been produced if it had not been clearly felt as a way of honouring Albert Einstein not only as a persistent, fully committed truth seeker, but as a person combining this, and the 'egocentricity' going with it, with perfect generosity. He used his name and his time to work for the persecuted, for emigrants, for the hungry. And, in addition, feeling the absurdity of the political developments, he partook in depressing world affairs, even compromising his deeply felt pacifism. And, last but not least, he retained a sense of humour, and even as a superstar, was unaware of his outer appearance to the extent of neglecting to keep his worn trousers properly shut when lecturing. He would perhaps laugh if he got to know that this text is an expression of personal devotion.

Preface by Øyvind Grøn

One day, early in the Autumn 1985, the seventy three year old philosopher Arne Næss appeared at my graduate course on general relativity. He immediatly decided that a new type of introduction to the general theory of relativity is needed; an introduction designed to meet the requirements of non-science educated people wanting to get a thorough understanding of this, most remarkable, theory.

The present text is the result of our efforts to provide a useful book for these people. It is neither a popular nor a semi-popular account of the theory. The book requires a rather large amount of patience from the reader, but nearly no previous knowledge of physics and mathematics. Our intention is to give an introduction that leads right up to Einstein's field equations and their most important consequences, starting at a lower level than what is common. The mathematical deductions are made with small steps so that the mathematically inexperienced reader may follow what happens. The meaning of the concepts that appear are explained and illustrated. And in some instances we mention points of a more philosophical character.

We devote a whole chapter to each of the topics 'vectors', 'differentiation', 'curves' and 'curved coordinate systems'. Tensors are indispensable tools in a formulation of the general theory of relativity intended to give the reader the possibility to apply the theory, at least to some simple, but nonetheless non-trivial, problems. The metric tensor is introduced in chapter 5, which also provides a most important discussion of the kinematic interpretation of the spacetime line element.

Albert Einstein demanded from his theory that no coordinate system is privileged. And in general curved coordinate systems are needed to describe curved spaces. In such coordinate systems the basis vectors vary with the position. This is described by the Christoffel symbols. In chapter 6 we give a thorough discussion, of a geometrical character, showing how the basis vectors of plane polar coordinates change with position, and relate this to the Christoffel symbols. Having worked yourself through this chapter, it is our hope that you have obtained a high degree of familiarity with the Christoffel symbols.

'Covariant differentiation', 'geodesic curves' and 'curvature' are important topics forming a 'package' of prerequisites necessary in order to be able to appreciate fully Einstein's geometric conception of gravity. These topics are treated in chapters 7, 8 and 9, respectively. The expression for the Riemann curvature tensor is deduced by utilising Green's theorem connecting circulation and curl.

Einstein's law of gravitation is formulated mathematically in his field equations. They tell how matter curves spacetime. The field equations require an appropriate tensor representation of some properties of matter, i.e. of density, stress and motion. The usual representation of these properties is motivated and explained in chapter 10, where the basic conservation laws of classical fluid mechanics are expressed in tensor form. In chapter 11 the expression for Einstein's divergence free curvature tensor is deduced. With this chapter our preparation for a presentation of general relativity has been fulfilled.

Einstein's general theory of relativity is presented in chapter 12. Here we discuss the conceptual contents of the general theory of relativity. We consider the Newtonian limit of the theory, and we give an elementary demonstration of the following theorem: From the general theory of relativity and the assumption that it is impossible to measure velocity relative to vacuum, it follows that vacuum energy acts upon itself with repulsive gravitation.

In chapters 13 and 14 we deduce some consequences of the theory. In particular we discuss the gravitational time dilation, the deflection of light, the relativistic contribution to Mercury's perihelion precession, and we give a detailed explanation of how the theory predicts the possible existence of black holes. Finally the most important relativistic universe models, including the so-called inflationary universe models, are discussed.

Detailed calculations of the form of the Laplacian differential operator in spherical coordinates, needed in chapter 12, and of the components of the Ricci curvature tensor, needed to write down Einstein's field equations for the applications in chapters 13 and 14, are presented in appendices A, B, and C, respectively.

Looking at the stars with Einstein's theory in mind, you may feel it is like a wonder that Einstein managed to reveal such deep secrets of cosmos to us. People do in fact search for black holes now. And there are strong indications that at least a few have been found. What mysterious connection is there that makes nature 'obey' Einstein's great mental construction - the general theory of relativity?

Acknowledgements

The authors would like to thank Henning R. Bråten, Jon M. Bjerkholt, Trond Bjerkholt, Jan Frøyland and Henning Knutsen for being our first readers and for giving several useful suggestions of a pedagogical character.

Most of this book was written in the home of Kit Fai and Arne. We want to thank Kit Fai for opening her and Arne's home for us in a very friendly manner which contributed much to the pleasure we had in the process of writing this book.

Jan Frøyland
Henning Knutsen
Harald Soleng
Jon Døhl

Contents

List of Figures

Chapter 1
Vectors

1.1 Introduction

Mathematically the general theory of relativity is a theory of vectors and quantities that generalize vectors, namely tensors. If one wants to master, or at least obtain some familiarity, with the mathematical apparatus of this theory, one should manage to have intercourse with vectors as dear friends. It is wise from the very start to listen to them with benevolence and patience. They have some rather strange habits which we should be well aware of, in order to be able to understand their message without a misinterpretation.

In what follows we shall not assume that the reader knows vectors. We shall give an heuristic introduction that suffices for our goal. More formal and general introductions of the calculus of vectors are found in mathematics textbooks. We urge the interested reader to consult such books.

The primary aim of this chapter is to introduce those parts of the calculus of vectors which will be needed in order to understand the mathematics used in general relativity. The secondary aim is to prepare the readers for more advanced topics, mathematical, physical and philosophical.

1.2 Vectors as arrows

Vectors can be imagined as arrows. Like an arrow a vector points in a certain direction. Which direction? It is important to note that the answer can be given without introducing a coordinate system.

We have arrows in a room or outdoors. The arrows can point towards the roof, a door, one's grandmother and so forth. The direction of the arrow is given by concrete things.

Thinking of the theory of relativity it is wise immediately to note that the length and direction of an arrow can be related to other arrows. They can be compared.

Ø. Grøn and A. Næss, *Einstein's Theory: A Rigorous Introduction for the Mathematically Untrained*, DOI 10.1007/978-1-4614-0706-5_1,

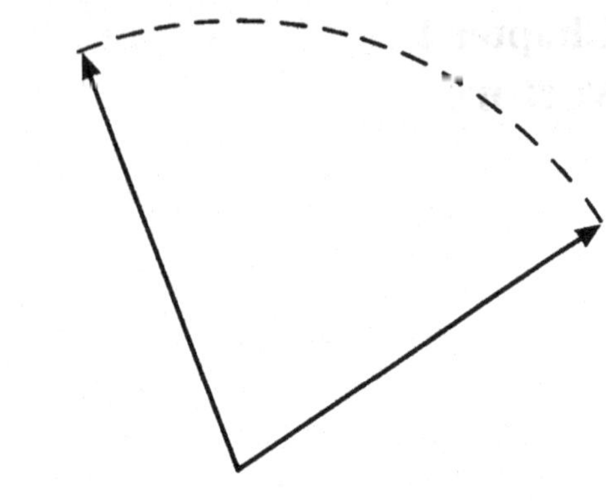

Fig. 1.1 Vectors with different directions

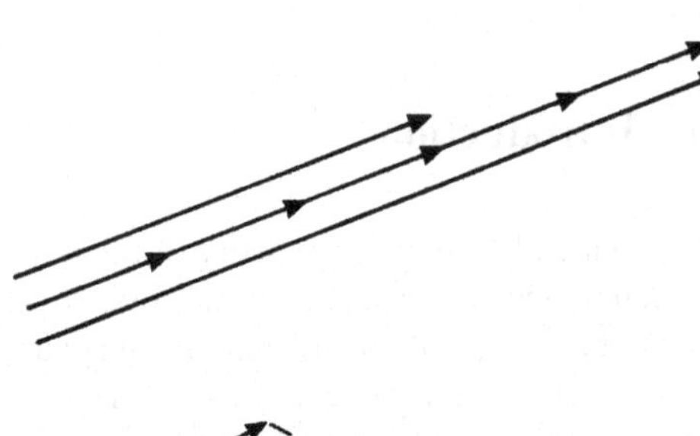

Fig. 1.2 Vectors with different magnitudes

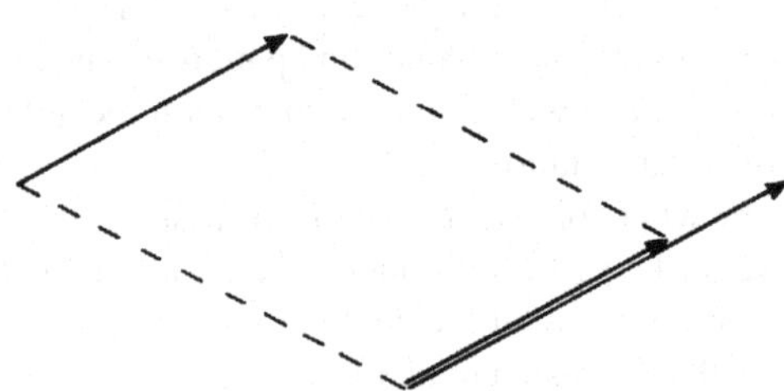

Fig. 1.3 Parallel transportation of a vector

We may move one vector to another without changing its direction, so that the starting points (roots) of the vectors touch each other. If the vector-arrows then lie along the same line, the vectors are said to have the same direction. If the vectors do not lie along the same line, we use the geometry of 'daily life' (see Ch. 4) to specify the difference of direction. In Fig. 1.1 is shown two vectors with equal magnitude, but different directions.

The difference of length between two vectors may be given by making small arrows which we call 'unit vectors', and have a magnitude or length equal to one. A line with unit vectors, and two vectors with magnitude 3 and 5, respectively, are drawn in Fig. 1.2.

If we find an arrow at one place and want to compare its direction with that of an arrow at another place, we must perform a difficult operation. We must 'parallel transport' to the one we want to compare it to (see Fig. 1.3). In daily life this can be done for example by means of straight threads along which we slide the ends of the arrows. Such procedures work well on a flat surface. However, difficulties will appear when we try to perform such parallel transport over great distances on the Earth, as the curvature of the Earth's spherical surface will disturb the result (see Fig. 9.2).

1.3 Vector fields

From now on vectors are no longer material arrows, but arrows in a purely abstract sense. In what follows a 'vector' is a geometric quantity with geometrically defined 'magnitude' and 'direction'. However, just as there is a correspondence between ordinary numbers and certain physical quantities, such as for example temperature (number of degrees) and length (number of centimetres), there is a correspondence between vectors and a special class of physical quantities; namely, those having a direction.

When we are outdoors in the wind, the moving air fills the region around us. There is a measurable velocity of the air everywhere in the region. In Fig. 1.4 we have shown, as an illustration, a weather map with the wind velocity field over Europe on February 3, 1988. The velocity has a magnitude and a direction. It is a vector. Thus a velocity field is linked conceptually with every point of the region. These abstract vectors are everywhere. If one can think of God as omnipresent, then one might also be able to think of vectors as omnipresent. In such a region there is said to be a vector field.

In order to be able to calculate with vectors, we must have a practical means of specifying their directions and magnitudes. Let us consider vectors on a flat surface, say on the sheet of paper you are writing on. Then we place unit vectors along the horizontal edge at the bottom and along the vertical left-hand edge of the sheet. These are reference vectors.

Drawing equidistant parallels with unit distance between each, we obtain a grid (see Fig. 1.5). Each corner is given a number containing figures; the first figure gives the distance of the point from the left edge of the sheet, and the second figure its distance from the downward edge. Imagine that each square is divided into four smaller squares. Again each corner is given a two-figured number. (The figures are not generally integer numbers.) If this division is proceeded indefinitely, we can associate a number with every point of our sheet. Note also that each point have different numbers. Such a set of numbers, by which we may specify the positions of arbitrary points in a region, is called a coordinate system. There is no reference to the grid we introduced originally, in the definition of a coordinate system. The grid which is usually drawn on maps, or a globe, for example, is only of heuristic help, making it easier to read off the position of a point in the coordinate system.

As is well known by users of atlases, one can map a region by means of different sorts of coordinate systems. The particular coordinate system that we have described above, with vertical and horizontal straight axes, is called a Cartesian coordinate system, in honour of the great mathematician René Descartes (1596–1650).

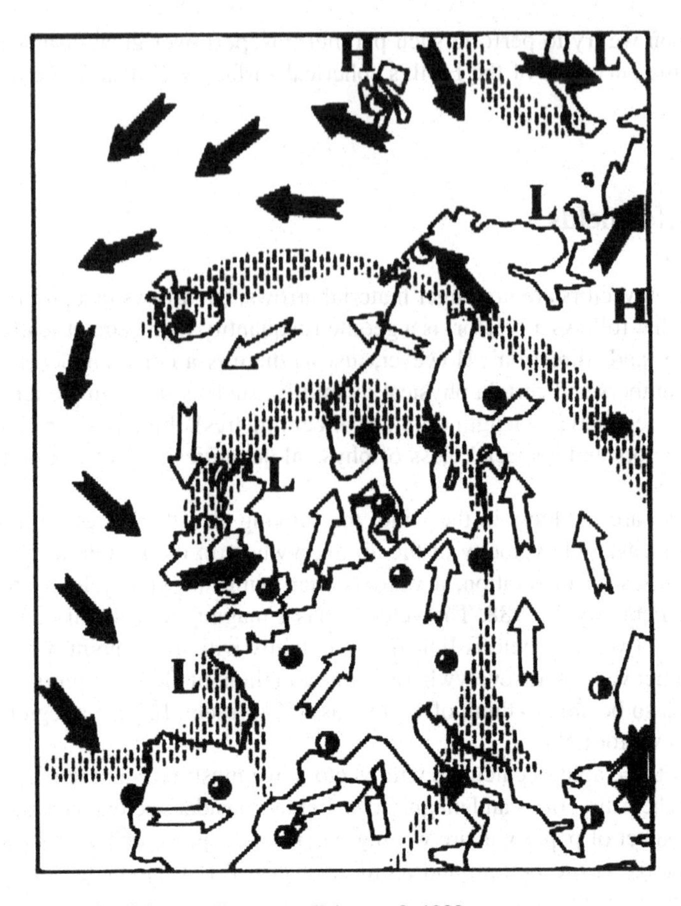

Fig. 1.4 Wind velocity field over Europe on February 3, 1988

The horizontal axis is usually called the x-axis, and the vertical axis the y-axis. The position of a point is given by a number (x, y) relative to a chosen origin, which has coordinates $(0, 0)$.

By means of our coordinate-system we can specify the magnitudes of the vectors in our vector-field. In order to be able to write a vector as the sum of 'vector components' (see below) along the coordinate lines we introduce reference vectors (see above) along these lines. They are called basis vectors. The basis vectors are introduced at every point in the region of interest. Generally the basis vectors may have unequal directions and magnitudes at different positions (see for example section 4.1.1, where a system of plane polar coordinates is considered). Only in the special case of a Cartesian coordinate system (in a flat space) do they have the same magnitude and direction all over space.

At every point of an n-dimensional space there are n basis vectors. There are, for example, two basis vectors at every point of a surface. Having a coordinate system

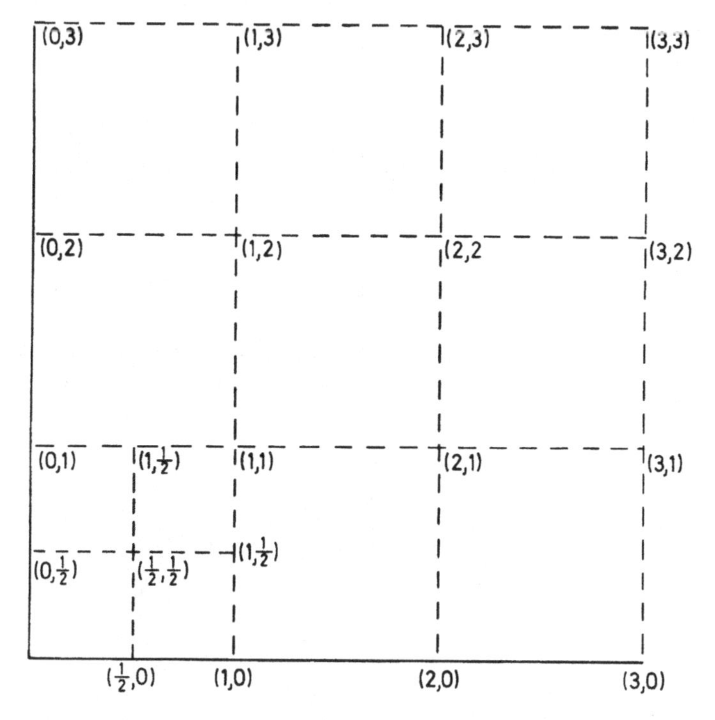

Fig. 1.5 Grid

and the associated basis vector field at our disposal, we can specify the components of, for example, a velocity vector field at every point in a region. By this means we shall be able to develop the calculus of vectors.

1.4 Calculus of vectors. Two dimensions

As to notation, vectors shall be denoted by placing an arrow above the letter. Thus, 'A-vector' is denoted by $\vec{A}$. Basis vectors are denoted by a the letter 'e' with an arrow and a subscript, which tells which basis vector we are talking about, for example $\vec{e}_x$.

Consider a vector $\vec{A}$. In Fig. 1.6 this vector is placed in a two-dimensional coordinate-system. The coordinate axes are straight lines. The difference between their directions amounts to 90°. They are said to be *orthogonal* to each other. (This term will be used frequently in the present text.) We have a Cartesian coordinate system. The basis vectors along the x-axis and y-axis, are written $\vec{e}_x$ and $\vec{e}_y$, respectively. The magnitude of a vector $\vec{A}$ is designated by $|\vec{A}|$. The basis vectors $\vec{e}_x$ and $\vec{e}_y$ have per definition magnitude equal to 1, $|\vec{e}_x| = |\vec{e}_y| = 1$.

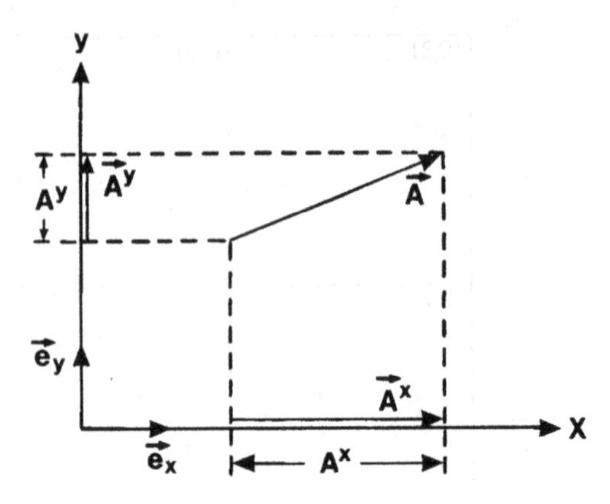

Fig. 1.6 Basis vectors and vector components

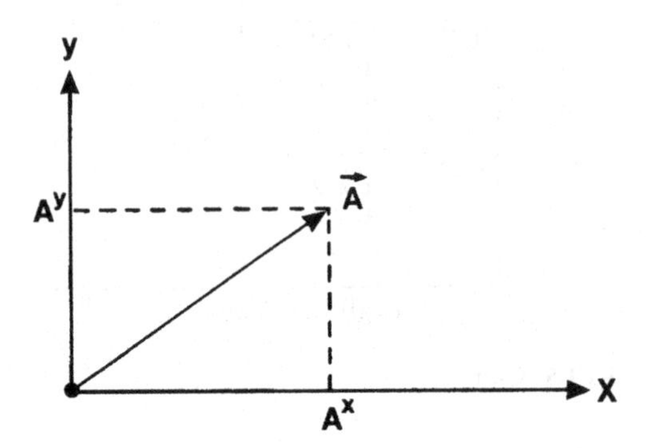

Fig. 1.7 Vector components

The quantities A^x and A^y in Fig. 1.6 will be called the components of the vector $\vec{A}$. (Note that in the calculus of vectors it is usual to use the upper right suffix to select a component rather than to indicate the exponent of a power. Even if this may seem confusing at the beginning, this notation is indispensable when one calculates with vector components, and after a while its use will be a matter of routine.) The vector components are defined as follows. Let a vector $\vec{A}$ be situated with its initial point at the origin, as shown in Fig. 1.7.

Then the components A^x and A^y are the coordinates of the terminal point of $\vec{A}$. They are not vectors, but ordinary numbers. Such numbers are called scalars. We might have called them 'scalar components'. Note that the scalar components of a vector might be negative (for example $A^x < 0$, if the vector points in the negative x direction).

The vector quantities $\vec{A}^x$ and $\vec{A}^y$ (as shown in Fig. 1.6) are termed the component vectors of $\vec{A}$. In order to define these we must first consider products of scalars and vectors.

If the scalar is equal to a number k, the product $k\vec{A}$ is defined as a vector with magnitude $k|\vec{A}|$, and with the same direction as $\vec{A}$. With, for example $k = 2$, the magnitude of the vector is doubled. If the scalar is a negative number, $-k$, the

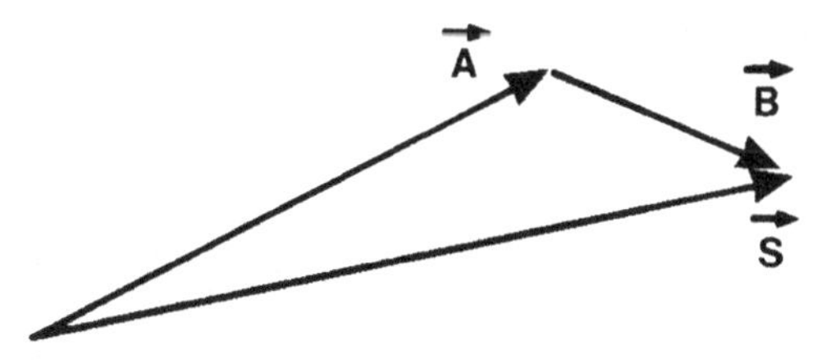

Fig. 1.8 Vector addition

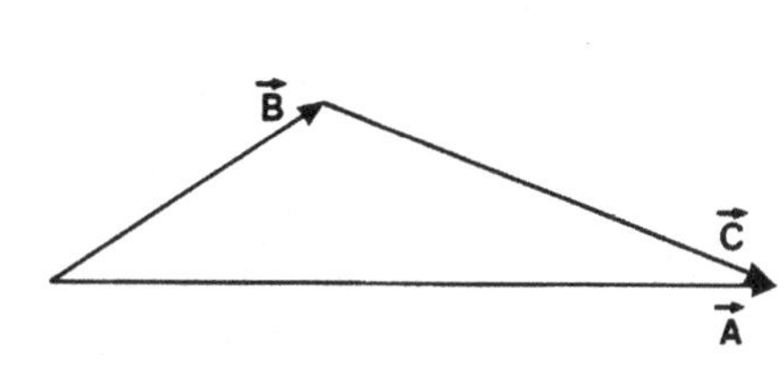

Fig. 1.9 Vector subtraction

product $-k\vec{A}$ is defined as a vector with magnitude $k|\vec{A}|$, and with opposite direction to that of $\vec{A}$. If $k = 0$, we write $k\vec{A} = \vec{0}$ and call $\vec{0}$ the null vector. The magnitude of the null vector is zero.

If two vectors $\vec{A}$ and $\vec{B}$ are parallel, there exists a scalar k so that $\vec{B} = k\vec{A}$. Thus $\vec{A} = (1/k)\vec{B}$. One might also be tempted to put $\vec{B}/\vec{A} = k$. But division by a vector is not defined within the calculus of vectors.

We now have the necessary tools for defining the component vectors of $\vec{A}$. The definition may be stated as follows: the component vectors of $\vec{A}$ are the products of the components of $\vec{A}$ and the basis vectors. In the special case considered above, with the vector $\vec{A}$ placed in a two-dimensional Cartesian coordinate-system, we have

$$\vec{A}^x = A^x\,\vec{e}_x, \qquad \vec{A}^y = A^y\,\vec{e}_y. \tag{1.1}$$

These equations, together with figures 1.6 and 1.7, offer a clear picture of the link between the component vectors $\vec{A}^x$, $\vec{A}^y$ and the (scalar) components A^x, A^y of a vector $\vec{A}$.

We will now define addition and subtraction of vectors. Consider two vectors $\vec{A}$ and $\vec{B}$. The vector-sum of $\vec{A}$ and $\vec{B}$ is a new vector, which we shall designate by $\vec{C}$. The sum of two vectors is written in a similar way as the sum of two real numbers,

$$\vec{C} = \vec{A} + \vec{B}. \tag{1.2}$$

However, the meaning of '+' is different. Vector-addition is defined as follows. Let the initial point of $\vec{B}$ be positioned at the terminal point of $\vec{A}$. Then $\vec{C}$ is the vector with initial point at $\vec{A}$'s initial point and terminal point at $\vec{B}$'s terminal point. This is illustrated in Fig. 1.8.

Subtraction of two vectors is defined as follows: $\vec{A} - \vec{B}$ is that vector $\vec{C}$ which gives $\vec{A}$ when it is added to $\vec{B}$. This is illustrated in Fig. 1.9.

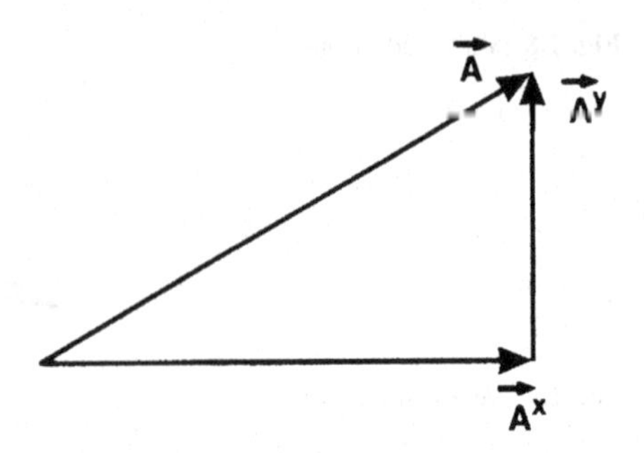

Fig. 1.10 Decomposition of a vector

Let us now turn to Fig. 1.6 again. The component vectors of $\vec{A}$ may be parallel transported so that we obtain the figure 1.10. (The coordinate system is not drawn, only the vectors.)

The figure shows that

$$\vec{A} = \vec{A}^x + \vec{A}^y. \tag{1.3}$$

The vector $\vec{A}$ is equivalent to the vector sum of $\vec{A}^x$ and $\vec{A}^y$. This is expressed by saying that $\vec{A}$ can be decomposed into the component vectors $\vec{A}^x$ and $\vec{A}^y$. From eqs. (1.1) and (1.3) follow

$$\vec{A} = A^x\vec{e}_x + A^y\vec{e}_y \tag{1.4}$$

Note that the sum of the (scalar) components, $A^x + A^y$, is different from the magnitude of $\vec{A}$: $|\vec{A}| \neq A^x + A^y$.

Applying the well-known result of Pythagoras, and of less well-known Indian and Chinese mathematicians, to figure 1.7, we get

$$|\vec{A}|^2 = |\vec{A}^x|^2 + |\vec{A}^y|^2 = (A^x)^2 + (A^y)^2 \tag{1.5}$$

(here the superscripts x and y denote 'which component', and the superscripts 2 are exponents). Taking the square root of each side of Eq. (1.5), we get the magnitude of $\vec{A}$ in terms of its components

$$|\vec{A}| = \sqrt{(A^x)^2 + (A^y)^2} \tag{1.6}$$

The addition of two vectors, $\vec{A} + \vec{B}$, may be done by adding their components. This is illustrated in Fig. 1.11, which looks complicated, but has a fairly simple meaning.

In mathematical form

$$\vec{C} = \vec{A} + \vec{B} \Longleftrightarrow C^x = A^x + B^x, \quad C^y = A^y + B^y, \tag{1.7}$$

where the symbol $\Longleftrightarrow$ means mutual implication (equivalence), and the comma symbolizes 'and'.

There exist several different 'products' of vectors. In the following we will only need the so-called dot product or scalar product of two vectors. (A generalized

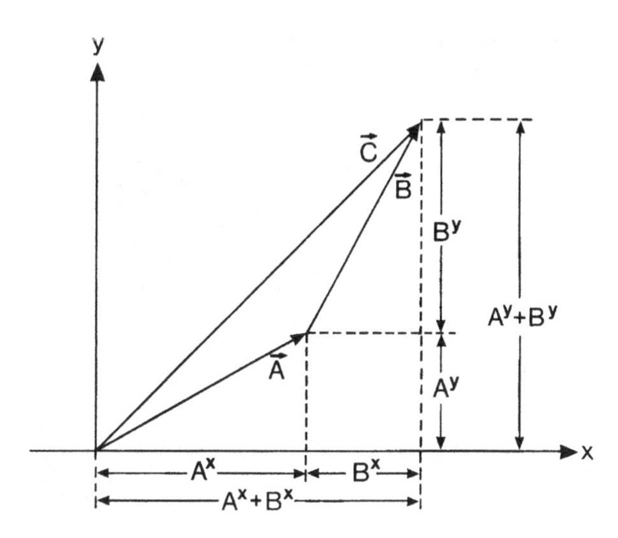

Fig. 1.11 Vector addition

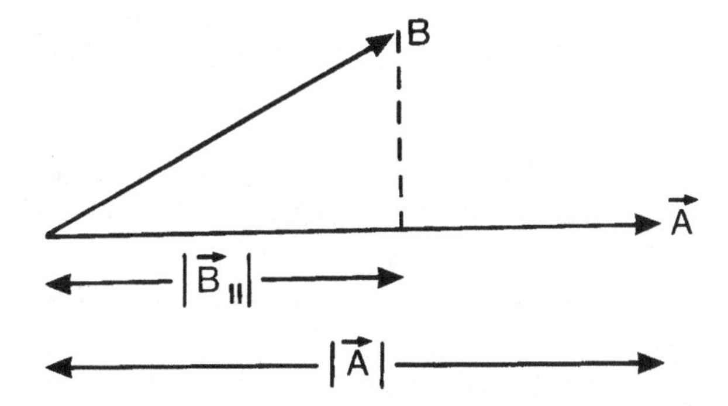

Fig. 1.12 Vector projection

version of the dot product is involved in the relativistic expression for distance, which we shall consider in great detail in Ch. 5.) The dot product of $\vec{A}$ and $\vec{B}$ is denoted by $\vec{A} \cdot \vec{B}$ and defined as the magnitude of $\vec{A}$ times the magnitude of $\vec{B}$'s projection onto $\vec{A}$. This is illustrated in Fig. 1.12.

The magnitude of $\vec{B}$'s projection onto $\vec{A}$ is denoted by $|\vec{B}_{\parallel}|$. The dot product of $\vec{A}$ and $\vec{B}$ may thus be written

$$\vec{A} \cdot \vec{B} = \pm|\vec{A}| \cdot |\vec{B}_{\parallel}| \tag{1.8}$$

(See points 5 and 6 below as to whether + or − is to be used.) Some properties of this product should be noted.

1. The product is a scalar quantity.
2. $\vec{B} \cdot \vec{A} = \vec{A} \cdot \vec{B}$.

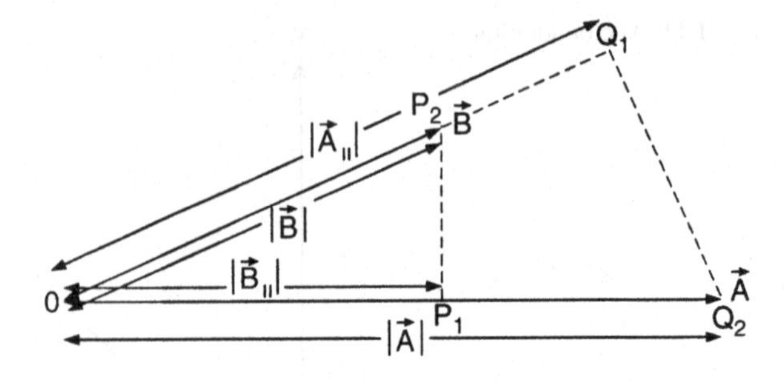

Fig. 1.13 Commutation of the dot product

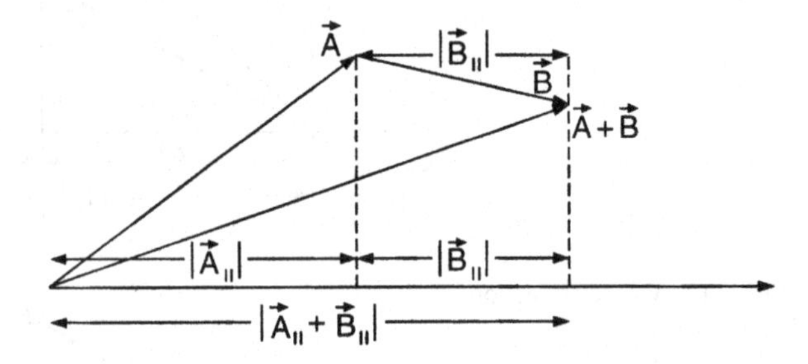

Fig. 1.14 Associative rule

3. For vectors of given magnitude, the product has a maximal value, equal to the magnitude of $\vec{A}$ times the magnitude of $\vec{B}$, if the vectors have the same direction.
4. The value of the dot product is zero if the vectors are orthogonal.
5. The dot product is positive if the projection of $\vec{B}$ onto $\vec{A}$ has the same direction as $\vec{A}$.
6. The product is negative if the projection of $\vec{B}$ onto $\vec{A}$ is oppositely directed to $\vec{A}$.
7. $\vec{D} \cdot (\vec{A} + \vec{B}) = \vec{D} \cdot \vec{A} + \vec{D} \cdot \vec{B}$.

Property 1 follows immediately from the expression (1.8). Property 2 is shown by considering Fig. 1.13.

The triangles OP_1P_2 and OQ_1Q_2 have the same shape. It follows that $|\vec{B}_\parallel|/|\vec{B}| = |\vec{A}_\parallel|/|\vec{A}|$. Hence $|\vec{B}| \cdot |\vec{A}_\parallel| = |\vec{A}| \cdot |\vec{B}_\parallel|$ or $\vec{B} \cdot \vec{A} = \vec{A} \cdot \vec{B}$.

Properties 3 and 4 follow immediatley from Eq. (1.8) and Fig. 1.12. Properties 5 and 6 are definitions. Property 7 is shown by considering Fig. 1.14. From the figure is seen that $|(\vec{A} + \vec{B})_\parallel| = |\vec{A}_\parallel| + |\vec{B}_\parallel|$. Multiplying each term by the number $|\vec{D}|$ gives $|\vec{D}| \cdot |(\vec{A} + \vec{B})_\parallel| = |\vec{D}| \cdot |\vec{A}_\parallel| + |\vec{D}| \cdot |\vec{B}_\parallel|$. Hence $\vec{D} \cdot (\vec{A} + \vec{B}) = \vec{D} \cdot \vec{A} + \vec{D} \cdot \vec{B}$. Due to the property 2 it also follows that $(\vec{A} + \vec{B}) \cdot \vec{D} = \vec{A} \cdot \vec{D} + \vec{B} \cdot \vec{D}$.

Since $\vec{e}_x$ and $\vec{e}_y$ are unit vectors and orthogonal to each other, it follows from properties 2, 3 and 4 that

$$\vec{e}_x \cdot \vec{e}_x = 1, \quad \vec{e}_y \cdot \vec{e}_y = 1, \quad \vec{e}_x \cdot \vec{e}_y = 0, \quad \vec{e}_y \cdot \vec{e}_x = 0 \tag{1.9}$$

We can now express the dot product of $\vec{A}$ and $\vec{B}$ by the components of $\vec{A}$ and $\vec{B}$. From equations (1.4), (1.9), the rule 7, and the product rule $(a+b)(c+d) = ac+ad+bc+bd$ which is valid for vectors as well as for scalars, we get (substitute $A^x\vec{e}_x$ for a, $A^y\vec{e}_y$ for b, $B^x\vec{e}_x$ for c and $B^y\vec{e}_y$ for d)

$$\begin{aligned}\vec{A}\cdot\vec{B} &= (A^x\vec{e}_x + A^y\vec{e}_y)\cdot(B^x\vec{e}_x + B^y\vec{e}_y)\\ &= A^x B^x\ \vec{e}_x\cdot\vec{e}_x + A^x B^y\ \vec{e}_x\cdot\vec{e}_y\\ &\quad + A^y B^x\ \vec{e}_y\cdot\vec{e}_x + A^y B^y\vec{e}_y\cdot\vec{e}_y\\ &= (A^x B^x)\ 1 + (A^x B^y)\ 0\\ &\quad + (A^y B^x)\ 0 + (A^y B^y)\ 1. \end{aligned} \tag{1.10}$$

Since a number, such as $A^y B^x$, times zero is equal to zero, this leads to

$$\vec{A}\cdot\vec{B} = A^x B^x + A^y B^y. \tag{1.11}$$

The expression (1.11) is not valid in arbitrary coordinates. It is specific of Cartesian coordinates. In general the dot products of the basis vectors must be included in the component expression for the dot product of two vectors (see Eq. (1.26)).

1.5 Three and more dimensions

Let us consider a three-dimensional generalization of Fig. 1.6.

The vector $\vec{A}$ is now placed in a three-dimensional Cartesian coordinate system, as shown in fig 1.15. Here $\vec{B}$ is the projection of $\vec{A}$ in the (x, y)-plane. It will be useful, in order to see more easily the geometrical contents of the deductions below, to draw a new figure with coordinate axes parallel to the old ones, but now through the root of the vector $\vec{A}$.

As in the two-dimensional case, the vector components are the coordinates of the terminal point of $\vec{A}$ when it is situated with its initial point at the origin (see Fig. 1.16).

The component vectors of $\vec{A}$ are

$$\vec{A}^x = A^x\vec{e}_x, \quad \vec{A}^y = A^y\vec{e}_y, \quad \vec{A}^z = A^z\vec{e}_z. \tag{1.12}$$

From Fig. 1.15 is seen that

$$\vec{B} = \vec{A}^x + \vec{A}^y, \quad \vec{A} = \vec{B} + \vec{A}^z. \tag{1.13}$$

Thus

$$\vec{A} = \vec{A}^x + \vec{A}^y + \vec{A}^z \tag{1.14}$$

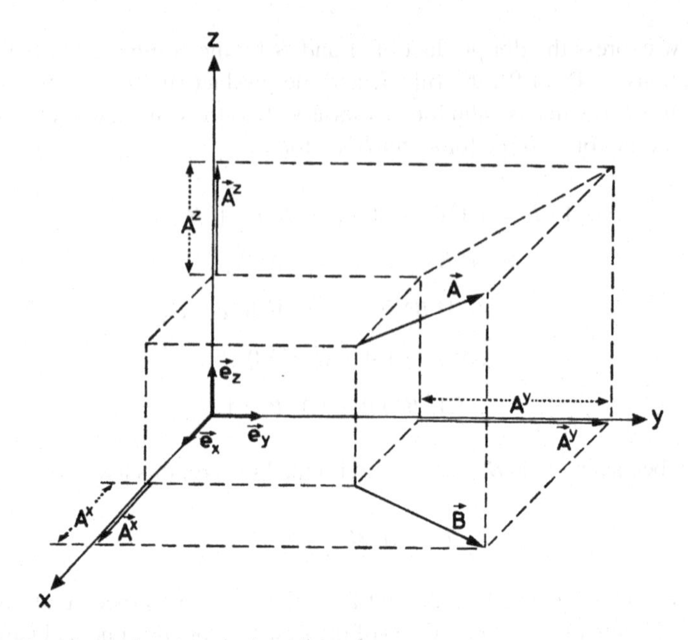

Fig. 1.15 Cartesian basis vectors in three dimensions

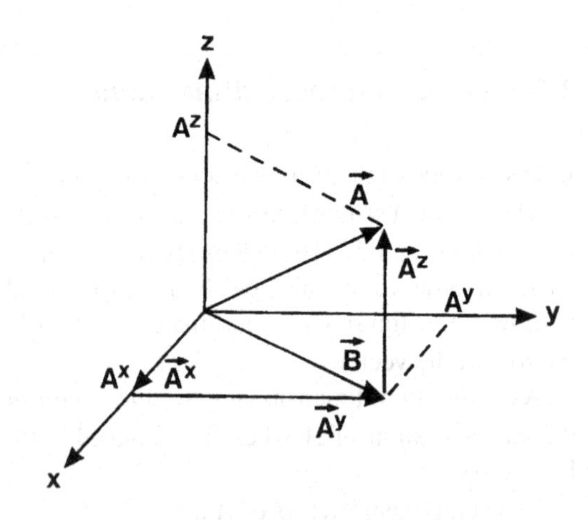

Fig. 1.16 Vector components in three dimensions

Inserting the expression (1.12) we get the component form of the vector $\vec{A}$

$$\vec{A} = A^x\vec{e}_x + A^y\vec{e}_y + A^z\vec{e}_z. \tag{1.15}$$

Also from Fig. 1.15, and using the Pythagorean theorem as we did in the two-dimensional case in Eq. (1.5), it is seen that

$$|\vec{B}|^2 = |\vec{A}^x|^2 + |\vec{A}^y|^2, \quad |\vec{A}|^2 = |\vec{B}|^2 + |\vec{A}^z|^2, \tag{1.16}$$

so that

$$|\vec{A}|^2 = |\vec{A}^x|^2 + |\vec{A}^y|^2 + |\vec{A}^z|^2 = (A^x)^2 + (A^y)^2 + (A^z)^2 \tag{1.17}$$

Taking the square root of both sides of Eq. (1.17), we get the magnitude of $\vec{A}$ in terms of its components

$$|\vec{A}| = \sqrt{(A^x)^2 + (A^y)^2 + (A^z)^2}. \tag{1.18}$$

If two vectors $\vec{A}$ and $\vec{B}$ in a three-dimensional space are added to give a third vector $\vec{C}$, we obtain a projection onto the (x, y) plane just like Fig. 1.8. The projection onto the (x, z) and (y, z) planes are similar, and we obtain

$$\vec{C} = \vec{A} + \vec{B} \Longleftrightarrow \begin{cases} C^x = A^x + B^x, \\ C^y = A^y + B^y, \\ C^z = A^z + B^z. \end{cases} \tag{1.19}$$

This rule is valid in general: two vectors may be added by adding their components.

Applying the multiplication rule $(a+b+c)(d+e+f) = ad+ae+af+bd+be+bf+cd+ce+cf$, we find the expression of the dot product of two vectors in a three-dimensional space, in terms of their components in a Cartesian coordinate system (compare the expression for the two-dimensional case in Eq. (1.10)

$$\begin{aligned}
\vec{A} \cdot \vec{B} &= (A^x\vec{e}_x + A^y\vec{e}_y + A^z\vec{e}_z) \\
&\quad \cdot (B^x\vec{e}_x + B^y\vec{e}_y + B^z\vec{e}_z) \\
&= A^x B^x\vec{e}_x \cdot \vec{e}_x + A^x B^y\vec{e}_x \cdot \vec{e}_y + A^x B^z\vec{e}_x \cdot \vec{e}_z \\
&\quad + A^y B^x\vec{e}_y \cdot \vec{e}_x + A^y B^y\vec{e}_y \cdot \vec{e}_y + A^y B^z\vec{e}_y \cdot \vec{e}_z \\
&\quad + A^z B^x\vec{e}_z \cdot \vec{e}_x + A^z B^y\vec{e}_z \cdot \vec{e}_y + A^z B^z\vec{e}_z \cdot \vec{e}_z \\
&= A^x B^x 1 + A^x B^y 0 + A^x B^z 0 \\
&\quad + A^y B^x 0 + A^y B^y 1 + A^y B^z 0 \\
&\quad + A^z B^x 0 + A^z B^y 0 + A^z B^z 1 \\
&= A^x B^x + A^y B^y + A^z B^z.
\end{aligned} \tag{1.20}$$

It is easily seen that the *form* of the equations characteristic of three-dimensional vector calculus is exactly equal to those of two dimensions (see Eq. 1.10)).

The generalization from three to n dimensions is simple as far as the mathematical expressions are concerned. But convenient figures like 1.8 and 1.15 cannot be drawn for the higher-dimensional cases. Intuitive presentation is problematic, except as intuition of algebraic forms, where this kind of computation, leading to Eq. (1.20), is easy also in four or five dimensions. Any error is readily *seen* by looking at the formal sequences of letters and signs. Goethe said: "The value of mathematics is nothing but that of forms" ("[...] nichts wert als die Form.")

Looking at Eq. (1.20), with its eight lines full of symbols, it will be reasonable if you get a slight feeling of terror. We would like to make a digression here, telling you about the intention of this text. We wish to lay before you a text where the mathematical deductions are presented in such detail that you may see what has been done mathematically in each step of the calculations. What you see depends upon what you know. And as the text proceeds, you will often find references to equations earlier in the book, as a reminder of the particular mathematical rule that has been applied in a certain step.

1.6 The vector product

In chapter 9 we need the concept of the vector product.

Two arbitrary vectors $\vec{A}$ and $\vec{B}$ define a plane, that of the parallelogram between them. Let us choose a three-dimensional coordinate system so that the vectors are lying in the (x, y)-plane. This is drawn in perspective in Fig. 1.17. The vector product of $\vec{A}$ and $\vec{B}$ is denoted by $\vec{A} \times \vec{B}$, and is defined as a vector, the 'vector product vector', whose magnitude is $|\vec{A}||\vec{B}_\perp|$, where $\vec{B}_\perp$ is the component of $\vec{B}$ perpendicular to $\vec{A}$. The direction of $\vec{A} \times \vec{B}$ is normal to the plane containing $\vec{A}$ and $\vec{B}$, with positive sense such that $\vec{A}$, $\vec{B}$ and $\vec{A} \times \vec{B}$ is oriented, as shown in Fig. 1.17. This product is antisymmetric, meaning that the sign of the product changes if the succession of the vectors changes, $\vec{B} \times \vec{A} = -\vec{A} \times \vec{B}$. If $\vec{A}$ and $\vec{B}$ are parallel, $\vec{B}$ has no component perpendicular to $\vec{A}$, and $\vec{A} \times \vec{B} = \vec{0}$. In particular it follows that the vector product of a vector with itself is equal to the null vector. Also, there are no vector products in a two-dimensional space.

In Fig. 1.18 we have drawn the parallelogram defined by the vectors $\vec{A}$ and $\vec{B}$. We have also drawn a dotted line from the end-point of vector $\vec{B}$ and perpendicular to the vector $\vec{A}$. If the triangle OQB is moved to the other side of the parallelogram, a rectangle $QPRB$ is formed with the same area as the triangle. The area of the rectangle, and thus of the parallelogram, is the length of $\vec{A}$ times the height QB. This height is just the component of $\vec{B}$ perpendicular to $\vec{A}$. It follows that the magnitude of the vector product $\vec{A} \times \vec{B}$ is equal to the area of the parallelogram defined by these vectors. The vector product $\vec{A} \times \vec{B}$ can therefore be said to represent the area as a vector quantity.

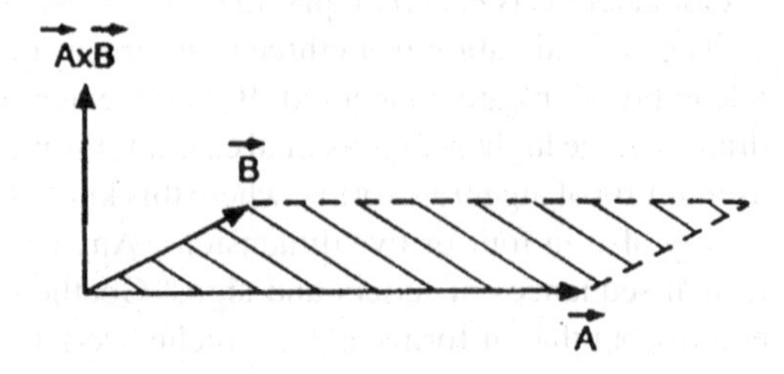

Fig. 1.17 The vector product of $\vec{A}$ and $\vec{B}$

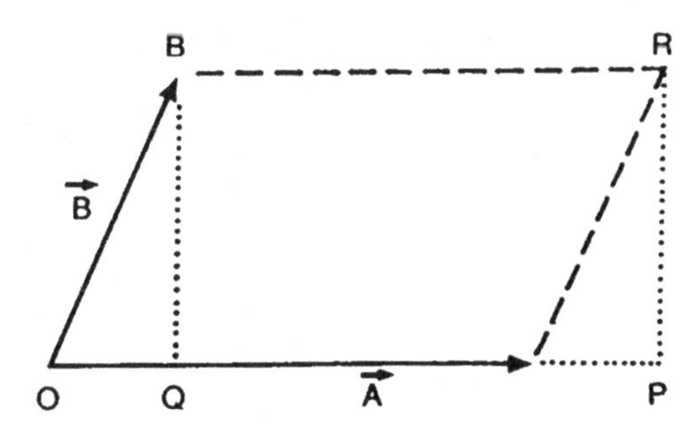

Fig. 1.18 Parallelogram extended by $\vec{A}$ and $\vec{B}$

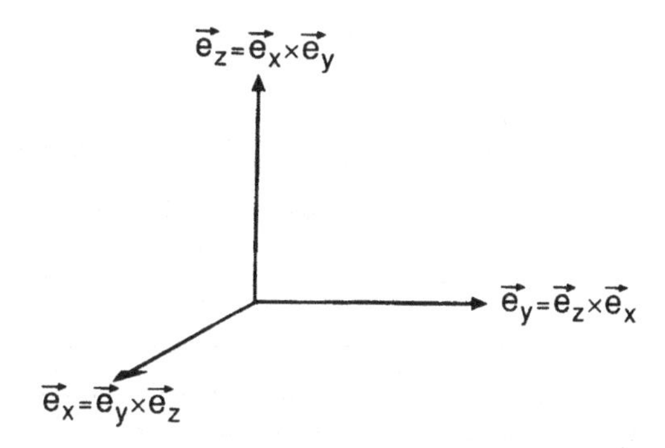

Fig. 1.19 Vector products of basis vectors in Cartesian coordinates

From the definition of the vector product follows that the non-vanishing vector products of the basis vectors in a Cartesian coordinate system, are (see Fig. 1.19)

$$\left.\begin{aligned} \vec{e}_x \times \vec{e}_y &= -\vec{e}_y \times \vec{e}_x = \vec{e}_z, \\ \vec{e}_y \times \vec{e}_z &= -\vec{e}_z \times \vec{e}_y = \vec{e}_x, \\ \vec{e}_z \times \vec{e}_x &= -\vec{e}_x \times \vec{e}_z = \vec{e}_y. \end{aligned}\right\} \tag{1.21}$$

Usually when we apply the calculus of vectors to physical problems, we concentrate on one direction at a time. We calculate with components. This will be the case for example in chapter 9 where we shall apply the vector product in connection with the curvature of spacetime. Therefore we shall need the component expression of the vector product.

In our Cartesian coordinate system the component forms of the vectors $\vec{A}$ and $\vec{B}$ are

$$\vec{A} = A^x\vec{e}_x + A^y\vec{e}_y + A^z\vec{e}_z, \quad \vec{B} = B^x\vec{e}_x + B^y\vec{e}_y + B^z\vec{e}_z. \tag{1.22}$$

Making a calculation similar to that in Eq. (1.20), but using now, Eq. (1.21) and the rule $\vec{A} \times (\vec{B} + \vec{C}) = \vec{A} \times \vec{B} + \vec{A} \times \vec{C}$, we arrive at

$$\begin{aligned} \vec{A} \times \vec{B} = {} & (A^x\vec{e}_x + A^y\vec{e}_y + A^z\vec{e}_z) \\ & \times (B^x\vec{e}_x + B^y\vec{e}_y + B^z\vec{e}_z) \end{aligned}$$

or, after expanding the right-hand side,

$$
\begin{aligned}
\vec{A} \times \vec{B} &= A^x B^x\, \vec{e}_x \times \vec{e}_x + A^x B^y\, \vec{e}_x \times \vec{e}_y \\
&\quad + A^x B^z\, \vec{e}_x \times \vec{e}_z + A^y B^x\, \vec{e}_y \times \vec{e}_x \\
&\quad + A^y B^y\, \vec{e}_y \times \vec{e}_y + A^y B^z\, \vec{e}_y \times \vec{e}_z \\
&\quad + A^z B^x\, \vec{e}_z \times \vec{e}_x + A^z B^y\, \vec{e}_z \times \vec{e}_y \\
&\quad + A^z B^z\, \vec{e}_z \times \vec{e}_z \\
&= A^x B^x\, \vec{0} + A^x B^y\, \vec{e}_z + A^x B^z\, (-\vec{e}_y) \\
&\quad + A^y B^x\, (-\vec{e}_z) + A^y B^y\, \vec{0} + A^y B^z\, \vec{e}_x \\
&\quad + A^z B^x\, \vec{e}_y + A^z B^y\, (-\vec{e}_x) + A^z B^z\, \vec{0} \\
&= (A^y B^z - A^z B^y)\, \vec{e}_x + (A^z B^x - A^x B^z)\, \vec{e}_y \\
&\quad + (A^x B^y - A^y B^x)\, \vec{e}_z. \qquad (1.23)
\end{aligned}
$$

Note that the components of the vector product are *antisymmetric* in the indices, meaning that if the succession of the indices are reversed, the components change sign (due to the minus signs in the expression). The same happens if the components of $\vec{A}$ and $\vec{B}$ are exchanged. The geometrical meaning of this is that $\vec{B}\times\vec{A} = -\vec{A}\times\vec{B}$, i.e. exchanging the succession of the vectors reverses the direction of the vector product vector.

1.7 Space and metric

We shall anticipate a little, and give you a glimpse of topics that will be treated at length later in this text. In order to be able to use the mathematical machinery that have been developed so far, we shall in this section assume that space is flat (as has been assumed also in the earlier sectons). However we shall now permit the use of arbitrary coordinates. Concrete examples will be thoroughly treated in chapters 4 and 6. The generalization to a mathematical structure which may be applied to curved space will be given in Ch. 5.

The general concept of space is such that we, with a suitable metaphor, can put different kinds of coordinate systems *into* a space. However, not any coordinate system! For example you cannot cover the spherical surface of the Earth by a two-dimensional Cartesian coordinate system in which the coordinate curves are everywhere straight lines orthogonal or parallel to each other. If you try to construct straight, parallel coordinate lines on a sphere, you will find that they will cross each other at two points, for example at the poles. The conclusion is: *you need curved coordinate systems to cover curved surfaces.*

In the general theory of relativity gravitation is described in terms of curved four-dimensional spacetime. So in this theory curved coordinate systems are inevitable.

It will be useful to introduce some notational conventions. Coordinates of spacetime are written x^μ, i.e. with Greek indices, and spatial coordinates are written x^i, i.e. with Latin indices. Here μ and i are not exponents, but labels indicating the coordinate in question. The three spatial coordinates are indicated by $x^i, i = 1, 2, 3$. In a Cartesian coordinate system this means $x^1 = x, x^2 = y$, and $x^3 = z$. The four coordinates of spacetime are indicated by $x^\mu, \mu = 1, 2, 3, 4$. Again, in a Cartesian coordinate system this means $x^1 = x, x^2 = y, x^3 = z$, and $x^4 = ct$, where c is the velocity of light. Note that x^4 is a time coordinate representing time as measured in a unit of length.

In Sect. 1.4 we introduced the basis vectors $\vec{e}_x$ and $\vec{e}_y$ of a two-dimensional Cartesian coordinate system. The basis vectors of an arbitrary coordinate system, $\{x^\mu, \mu = 0, 1, 2, \ldots\}$ or $\{x^i, i = 1, 2, \ldots\}$, are denoted by $\vec{e}_\mu$ or $\vec{e}_i$, i.e. the index of an arbitrary basis vector is just the number of the coordinate and not the coordinate itself. However, when we specialize to a certain basis vector in a specified coordinate system, the coordinate is inserted as index of the basis vector. For example, if we consider $\vec{e}_2$ with $x^2 = \theta$, we insert $\vec{e}_\theta$ for $\vec{e}_2$.

The basis vectors of an arbitrary coordinate system are neither orthogonal to each other, nor are they unit vectors. It follows that the dot products of the basis vectors are arbitrary functions of position. The dot products of basis vectors in spacetime are termed $g_{\mu\nu}$ and in space g_{ij}. We may then write, for the dot products of any two basis vectors $\vec{e}_\mu$ and $\vec{e}_\nu$ in four-dimensional spacetime

$$g_{\mu\nu} \equiv \vec{e}_\mu \cdot \vec{e}_\nu, \ \mu = 1, 2, 3, 4 \quad \nu = 1, 2, 3, 4, \tag{1.24}$$

where the symbol $\equiv$ means 'defined as'. From the generalization of Eq. (1.9) to three dimensions we understand that a three-dimensional Cartesian system is characterized by

$$\begin{aligned} g_{11} &= g_{22} = g_{33} = 1, \\ g_{12} &= g_{13} = g_{23} = g_{21} = g_{31} = g_{32} = 0. \end{aligned} \tag{1.25}$$

Hence, for all Cartesian coordinate systems $g_{ii} = 1$, which means that the basis vectors have unit length, and $g_{ij} = 0$ for $i \neq j$ (where $\neq$ means 'different from'), which says that all the dot products of two different basis vectors are zero, meaning that the vectors are orthogonal to each other. In general $g_{\mu\nu}$ or g_{ij} characterizes the angles of all pairs of basis vectors and their magnitudes in a coordinate system of a space. The quantities $g_{\mu\nu}$ and g_{ij} are said to tell us the metric of the spacetime and space, respectively. So you hear professionals frequently ask each other: "Well, but what is the metric of the spacetime you now are talking about?"

The general expression of the dot product of two vectors in three-dimensional space is

$$\vec{A} \cdot \vec{B} = A^i \vec{e}_i \cdot B^j \vec{e}_j = \vec{e}_i \cdot \vec{e}_j A^i B^j = g_{ij} A^i B^j, \tag{1.26}$$

where one is to summarize over the range of indices that appear in products as a subscript in one factor and as a superscript in the other. This is Einstein's summation convention.

The efficiency of Einstein's summation convention can be seen by applying it to calculate the dot product of two vectors in a three-dimensional Cartesian coordinate system, for instance,

$$\begin{aligned} \vec{A} \cdot \vec{B} &= g_{11} A^1 B^1 + g_{12} A^1 B^2 + g_{13} A^1 B^3 \\ &\quad + g_{21} A^2 B^1 + g_{22} A^2 B^2 + g_{23} A^2 B^3 \\ &\quad + g_{31} A^3 B^1 + g_{32} A^3 B^2 + g_{33} A^3 B^3. \end{aligned} \tag{1.27}$$

Using Eq. (1.27) and inserting $1 = x$, $2 = y$, and $3 = z$, for the indices, we get

$$\begin{aligned} \vec{A} \cdot \vec{B} &= 1\,(A^x B^x) + 0\,(A^x B^y) + 0\,(A^x B^z) \\ &\quad + 0\,(A^y B^x) + 1\,(A^y B^y) + 0\,(A^y B^z) \\ &\quad + 0\,(A^z B^x) + 0\,(A^z B^y) + 1\,(A^z B^z) \\ &= A^x B^x + A^y B^y + A^z B^z \end{aligned} \tag{1.28}$$

as in Eq. (1.20).

The distance vector between two points that are infinitesimally close to each other is denoted by $d\vec{r}$. (Further explanation of this notation is given in the next chapter.) The components of $d\vec{r}$ are called coordinate differentials and are written dx^i. If $i = 1, 2, 3$ then

$$d\vec{r} = dx^i \vec{e}_i = dx^1 \vec{e}_1 + dx^2 \vec{e}_2 + dx^3 \vec{e}_3. \tag{1.29}$$

(Note that here dx^2 is not dx squared, but refers to the second component of $d\vec{r}$.) The squared distance between the points is denoted by ds^2. Using Eqs. (1.28) and (1.24) we get

$$\begin{aligned} ds^2 &= d\vec{r} \cdot d\vec{r} = dx^i \vec{e}_i \cdot dx^j \vec{e}_j = \vec{e}_i \cdot \vec{e}_j dx^i dx^j \\ &= g_{ij} dx^i dx^j. \end{aligned} \tag{1.30}$$

The corresponding expression for the squared 'distance' between two points in spacetime is

$$ds^2 = g_{\mu\nu} dx^\mu dx^\nu. \tag{1.31}$$

This is a highly condensed and very famous formula. If μ and ν go from 1 to 4 the far right-hand side is the sum of $4 \times 4 = 16$ terms. Equations (1.30) and (1.31) show that the general expression for distance in arbitrary coordinates involves the components, g_{ij} or $g_{\mu\nu}$, of the metric. That is the main reason for the name 'metrical'—it 'measures'.

Equation (1.31) generalizes the Pythagorean theorem to a form which is valid in an arbitrary coordinate system in a space with an arbitrary number of dimensions. The most simple application of this equation is to calculate the distance of two nearby points on a (two-dimensional) plane in terms of the coordinate differentials dx and dy in a Cartesian coordinate system. In this case

$$g_{11} = g_{22} = 1, \qquad g_{12} = g_{21} = 0. \tag{1.32}$$

From Eq. (1.30) we then get

$$\begin{aligned} ds^2 &= g_{ij} dx^i dx^j \\ &= g_{11} dx^1 dx^1 + g_{12} dx^1 dx^2 \\ &\quad + g_{21} dx^2 dx^1 + g_{22} dx^2 dx^2 \\ &= (dx^1)^2 + (dx^2)^2. \end{aligned} \tag{1.33}$$

Using that $dx^1 = dx$ and $dx^2 = dy$ we get

$$ds^2 = dx^2 + dy^2. \tag{1.34}$$

This, we see, is in fact the usual form of the Pythagorean theorem!

In what is called a 'flat' space one may define distance vectors $\vec{r}$ of finite magnitude. This is not possible in a so-called curved space. The surface of a sphere is a two-dimensional curved space in three-dimensional Euclidean (flat) space. Let us jump from geometry to physics and 'apply' the geometry of spheres to a physical object, for instance the Earth, with its physical surface. Consider a distance vector from, say, the North Pole to a point on the Equator. The vector-arrow, which is straight in the usual Euclidean meaning of this word, will go through the Earth, beneath the surface. For two-dimensional creatures, whose universe is the surface of the sphere, this distance vector simply does not exist. Therefore, in general there will exist only infinitesimally short distance vectors in a curved space.

Three points may be noted. Firstly, the picture of vectors as straight arrows is a metaphor of great heuristic help, but is not needed in the calculus of vectors. Secondly, in connection with curved spaces vectors are defined in tangent spaces (see Ch. 3). The tangent space of the North Pole of the Earth, for example, is an Euclidean plane touching the North Pole. And a vector representing, say, the wind velocity at the North Pole may be thought of as an arrow in this tangent plane.

Thirdly, finite distances can be defined in curved spaces. They are calculated by adding infinitely many infinitesimally small distances along a curve between the point whose distance we are going to find. This is performed by integration, which will be introduced in section 3.5.

Chapter 2
Differential calculus

This chapter is written for those people who have the courage to approach the mathematics of general relativity without being familiar with differential calculus. The use of this fabulous creation by Newton and Leibniz is essential and omnipresent on our way to Einstein's field equations.

2.1 Differentiation

In Fig. 2.1 we have drawn Cartesian coordinates and a curve. Five lines lead from a point P on the curve to points Q_1, Q_2, Q_3, Q_4, Q_5, also on the curve. If we continue plotting lines like that, the Q's approach P indefinitely. They are said to approach the tangent line at P. The lines PQ_1, PQ_2, PQ_3, PQ_4, and PQ_5 have different slopes in relation to the x-axis. The slope of the curve at a point P is defined as the slope of the tangent line at P. Knowing the slope of a curve at any point, and the value of the function at one point, we can plot the curve (Fig. 2.2). If the the tangent lines are close enough the curve 'plots itself'. Could we find simple expressions for the slope of those lines, that is, for the slope of the curve itself? The slope k of a line is quantitatively expressed by what we shall call the 'slope quotient' defined as 'increment in y-direction divided by increment in x-direction' (see Fig. 2.3).

$$k = \frac{y_2 - y_1}{x_2 - x_1}. \tag{2.1}$$

In Fig. 2.2 the slope quotient at P is $1/5 = 0.20$. The slopes of the series of lines $PQ_1, PQ_2, PQ_3, \ldots$ in Fig. 2.2 are expressed by the same sort of quotient as that of the line pictured in Fig. 2.3. Their slope quotients approach that of the tangent line, as Q_n approaches P.

Ø. Grøn and A. Næss, *Einstein's Theory: A Rigorous Introduction for the Mathematically Untrained*, DOI 10.1007/978-1-4614-0706-5_2,

Fig. 2.1 A curve

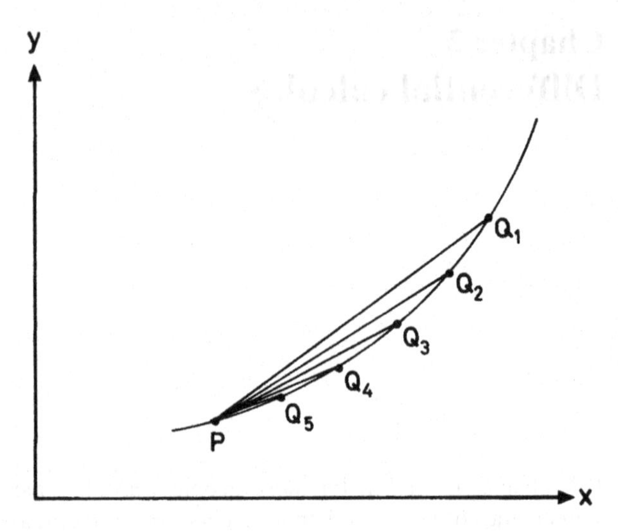

Fig. 2.2 A curve with tangent lines

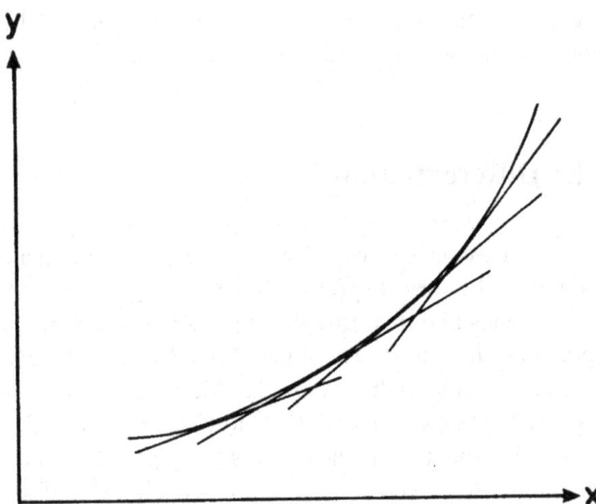

Fig. 2.3 The slope of the tangent line

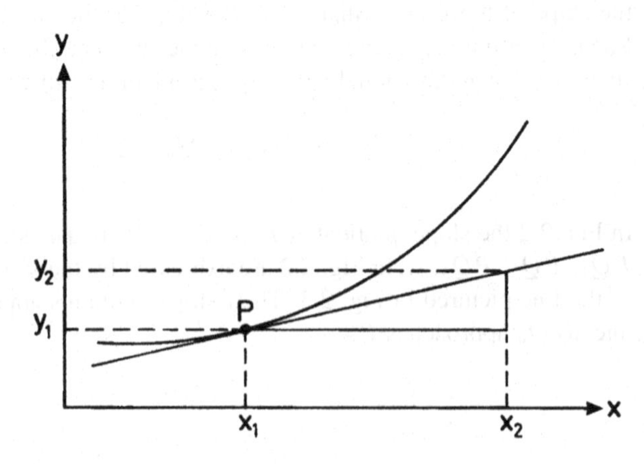

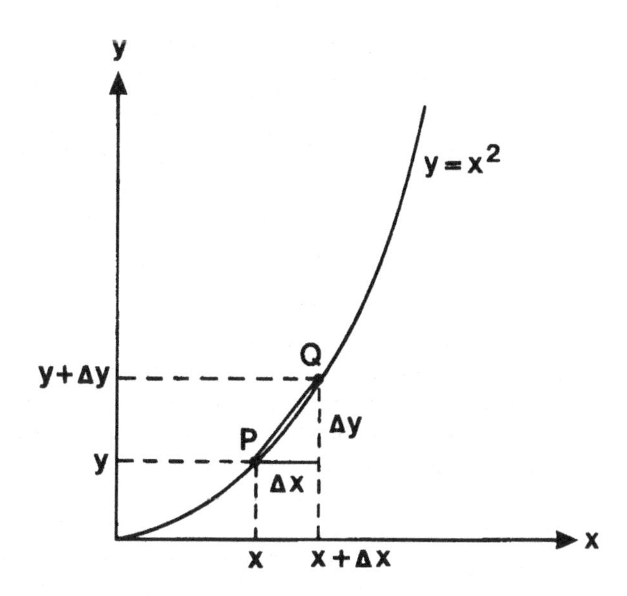

Fig. 2.4 The parabola $y = x^2$

The slope quotient of the curve on Fig. 2.2 increases continuously, from less than $1/5$ to a fairly big number. That of Fig. 2.3 seems to start from zero and remain there for a while, like a straight line parallel to the x-axis.

A word of caution: not all curves have a tangent at every point. If the curve has a sharp corner at a point it does not have a tangent at that point. Also, if the curve has a vertical step (the vertical interval is not reckoned as part of the curve), then it is disconnected for a certain value of x, say $x = x_1$. This is called a discontinuity. If the curve is the graph of a function, the function is said to have a discontinuity at $x = x_1$. A curve does not have a tangent at a discontinuity. We shall assume that the functions we need to consider, are such that their graphs have one definite tangent line at every point, i.e. that the functions are continuous and their graphs are without sharp corners or discontinuous steps. If you inspect Fig. 2.4 you will see that a new notation has been introduced. The figure suggests that we move along a curve, called a parabola, with equation $y = x^2$, a short distance from P to Q. If P is an arbitrary point on the curve, we denote the coordinates of P by x and y, and Q has coordinates $x + \Delta x$ and $y + \Delta y$, where Δx ('delta x') is a small increment added to x, and Δy a corresponding (i.e. related to Δx and the steepness of the curve, see Fig. 2.4) small increment of y.

In the new notation we may now express in a general way the slope quotient of the curve making use of Eq. (2.1), one page 21:

$$\frac{(y + \Delta y) - y}{(x + \Delta x) - x} = \frac{\Delta y}{\Delta x}. \tag{2.2}$$

However close the points P and Q on a curve may be, there is a definite quotient $\Delta y/\Delta x$. When points on the fairly simple curves (graphs of continuous and singularity free functions) we shall study are brought closer and closer, the quotient approaches a definite limit, which represents the slope of the curve at the point P. There is a symbol for the approach towards this 'limit' which some of us find quite elegant:

$$\lim_{\Delta x \to 0} \frac{\Delta y}{\Delta x}. \tag{2.3}$$

Let us now consider the curve in Fig. 2.4 as the graph of a function $y = f(x)$. The limit (2.3) changes when x varies except when the curve, in our wide sense of the term, is a straight line. Therefore the limit is evidently a function of x, and is called *the derivative* of $f(x)$. Note that $f(x)$ and the derivative of $f(x)$ are two different functions. The process of finding the derivative is not called derivation but 'differentiation'. As could be expected there are many symbols expressing this crucial notion. The most intuitively powerful is perhaps dy/dx. Since $y = f(x)$ we may also write $df(x)/dx$ and df/dx. The shortest notation is $f'(x)$ or y'.

It was the great Gottfried Wilhelm Leibniz who first used this elegant notation:

$$f'(x) \equiv \frac{dy}{dx} \equiv \lim_{\Delta x \to 0} \frac{\Delta y}{\Delta x} = \lim_{\Delta x \to 0} \frac{f(x + \Delta x) - f(x)}{\Delta x}. \tag{2.4}$$

Note that the limit in Eq. (2.4) is well-defined only if the curve $y = f(x)$ is continuous. Functions can only be differentiated where they are continuous.

The material world is not continuous. At small distances the discontinuity of the atomic world appears. Such discontinuities are neglected when we apply the differential calculus to the description of the world, which we do in the general theory of relativity. Then we idealize the world as a continuum. This is, however, an eminently adequate idealization for the purpose of describing most macroscopic phenomena.

The derivative, as defined in Eq. (2.4), is simply an expression of the slope of a curve. Since this is identical to the direction of the tangent at any point of a curve, the derivative is an expression of the slope of the tangent line at any point of the curve.

We shall now introduce a new concept called the differential.

In Fig. 2.5 we have drawn the graph of a function $y = f(x)$ and a tangent line of the curve in the neighbourhood of a point P with x-coordinate x_0. The tangent, which is called the linearization $L(x)$ of f, has constant slope quotient, $k = \Delta L/\Delta x$. Then $k = \Delta L/\Delta x = f'(x_0)$, since the slope of the tangent is equal to the slope of the curve at x_0. It follows that the increment of the linearization of f as x increases by Δx is $\Delta L = f'(x_0)\Delta x$. The quantity ΔL is called the differential of f and will be denoted by Df. Hence

$$Df = f'(x_0)\Delta x. \tag{2.5}$$

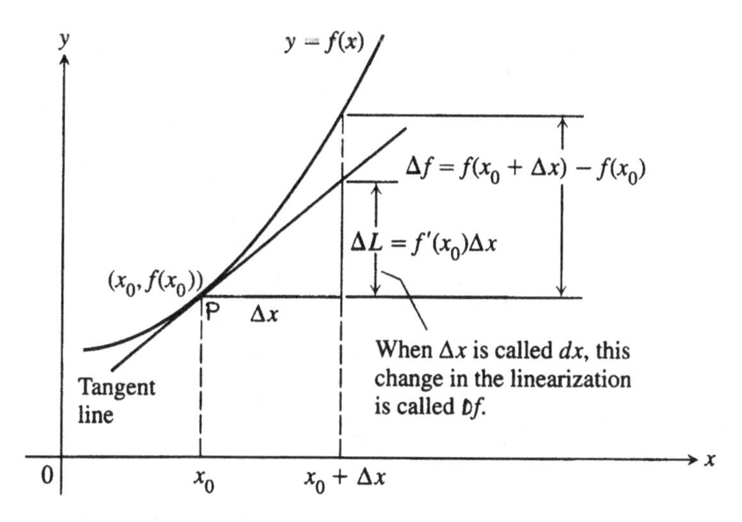

Fig. 2.5 The tangent line at P

The increment Δf of f as x increases by Δx is $\Delta f = f(x_0 + \Delta x) - f(x_0)$. Later (see Eq. 2.93) we shall show that if Δx is small, the difference between the differential Df and the increment Δf of f is given by a sum of terms proportional to increasing powers of Δx, starting by $(\Delta x)^2$. This means that when we calculate to first order in Δx, this difference may be neglected.

2.2 Calculation of slopes of tangent lines

Let us as a first example consider the function

$$y \equiv f(x) = x^2. \tag{2.6}$$

Writing x_1 for $x + \Delta x$ and y_1 for $y + \Delta y$ the expression for the slope quotient takes the form

$$\frac{\Delta y}{\Delta x} = \frac{y_1 - y}{x_1 - x}. \tag{2.7}$$

According to Eq. (2.6), $y = x^2$, and $y_1 = x_1^2$. Hence we can do a first calculation, using the rule $a^2 - b^2 = (a - b)(a + b)$

$$\frac{y_1 - y}{x_1 - x} = \frac{{x_1}^2 - x^2}{x_1 - x} = \frac{(x_1 - x)(x_1 + x)}{x_1 - x} = x_1 + x. \tag{2.8}$$

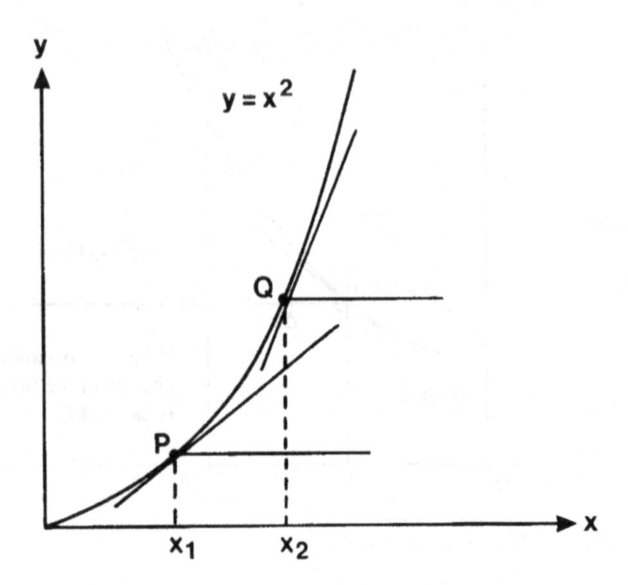

Fig. 2.6 The slope of the curve $y = x^2$

When x_1 approaches x, the quantity $x_1 + x$ approaches $2x$. In symbols (with $y = x^2$)

$$\lim_{\Delta x \to 0} \frac{\Delta y}{\Delta x} = \lim_{x_1 \to x} \frac{y_1 - y}{x_1 - x} = \lim_{x_1 \to x} (x_1 + x) = 2x. \tag{2.9}$$

From Eq. (2.9) we conclude

$$\text{if} \quad y \equiv f(x) = x^2 \quad \text{then} \quad \frac{dy}{dx} = 2x. \tag{2.10}$$

The derivative of $y = x^2$ is simply $2x$. Geometrically this means that at any point (x_1, y_1) the slope of the curve increases proportionally to x_1, as is illustrated in Fig. 2.6. Note that the derivative of a function is itself a function. When x changes, the derivative changes accordingly.

Our next example involves a little more calculation with fractions, but will be needed below. We shall find the derivative of the function $f(x) = 1/x$.

$$\begin{aligned} \left(\frac{1}{x}\right)' &= \lim_{\Delta x \to 0} \frac{\frac{1}{x+\Delta x} - \frac{1}{x}}{\Delta x} = \lim_{\Delta x \to 0} \frac{\frac{x-(x+\Delta x)}{(x+\Delta x)x}}{\Delta x} \\ &= \lim_{\Delta x \to 0} \frac{\frac{-\Delta x}{(x+\Delta x)x}}{\Delta x}. \end{aligned} \tag{2.11}$$

Dividing by Δx in the upper numerator and the denominator, gives

$$\left(\frac{1}{x}\right)' = \lim_{\Delta x \to 0} \frac{-1}{(x + \Delta x)x} = -\frac{1}{x^2}. \tag{2.12}$$

As a third example let us take the derivative of the derivative of $f(x) = x^2$. This derivative is called a derivative of the second order with respect to x^2, or the second derivative of $f(x)$, and is denoted by d^2f/dx^2. This time we get

$$\frac{d^2 f}{dx^2} = \lim_{\Delta x \to 0} \frac{2\left[(x + \Delta x) - x\right]}{\Delta x} = \frac{2\Delta x}{\Delta x} = 2. \tag{2.13}$$

The geometrical meaning of this result will be made clear in section 2.5. Note, however, that a 'constant function', $f(x) = k$, for instance where $k = 2$, corresponds geometrically to a horizontal straight line. There is no slope; the derivative of a constant function is zero. In the section on series expansions we shall need the following consequence of this: The third derivative of x^2 is zero, and so on for the fourth, the fifth and sixth derivative and so on.

We shall often use a couple of elementary rules:

1. The derivative of the sum of two or more functions is equal to the sum of their derivatives.
2. The derivative of the product of a constant and a function is equal to the constant times the derivative of the function.

Example 2.1. Using the results that the derivative of x^2 is $2x$, the derivative of x is 1, and the derivative of a constant is zero, we obtain: if $y = ax^2 + bx + c$, then $dy/dx = 2ax + b$.

2.3 Geometry of second derivatives

Each curve we talk about in this section is assumed to be the graph of a function $f(x)$. The function value $f(x)$ gives the height y above the x-axis of a point on the curve $y = f(x)$. The first derivative gives the change of height per unit distance in the x-direction. Consider the function $y = 2x$. For $x = 1$ we then have $y = 2$, for $x = 2$ we have $y = 2 \times 2 = 4$, for $x = 3$ we have $y = 2 \times 3 = 6$, and so on. The change of y is the double of the change of x. The rate of change is constant. The derivative is a constant, namely 2. It determines the slope of the curve.

The second derivative gives the rate of change of the slope with distance in the x-direction. The second derivative of $y = 2x$, that is, the first derivative of 2, is zero, as it should be. But for $y = x^2$, the second derivative is not zero, but is a constant: $y' = 2x$, $y'' = 2$. The slope of the curve changes with x, and the value '2' is a measure of the change.

A straight line, i.e. a line which is not curved, has a constant slope. The more a curve curves, the faster its slope changes, and the larger is the value of the second derivative of the corresponding function $f(x)$. There is a close relation between the second derivative of a function and the curvature of its graph. This will be discussed in more detail in chapter 9.

2.4 The product rule

We shall now find a formula showing how the slope changes along the graph of a function which is the product of two other functions. Let

$$y = f(x)g(x), \quad y + \Delta y = f(x + \Delta x)g(x + \Delta x). \tag{2.14}$$

We can get an expression for Δy by subtracting y from (2.14) i.e. $\Delta y = (y+\Delta y)-y$

$$\Delta y = f(x + \Delta x)g(x + \Delta x) - f(x)g(x). \tag{2.15}$$

In order to arrive at a limit analogous to that of Eq. (2.4) we use a trick, adding and subtracting in Eq. (2.15) the same expression, that is since $-f(x)g(x + \Delta x) + f(x)g(x + \Delta x) = 0$, the right-hand side of Eq. (2.15) can be written as

$$\begin{aligned}\Delta y =& f(x \overset{1}{+} \Delta x)g(x \overset{2}{+} \Delta x) - \overset{3}{f}(x)g(x \overset{4}{+} \Delta x)\\ &+ \overset{5}{f}(x)g(x \overset{6}{+} \Delta x) - \overset{7}{f}(x)\overset{8}{g}(x).\end{aligned} \tag{2.16}$$

From the terms marked 1, 2, 3, and 4 we form the expression $g(x + \Delta x)[f(x + \Delta x) - f(x)]$, and from the functions 5, 6, 7, and 8 we form $f(x)[g(x+\Delta x)-g(x)]$. Dividing by Δx we get the formula

$$\begin{aligned}\frac{\Delta y}{\Delta x} =& g(x + \Delta x)\frac{f(x + \Delta x) - f(x)}{\Delta x}\\ &+ f(x)\frac{g(x + \Delta x) - g(x)}{\Delta x}.\end{aligned} \tag{2.17}$$

From now on it is getting easier. We proceed to the limit where $\Delta x \to 0$. In the first term at the right-hand side we can then use the limiting value

$$\lim_{\Delta x \to 0} g(x + \Delta x) = g(x). \tag{2.18}$$

From the definition (2.4) we get

$$\lim_{\Delta x \to 0} \frac{f(x + \Delta x) - f(x)}{\Delta x} = \frac{df(x)}{dx} \tag{2.19}$$

and

$$\lim_{\Delta x \to 0} \frac{g(x + \Delta x) - g(x)}{\Delta x} = \frac{dg(x)}{dx}. \tag{2.20}$$

Consequently, looking at Eqs. (2.17) and (2.18), and then at (2.17) together with (2.19) and (2.19), we find

$$\frac{d\,[f(x)g(x)]}{dx} = g(x)\frac{df(x)}{dx} + f(x)\frac{dg(x)}{dx}. \tag{2.21}$$

Perceptually clearer

$$[f(x)\,g(x)]' = g(x)f'(x) + f(x)g'(x). \tag{2.22}$$

Perceptually? Visually? What is the relevance in mathematics? Ideally no relevance, in practice quite central. A professional glances half a second at a formidable formula with 1000 signs, calmly announcing: "There is a mistake—*here*". What a formidable intellect, what a deep understanding, we are apt to think. But the expert is likely only to have activated his perceptual apparatus, nothing more. (His 'gestalt vision' I (A.N.) would say, as a philosopher.) The equation (2.21) has nearly half a hundred separate meaningful signs, Eq. (2.22) has 28. We are now offering a version of the product rule with only 12 signs. Let us pose

$$u \equiv f(x), \quad v \equiv g(x). \tag{2.23}$$

From this emerges supreme simplicity and surveyability:

$$(uv)' = vu' + uv'. \tag{2.24}$$

In words: The derivative of the product of two functions is equal to the second function multiplied by the derivative of the first function plus the first function multiplied by the derivative of the second.

Example 2.2. Find the derivative of $y = x^3$.

Writing u for x^2 and v for x, that is uv for x^3, the use of Eqs. (2.24) and (2.10) gives us

$$(x^3)' = (x^2 \times x)' = x(x^2)' + x^2 \times x' = x \times 2x + x^2 \times 1 = 3x^2.$$

2.5 The chain rule

What is the derivative of the more complicated function $y = (x^2 + 3)^3$? Here, too, we can find a simple rule which is called 'the chain rule'. What is inside the parenthesis is itself a function of x, and we may easily perceive what is to be done by denoting it by a letter of its own, say, u. Accordingly, $y = u^3$ with $u = x^2 + 3$. The function y may then be written

$$y(u) = y[u(x)]. \tag{2.25}$$

We may think of this function in two ways; either as a function $y(u)$ of u, or as a *composite function* $y[u(x)]$ of x, i.e. a function of a function.

Let us now free ourselves of the particular example $y = (x^2 + 3)^3$ and consider two arbitrary functions $y(u)$ and $u(x)$. We shall deduce a formula for the derivative of the composite function $y[u(x)]$. Here y is called the *outer function* and u the *inner function*. According to Eq. (2.5) the differential of y is

$$Dy = y'(u) \times \Delta u, \tag{2.26}$$

where Δu is the increment of u as x increases by Δx, and the differential of the function $u(x)$ is

$$Du = u'(x) \times \Delta x. \tag{2.27}$$

As noted above the difference between Δu and Du is given by a sum of terms proportional to increasing powers of Δx, starting by $(\Delta x)^2$. We shall take the limits $\Delta u \to 0$ and $\Delta x \to 0$, so it will be sufficient to calculate to first order in Δu and Δx. Hence it is sufficient to use the approximations $Dy \approx \Delta y$ and $Du \approx \Delta u$. From Eqs. (2.26) and (2.27) we then get

$$\Delta y \approx y'(u) \times u'(x)\Delta x. \tag{2.28}$$

The increments Δy and Δx are finite quantities. Hence we may divide by Δx, and get

$$\frac{\Delta y}{\Delta x} \approx y'(u) \times u'(x). \tag{2.29}$$

Taking the limit $\Delta x \to 0$ we get

$$\lim_{dx \to 0} \frac{dy}{dx} = \lim_{dx \to 0} \frac{\Delta y}{dx} = y' = y'(u) \times u'(x). \tag{2.30}$$

This is the chain rule for differentiation of a composite function. It may be phrased as follows: The derivative of a composite function is equal to the derivative of the outer function with respect to the inner function multiplied by the derivative of the inner function.

Sometimes the notation with a quotient of two differentials is convenient when we write the derivative of a function. Using this notation the chain rule takes the form

$$\frac{dy}{dx} = \frac{dy}{du} \times \frac{du}{dx} = y'(u) \times u'. \tag{2.31}$$

As a simple application of the chain rule we shall differentiate $y = (x^2 + 3)^3$. Here $y = u^3$ and $u = x^2 + 3$. Using the result of the example above with x replaced by u, we may write

$$\frac{dy}{du} = 3u^2 \tag{2.32}$$

or

$$dy = 3u^2 du. \tag{2.33}$$

If we divide each of the differentials by dx we obtain

$$y' = \frac{dy}{dx} = 3u^2 \frac{du}{dx} = 3u^2 u'. \tag{2.34}$$

Inserting the expressions for y, u, and u' we finally get

$$\left[\left(x^2+3\right)^3\right]' = 3\left(x^2+3\right)^2 2x = 6x\left(x^2+3\right)^2. \tag{2.35}$$

2.6 The derivative of a power function

We can use the product rule and the chain rule to find the derivative of the function $f(x) = x^p$, where p is a real number.

Equation (2.10)) and Example 2.1 suggest that $(x^n)' = nx^{n-1}$ if n is an integer number. This can easily be proved by so-called mathematical induction. The formula is clearly correct for $n = 1$, in which case $(x^1)' = x^0 \equiv 1$.

If we now assume that the formula is valid for x^{n-1}, then $(x^{n-1})' = (n-1)x^{n-2}$. By means of the product rule we get, by setting $u = x$ and $v = x^{n-1}$

$$\begin{aligned}(x^n)' &= (x \times x^{n-1})' = x(n-1)x^{n-2} + 1 \times x^{n-1} \\ &= [(n-1)+1]x^{n-1} = nx^{n-1}.\end{aligned} \tag{2.36}$$

We now know

1. the rule is valid for $x = 1$, and
2. if the rule is valid for x^{n-1}, then it is also valid for x^n, with n as an integer number.

These are the two criteria for the proof by mathematical induction. We have now proved the rule for the case that p is an integer number, $p = n$.

Note that by inserting $n = 3$, we get the result of the above example. Furthermore the rule is correct for $n = 0$, in which case it gives $(x^0)' = 0$, which is obviously correct since $x^0 = 1$ is a constant function. Its graph is a horizontal line with vanishing slope, i.e. the derivative of a constant is zero. From Eq. (2.12), noting that $1/x = x^{-1}$ and $1/x^2 = x^{-2}$, follows that the rule is also valid for $n = -1$, which will be used in the next section.

Is it possible to prove that the rule is valid also if p is a fraction? Consider a function $u = x^{1/n}$, where n is an integer number. This function is defined in the following way

$$\left(x^{1/n}\right)^n = x. \tag{2.37}$$

Hence

$$u^n = x. \tag{2.38}$$

Here u is a function of x. Differentiation by means of the rule (2.36) and the chain rule leads to

$$nu^{n-1} \times u' = 1. \tag{2.39}$$

Dividing by nu^{n-1} on both sides gives

$$u' = \frac{1}{nu^{n-1}}. \tag{2.40}$$

Remember also that $u^n = x$ and $u = x^{1/n}$. Then we get

$$u' = \left(x^{1/n}\right)' = \frac{1}{nu^{n-1}} = \frac{u}{nu^n} = \frac{x^{1/n}}{nx} = \frac{1}{n}x^{(1/n)-1}. \tag{2.41}$$

Thus

$$(x^p)' = px^{p-1} \tag{2.42}$$

is valid also when p is a fraction of the form $p = 1/n$, which is sufficient for our applications later on.

Let us differentiate the square root of x, which is defined by

$$\sqrt{x} \equiv x^{1/2}.$$

Using the rule (2.42), we get

$$\left(\sqrt{x}\right)' = \left(x^{1/2}\right)' = \frac{1}{2}x^{-1/2} = \frac{1}{2\,x^{1/2}} = \frac{1}{2\,\sqrt{x}}.$$

2.7 Differentiation of fractions

In chapter 44we shall need to differentiate fractions of functions. By means of the product rule, the chain rule and the rule for differentiating power functions, we shall deduce the rule for differentiating such a fraction. Let $u(x)$ and $v(x)$ be functions of x. We first consider the function

$$y[v(x)] = 1/v = v^{-1}. \tag{2.43}$$

The chain rule, in the form (2.29) with u replaced by v, gives

$$y' = y'(v) \times v'. \tag{2.44}$$

From the rule (2.42) with $p = -1$ we get

$$y'(v) = (v^{-1})' = -v^{-2} = -\frac{1}{v^2}. \tag{2.45}$$

These equations lead to

$$\left(\frac{1}{v}\right)' = -\frac{1}{v^2}v'. \tag{2.46}$$

The product rule (2.22) with $f = u$ and $g = 1/v$ gives

$$\left(\frac{u}{v}\right)' = \left(u\frac{1}{v}\right)' = \frac{1}{v}u' + u\left(\frac{1}{v}\right)'. \tag{2.47}$$

Inserting Eq. (2.46) into the last term of Eq. (2.47) we find

$$\left(\frac{u}{v}\right)' = \frac{u'}{v} + u\left(-\frac{1}{v^2} \times v'\right) = \frac{u'}{v} - \frac{uv'}{v^2}. \tag{2.48}$$

Multiplying the first term at the far right-hand side by v in the numerator and the denominator, and putting the two terms on a common denominator, we finally arrive at

$$\left(\frac{u}{v}\right)' = \frac{u'v - uv'}{v^2}. \tag{2.49}$$

This is the rule for differentiating a fraction of two functions.

2.8 Functions of several variables

Geometrical pictures of functions of one variable, $y = f(x)$, are conveniently drawn as curves on an (x, y) plane. In this case the value of f is the distance (in the y direction) from the x axis to a point on the curve. But, inevitably, we have to proceed to functions of several variables. Functions with two variables, $z = f(x, y)$, can be illustrated on paper, but not so easily. They are pictured as surfaces in three-dimensional space, not curves. In this case the value of the function f is the height above the (x, y) plane of a point on the surface with coordinates (x, y, z) in the three-dimensional space (see Fig. 2.8 below).

In the above sections we have seen how the change of a function of one variable with position is described by the derivative of the function. The increase of a function $f(x, y)$ of *two* variables may be divided into two parts: The increase

$$\Delta_x f \equiv f(x + \Delta x, y) - f(x, y) \tag{2.50}$$

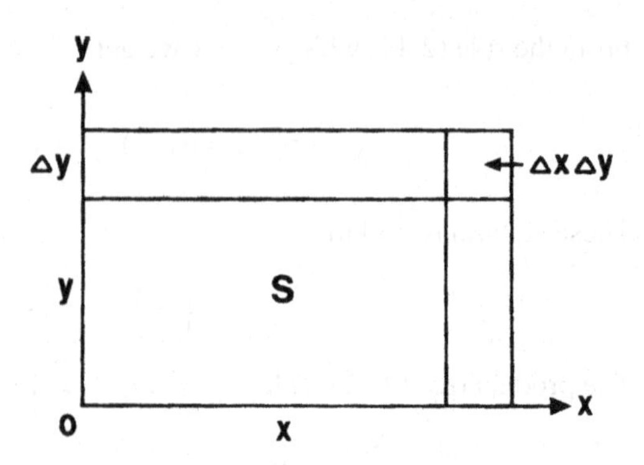

Fig. 2.7 A rectangular garden

that f gets by a displacement Δx along the x-axis, and the increase

$$\Delta_y f \equiv f(x, y + \Delta y) - f(x, y) \tag{2.51}$$

it gets by a displacement Δy along the y-axis. Mathematically such variations are described by what is called 'partial derivatives' of a function of several variables.

The partial derivatives of $f(x, y)$ with respect to x and y, respectively, are defined by

$$\begin{aligned} \frac{\partial f}{\partial x} \equiv \frac{\partial f(x, y)}{\partial x} &\equiv \lim_{\Delta x \to 0} \frac{f(x + \Delta x, y) - f(x, y)}{\Delta x} \\ &= \lim_{\Delta x \to 0} \frac{\Delta_x f}{\Delta x} \end{aligned} \tag{2.52}$$

and

$$\begin{aligned} \frac{\partial f}{\partial y} \equiv \frac{\partial f(x, y)}{\partial y} &\equiv \lim_{\Delta y \to 0} \frac{f(x, y + \Delta y) - f(x, y)}{\Delta y} \\ &= \lim_{\Delta y \to 0} \frac{\Delta_y f}{\Delta y}. \end{aligned} \tag{2.53}$$

The partial derivative of f with respect to x is calculated by differentiating f with respect to x while keeping y constant, and the partial derivative of f with respect to y is calculated by differentiating f with respect to y while keeping x constant.

We shall illustrate these new concepts by referring to an increase of the area of a rectangular garden. The Euclid-loving owner introduces a Cartesian coordinate system, as shown in Fig. 2.7. Let the function $f(x, y)$ represent the area of the garden,

$$f(x, y) = xy. \tag{2.54}$$

Let us consider two ways by which the owner may increase the area of the garden. Firstly he may extend it only in the x-direction; keeping y constant and increasing x by Δx. The corresponding increase of the area is given by Eq. (2.53),

$$\Delta_x f = (x + \Delta x) \times y - xy = xy + \Delta x \times y - xy = y\Delta x. \tag{2.55}$$

This is just the area of the column with width Δx and height y. Secondly he may extend the garden in the y-direction; keeping x constant and increasing y by Δy. Then the increase of area is

$$\Delta_y f = x(y + \Delta y) - xy = x\Delta y. \tag{2.56}$$

The partial derivatives of f are now found by dividing Eq. (2.50) by Δx and Eq. (2.51) by Δy, taking the limits $\Delta x \to 0$, $\Delta y \to 0$ and applying the definitions (2.52) and (2.53). The result is

$$\frac{\partial f}{\partial x} = y, \quad \frac{\partial f}{\partial y} = x \quad \text{with} \quad f(x, y) = xy. \tag{2.57}$$

The change of height of the surface by a small displacement in arbitrary direction, with component Δx along the x-axis and component Δy along the y-axis, is

$$\Delta f = f(x + \Delta x, y + \Delta y) - f(x, y). \tag{2.58}$$

Generalizing the definition (2.5) of the differential of a function of a single variable, we define the total differential of a function $f(x, y)$ of two variables, as

$$Df = \frac{\partial f}{\partial x}\Delta x + \frac{\partial f}{\partial y}\Delta y. \tag{2.59}$$

To first order in Δx and Δy there is no difference between the increment Δf and the differential Df of f. The total differential may therefore be used to calculate how the value of a function changes by small increments of the variables x and y. These increments are usually called *coordinate differentials*, and are denoted by dx and dy. With this notation the expression for a *total differential* takes the form

$$Df = \frac{\partial f}{\partial x}dx + \frac{\partial f}{\partial y}dy. \tag{2.60}$$

Using Einstein's summation convention as introduced in Eq. (1.26) in Ch. 1, this may be written

$$Df = \frac{\partial f}{\partial x^i}dx^i, \tag{2.61}$$

where $x^1 = x$ and $x^2 = y$ in the present case. (Remember that the numbers 1 and 2 are not exponents, but indices of coordinate axes.)

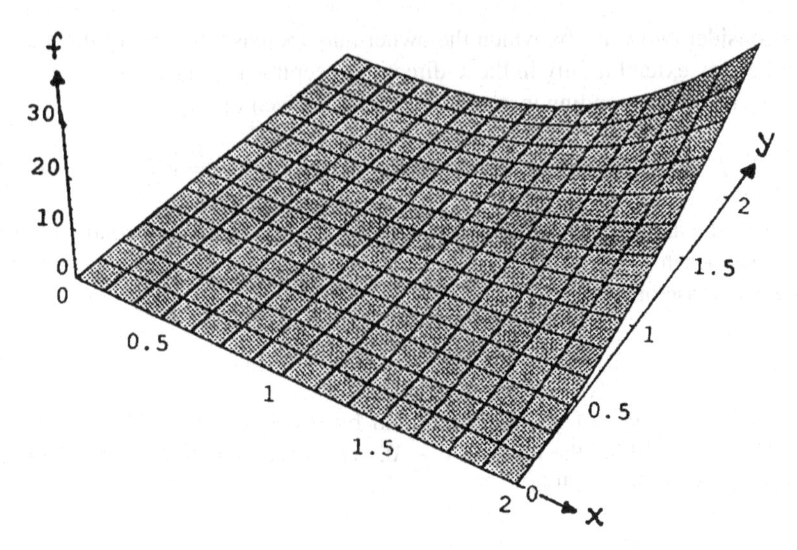

Fig. 2.8 The function $f(x, y) = x^3y^2$

Let us illustrate the concept 'total differential' by going back to the example with the garden. The finite increment of the area of the garden is

$$\begin{aligned} \Delta f &= (x + \Delta x)(y + \Delta y) - xy \\ &= xy + x\Delta y + y\Delta x + \Delta x\Delta y - xy \\ &= x\Delta y + y\Delta x + \Delta x\Delta y. \end{aligned} \tag{2.62}$$

Since Δx and Δy are small the product $\Delta x\Delta y$ is a 'small quantity of the second order'. In short, it is very, very small. When Δx and Δy are one to a million, $\Delta x\Delta y$ is one to a million millions. Geometrically it is the area of the small rectangle with sides Δx and Δy at the upper right-hand corner of the garden in Fig. 2.7. If we are interested in changes of the area to first order in the differentials, we can neglect products such as $\Delta x\Delta y$, which gives

$$Df = xdy + ydx. \tag{2.63}$$

Further examples:

Example 2.3. We consider a hill described by the function

$$f(x, y) = x^3y^2. \tag{2.64}$$

The surface of the hill is illustrated in Fig. 2.8.

Let us first differentiate f while y is kept constant. Then we get

$$\frac{\partial f}{\partial x} = 3x^2y^2.$$

Then we differentiate while x is kept constant,

$$\frac{\partial f}{\partial y} = x^3 2y = 2x^3 y.$$

Using Eq. (2.60) we get the total differential

$$Df = \frac{\partial f}{\partial x}dx + \frac{\partial f}{\partial y}dy = 3x^2y^2dx + 2x^3ydy.$$

Example 2.4. Let us, as a reasonably obvious generalization, consider an example with three variables.

$$g(x, y, z) = x^2y^3z.$$

Then

$$\frac{\partial g}{\partial x} = 2xy^3z, \quad \frac{\partial g}{\partial y} = 3x^2y^2z, \quad \frac{\partial g}{\partial z} = x^2y^3,$$

and

$$Dg = 2xy^3zdx + 3x^2y^2zdy + x^2y^3dz.$$

The above excessively complicated relations are found by easy, elegant manipulations of symbols. What does it mean, geometrically, or otherwise? "One cannot escape a feeling that these mathematical formulae have an independent existence and intelligence of their own, wiser than we are, ..." (Heinrich Hertz).

Einstein's field equations are 'second order partial differential equations' which contain partial derivatives of partial derivatives. Such second order partial derivatives are defined by

$$\frac{\partial^2 f}{\partial x^2} \equiv \frac{\partial}{\partial x}\left(\frac{\partial f}{\partial x}\right), \quad \frac{\partial^2 f}{\partial y \partial x} \equiv \frac{\partial}{\partial y}\left(\frac{\partial f}{\partial x}\right),$$
$$\frac{\partial^2 f}{\partial x \partial y} \equiv \frac{\partial}{\partial x}\left(\frac{\partial f}{\partial y}\right), \quad \frac{\partial^2 f}{\partial y^2} \equiv \frac{\partial}{\partial y}\left(\frac{\partial f}{\partial y}\right). \tag{2.65}$$

Just as the multiplication of 5 by 6 yields the same number as the multiplication of 6 by 5, successive differentiation with respect to x and then with respect to y, yields the same result as successive differentiation first with respect to y and then with respect to x. This is expressed mathematically by

$$\frac{\partial^2 f}{\partial y \partial x} = \frac{\partial^2 f}{\partial x \partial y}, \tag{2.66}$$

which means that different succession of differentiation does not affect the result. (The proof is of an advanced mathematical character and is not needed in our text.)

2.9 The MacLaurin and the Taylor series expansions

Most functions are more complicated and difficult to work with than power functions. It was therefore a very useful mathematical result, when the Scottish mathematician Colin MacLaurin (1698–1746) and the English mathematician Brook Taylor (1685–1731) made clear that most functions can be approximated by sums of power functions, called power series expansions. Such expansions will be applied in chapter 9 in our discussion of curvature.

Most anecdotes about the mathematical genius Carl Friedrich Gauss (1777–1855) are likely to require considerable mathematical knowledge, but one is rather innocent. Carl was a small boy when the tired teacher, hoping to get some rest, asked his pupils to for the sum of the first hundred numbers. To his annoyance Gauss practically at once raised his hand. He wrote from left to right $1, 2, 3, \ldots, 50$, then, on the next line, $100, 99, 98, \ldots, 51$, and on the third line he added the above, getting 50 equal sums $101, 101, 101, \ldots, 101$. Multiplying 101 with 50 he came up with the correct answer: 5050.

If 100 is replaced by an arbitrary natural number n, the formula Gauss found for the sum of the n first natural numbers, is

$$1 + 2 + 3 + \cdots + n = (1 + n)\frac{n}{2}. \tag{2.67}$$

Such a sum is called a *finite series*, meaning that it is a sum of a finite number of terms. When the number of terms increases indefinitely, the sum of this particular series does the same. The resulting *infinite series* is then said to *diverge*. There are, however, series with an infinitely large number of terms which have a definite sum. Even in the case that the sum is finite, in which case the series is said to *converge*, such a series is called an infinite series.

A particularly nice example of a convergent infinite series is $1+x+x^2+x^3+\cdots$ with $|x| < 1$. In this sum one gets the next term multplying the last one by x. Such a series with infinitely many terms is called an 'infinite geometrical series'. Note that for $|x| < 1$ each term is less than the foregoing. Let S_n denote the sum of the finite series,

$$S_n = 1 + x + x^2 + x^3 + \cdots + x^n. \tag{2.68}$$

Multiplying the left-hand side, and each term on the right-hand side, by x,

$$xS_n = x + x^2 + x^3 + \cdots + x^n + x^{n+1}. \tag{2.69}$$

Subtracting each side of (2.68) from each side of (2.69) leads to

$$S_n - xS_n = 1 - x^{n+1} \tag{2.70}$$

or

$$(1 - x)S_n = 1 - x^{n+1}. \tag{2.71}$$

Hence

$$S_n = \frac{1 - x^{n+1}}{1 - x}. \tag{2.72}$$

If $|x|<1$, each multiplication of x by itself gives a smaller number. Then $\lim_{n\to\infty} x^{n+1} = 0$, and the series with infinitely many terms has a finite sum equal to

$$S = \frac{1}{1 - x}. \tag{2.73}$$

The final result may be written as follows

$$1 + x + x^2 + x^3 + \cdots = \frac{1}{1 - x}, \quad |x| < 1. \tag{2.74}$$

Letting $x = 1/k$ we have an example of a convergent infinite series

$$1 + \frac{1}{k} + \frac{1}{k^2} + \cdots = \frac{1}{1 - 1/k}, \quad k > 1. \tag{2.75}$$

Here k is a real number. The left-hand side of Eq. (2.75) is meant to represent an infinite number of terms with values indicated by the first three ones that have been written down. The right-hand expression is the sum of all these terms. If $k = 2$, for example, the equation turns into

$$1 + \frac{1}{2} + \frac{1}{4} + \frac{1}{8} + \frac{1}{16} + \frac{1}{32} + \frac{1}{64} + \cdots$$
$$= \frac{1}{1 - 1/2} = \frac{1}{1/2} = 2. \tag{2.76}$$

Following MacLaurin and Taylor we now write a function $f(x)$ as a sum of power functions, i.e. as a *power series expansion*

$$f(x) = a_0 + a_1 x + a_2 x^2 + \cdots, \quad |x| < 1. \tag{2.77}$$

Here $a_0, a_1, a_2, \ldots$ are numbers which depend upon the function $f(x)$.

We shall now show how these numbers are determined by the values of the function and its derivatives at the point $x = 0$. Looking at Eq. (2.77) we see that if we insert the value $x = 0$ all terms on the right-hand side are equal to zero except the first one. This leads to

$$a_0 = f(0). \tag{2.78}$$

We now differentiate both sides of Eq. (2.77), using Eq. (2.42),

$$f'(x) = 1a_1 + 2a_2 x + 3a_3 x^2 + 4a_4 x^3 + \cdots. \tag{2.79}$$

Putting $x = 0$ we get

$$a_1 = f'(0). \tag{2.80}$$

The rest of the coefficients $a_2, a_3, \ldots$ can be determined in a similar way. Let us find the second and third derivatives of $f(x)$

$$f''(x) = 1 \times 2a_2 + 1 \times 2 \times 3a_3x + 1 \times 2 \times 3 \times 4a_4x^2 + \cdots . \tag{2.81}$$

$$f'''(x) = 1 \times 2 \times 3a_3 + 1 \times 2 \times 3 \times 4a_4x + \cdots . \tag{2.82}$$

Putting $x = 0$ in these equations give

$$a_2 = \frac{f''(0)}{1 \times 2}, \quad a_3 = \frac{f'''(0)}{1 \times 2 \times 3}. \tag{2.83}$$

Inserting these results in Eq. (2.77) we obtain a nice formula

$$f(x) = f(0) + \frac{f'(0)}{1}x + \frac{f''(0)}{1 \times 2}x^2 + \frac{f'''(0)}{1 \times 2 \times 3}x^3 + \cdots . \tag{2.84}$$

If we continue the differentiation we can get any of the coefficients a_i expressed through derivatives of $f(x)$ at $x = 0$. The resultant infinite series is called the *MacLaurin series.*

We shall consider, as an illustration, an example that shall be employed in chapter 9; the MacLaurin series of the function $f(x) = 1/\sqrt{1-x^2} = (1-x^2)^{-1/2}$. In order to differentiate this function we write it as a composite function, $f[g(x)]$, where $f(g) = g^{-1/2}$, and $g(x) = 1 - x^2$. From the rule (2.42) with $p = -1/2$, and the chain rule for differentiating composite functions, we deduce

$$\begin{aligned} f'(x) &= f'(g) \times g'(x) = -\left(\frac{1}{2}\right) g^{-3/2}(-2x) \\ &= x\left(1 - x^2\right)^{-3/2}. \end{aligned}$$

Differentiating once more, first using the product rule, and then the rule (2.42) with $p = -3/2$, we get

$$\begin{aligned} f''(x) &= \left(1 - x^2\right)^{-3/2} + x\left(-\frac{3}{2}\right)\left(1 - x^2\right)^{-5/2}(-2x) \\ &= \left(1 - x^2\right)^{-3/2} + 3x^2\left(1 - x^2\right)^{-5/2}. \end{aligned}$$

From these expressions follow

$$f(0) = 1, \quad f'(0) = 0, \quad f''(0) = 1.$$

Substituting this into Eq. (2.84) gives

$$\frac{1}{\sqrt{1-x^2}} = 1 + \frac{1}{2}x^2 + \cdots . \tag{2.85}$$

The third term of the series, which we have not written down, is proportional to x^4, the next one to x^6, and so on. For small x, say x smaller that 0.01, $x^4 < (0.01)^4 = 0.00000001$, showing that these higher order terms are then so small as to be negligible. For such small values of x we can use the approximation

$$\frac{1}{\sqrt{1-x^2}} \approx 1 + \frac{1}{2}x^2. \tag{2.86}$$

This approximate expression is useful because the right-hand side is easier to handle mathematically.

A second example is the MacLaurin series of the function $f(x) = \sqrt{1-x} = (1-x)^{1/2}$. Differentiation gives $f'(x) = (1/2)(1-x)^{-1/2}(-x') = (-1/2)(1-x)^{-1/2}$. Thus $f(0) = 1$ and $f'(0) = -1/2$, which gives

$$\sqrt{1-x} = 1 - \frac{1}{2}x + \cdots . \tag{2.87}$$

For small values of x we can apply the approximation

$$\sqrt{1-x^2} \approx 1 - \frac{1}{2}x^2. \tag{2.88}$$

The MacLaurin series is a series expansion of a function at the point $x = 0$, that converges for $|x| < 1$. This means that the value of the function at an arbitrary point inside the interval $-1 < x < 1$, can be found by adding terms with the values of the function and its derivatives calculated at the point $x = 0$. If, on the other hand, x is near a point outside this interval, for example $x = 2$, the MacLaurin series is of no use. Then one needs a generalization of the MacLaurin series. One needs an expansion about an arbitrary point, say $x = a$. This is called the *Taylor series*, and will be used later in our road to Einstein's field equations.

In order to arrive at the Taylor series, we introduce a function $F(x)$ defined by

$$F(x) = f(x_0 + x) \tag{2.89}$$

where x_0 is a fixed value of x. Differentiation gives

$$\begin{aligned} F'(x) &= f'(x_0 + x), \quad F''(x) = f''(x_0 + x), \\ F'''(x) &= f'''(x_0 + x), \quad \cdots . \end{aligned} \tag{2.90}$$

Inserting $x = 0$ leads to

$$F(0) = f(x_0), \quad F'(0) = f'(x_0), \quad F''(0) = f''(x_0),$$
$$F'''(0) = f'''(x_0), \quad \cdots. \tag{2.91}$$

The MacLaurin series for the function $F(x)$ is

$$F(x) = F(0) + \frac{F'(0)}{1}x + \frac{F''(0)}{1 \times 2}x^2 + \frac{F'''(0)}{1 \times 2 \times 3}x^3 + \cdots.$$

Substituting from Eq. (2.90), we get

$$f(x_0 + x) = f(x_0) + \frac{f'(x_0)}{1}x + \frac{f''(x_0)}{1 \times 2}x^2 + \frac{f'''(x_0)}{1 \times 2 \times 3}x^3 + \cdots.$$

Adding infinitely many terms we get the Taylor series.

Substituting a coordinate increment Δx for x, the corresponding increment $\Delta f = f(x_0 + \Delta x) - f(x_0)$ of the function f is

$$\Delta f = f'(x_0)\Delta x + \frac{1}{2}f''(x_0)(\Delta x)^2 + \frac{1}{6}f'''(x_0)(\Delta x)^3 + \cdots. \tag{2.92}$$

According to Eq. (2.5) the first term at the right-hand side is just the differential Df of f. Hence

$$\Delta f - Df = \frac{1}{2}f''(x_0)(\Delta x)^2 + \frac{1}{6}f'''(x_0)(\Delta x)^3 + \cdots \tag{2.93}$$

where the terms we have not written are of higher order in Δx.

Chapter 3
Tangent vectors

Curves of particular interest in physics are those representing paths of moving particles. Such curves may be described by giving the coordinates of the particle as functions of time. In the following we shall consider the path of a particle thrown horizontally, as an illustrating example (see Fig. 3.1).

3.1 Parametric description of curves

In chapter 2 (see Fig. 2.4) we became familiar with the parabola, given by

$$y = x^2. \tag{3.1}$$

Consider a similar parabola, curving downwards instead of upwards, see Fig. 3.1.

The equation of this parabola is

$$y = -x^2. \tag{3.2}$$

The path of a free-falling particle moving from the origin of the coordinate system with a horizontal initial velocity, has the shape of such a parabola.

In order to describe, and understand, the motion of particles, it is useful to describe the components of the motion in the x direction and the y direction separately. The curve followed by the particle is then described by giving, not y as a function of x, but by giving both x and y as functions of a variable t,

$$x = x(t), \quad y = y(t). \tag{3.3}$$

The variable t is termed a *parameter*. Describing a curve in this way, we give a 'parametric representation' of the curve.

Ø. Grøn and A. Næss, *Einstein's Theory: A Rigorous Introduction for the Mathematically Untrained*, DOI 10.1007/978-1-4614-0706-5_3,

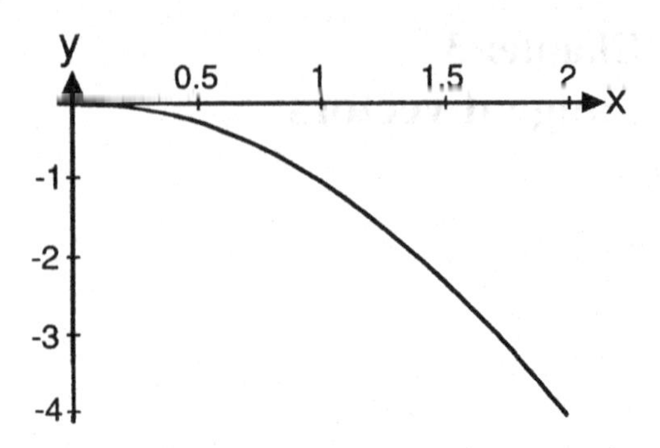

Fig. 3.1 Parabola

In the case of our particle moving horizontally away from the origin, the x and y coordinates can be given as functions of the parameter t as

$$x = t, \quad y = -t^2 \tag{3.4}$$

which represents the parabola given in Eq. (3.2).

The coordinates x and y tell the position, in the (x, y) plane, of points on the curve, and thus determine its shape. The parameter t may be thought of as a kind of coordinate along the curve. The value of t indicates where *on the curve* a certain point is. To every point on the curve there corresponds a certain value of t, and all points on the curve have different values of t.

A few more examples will make you a little more familiar with the 'parametric representation of curves'.

Consider a certain curve, given by the parametric form

$$x = t, \quad y = \sqrt{1 - t^2}. \tag{3.5}$$

In order to find the shape of the curve we can substitute different values of t, calculate the corresponding values of x and y, and then draw the curve through the points that have been found. Alternatively we may eliminate t from the parametric equations and thereby find an equation between x and y, from which one can identify the curve (if it is reasonably simple). Since

$$x^2 = t^2, \quad y^2 = 1 - t^2, \tag{3.6}$$

the curve represented by Eq. (3.5) has the equation

$$x^2 + y^2 = 1. \tag{3.7}$$

This equation describes a circle with centre at the origin and radius 1 (see Fig. 3.2). From the figure is seen that Eq. (3.7) is just an expression of the Pythagorean theorem. For an arbitrary curve the distance from the origin to a point on the curve, $r = \sqrt{x^2 + y^2}$, depends upon the position of the point. However, in the case of a

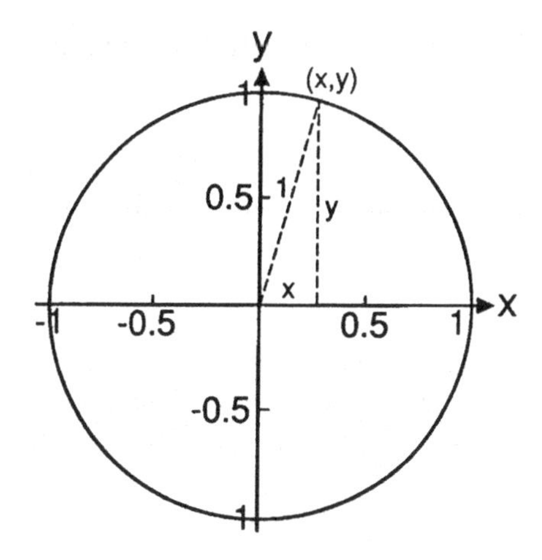

Fig. 3.2 Circle with unit radius centered at the origin

circle this distance is the same for all points on the curve, equal to the radius of the circle. In Fig. 3.2 we have drawn a circle with radius $r = 1$.

Subtracting x^2 from both sides of Eq. (3.7) gives

$$y^2 = 1 - x^2. \tag{3.8}$$

Taking the positive square root we find y as a function of x

$$y = \sqrt{1 - x^2}. \tag{3.9}$$

The graph of this function is only the upper half of the circle.

A whole circle cannot be represented as the graph of a single function. This is due to the fact that to every value of the variable x there corresponds only one value of the function y. So a graph of a single function cannot have two points with the same value of x and different values of y. However, by giving x and y as functions of a parameter, as in Eq. (3.6), one obtains a representation of the whole of an arbitrary curve, whether it is open or closed or even when it intersects itself.

Another example: A curve is given by the parametric representation

$$x = 7t - 3, \quad y = -49t^2 + 42t - 9. \tag{3.10}$$

Identify the curve! There is something suspicious about the expression for y: $49 = 7^2$, $9 = 3^2$, $42 = 2 \cdot 7 \cdot 3$. The expression for y is quadratic; $y = -(7t - 3)^2$. (Remember the rule $(a + b)^2 = a^2 + 2 \cdot a \cdot b + b^2$. Here $a = 7t$ and $b = -3$.) Eq. (3.10) implies that $x^2 = (7t - 3)^2$. Thus

$$y = -x^2,$$

which is again the parabola of Eq. (3.2). Evidently one and the same curve can be represented in infinitely many ways.

3.2 Parametrization of a straight line

A straight line L can be identified by giving its direction and a point P_0 that it passes through. Let the point P_0 have Cartesian coordinates (x_0, y_0, z_0), and let a certain vector, $\vec{v}$, specify the direction of the line (compare Eq. (1.15))

$$\vec{v} = v^x \vec{e}_x + v^y \vec{e}_y + v^z \vec{e}_z. \tag{3.11}$$

This is illustrated in Fig. 3.3. A vector from the origin to a point is called the *position vector* of the point. The position vector of P_0, and an arbitrary point P on the line L with coordinates (x, y, z) are, respectively

$$\vec{r}_0 = x_0 \vec{e}_x + y_0 \vec{e}_y + z_0 \vec{e}_z, \quad \vec{r} = x \vec{e}_x + y \vec{e}_y + z \vec{e}_z. \tag{3.12}$$

The vector from $P_0$to P is $t\vec{v}$. Here t is a scalar quantity which determines how far the point P is removed from P_0. The quantity t is the curve parameter. From Fig. 3.3 and the rule for vector addition (see Fig. 1.8) we get

$$\vec{r} = \vec{r}_0 + t\vec{v}. \tag{3.13}$$

Substitution from Eqs. (3.11) and (3.12) gives

$$\begin{aligned} \vec{r} &= x \vec{e}_x + y \vec{e}_y + z \vec{e}_z \\ &= x_0 \vec{e}_x + y_0 \vec{e}_y + z_0 \vec{e}_z + t v^x \vec{e}_x + t v^y \vec{e}_y + t v^z \vec{e}_z \\ &= (x_0 + t v^x) \vec{e}_x + (y_0 + t v^y) \vec{e}_y + (z_0 + t v^z) \vec{e}_z. \end{aligned} \tag{3.14}$$

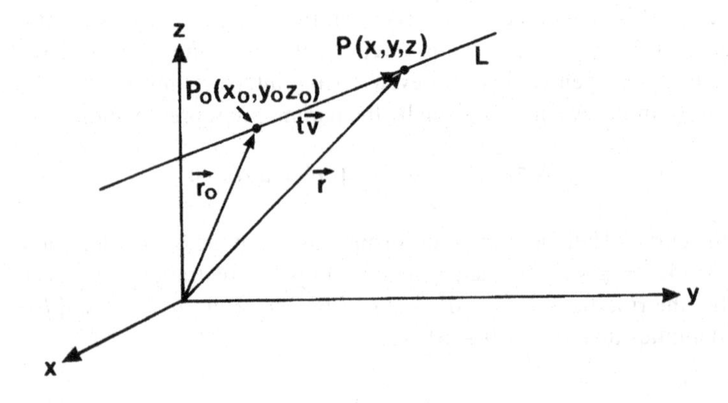

Fig. 3.3 Parametrisation of a line

This is the component form of Eq. (3.13) in three dimensions. If two vectors are equal, then all their components must be equal. Thus, in a three-dimensional space the vector equation (3.13) leads to the three scalar equations

$$x = x_0 + tv^x, \quad y = y_0 + tv^y, \quad z = z_0 + tv^z. \tag{3.15}$$

These are the equations of a straight line in parametric form. The usefulness of this form is easy to see. Think of the line L in Fig. 3.3 as the path of a particle moving without being acted upon by any forces. Then t is the time, and the particle passes P_0 at the time $t = 0$ and P at time t.

Let us for a moment go back to Eq. (3.13). It is a marvelously simple equation, especially when you realize that it is valid for lines in spaces of any dimension; it holds for vectors in 5 or 55 dimensions (see Ch. 9) as elegantly as on a plane. However, the component form of the vector equation for 55 dimensions is, if not complicated, at least rather long. On the right-hand side of the equivalent of Eq. (3.15) you find a sum of 55 terms.

3.3 Tangent vector fields

In section 1.3 we introduced the term 'vector field' representing the wind in a region, the air above Western Europe. Roughly a vector field is a continuum of vectors filling a region. The vectors representing the wind velocity was implicitly assumed to exist in a flat three-dimensional region. However, the spacetime of general relativity is curved. We need to become acquainted with vectors in curved spaces. A problem immediately appears. Vectors are drawn as arrows. And these arrows are straight in the usual Euclidean sense. They are usually drawn in a flat (x, y) plane.

Imagine instead that the vectors are drawn on a spherical surface. The expression 'two-dimensional curved space' means just such curved surfaces. If a vector-arrow could exist in such a curved two-dimensional space, it would have to curve in order not to point out of the space, at least if the vector arrow has a finite length. However, curved vector arrows are not allowed! The conclusion is that *vectors of finite length do not exist in curved spaces.* (Infinitely short vectors, however, are allowed as far as their departure from the curved space can be neglected.)

The simplest example of a curved space is a space of only one dimension: a curve. We cannot define finite vectors on a curve. But look at Fig. 2.2. There a curve is generated by a succession of tangent lines. Each of the tangent lines are straight. Such a line is called the *tangent space* of the curve at a point. The tangent space is flat. Even if we cannot define vectors *in* the one-dimensional space defined by the curve, we can do something which will turn out to be very useful: we can define a succession of tangent vectors along the curve. They can be drawn as straight arrows along the tangent lines. If we define a tangent vector at every point of the curve, we have what is a called a *tangent vector field* in a one-dimensional curved space. The vectors of the tangent vector field exist in a succession of different tangent spaces along the curve.

These concepts are readily generalized to spaces of higher dimensions. For example the tangent space of the surface of the Earth at the North Pole is a plane touching the North Pole. Now, think of a vector representing the velocity of the drifting ice at the North Pole. Imagine that no vertical extension exists. The world is just the two-dimensional surface of the Earth. And the ice exists on this surface. The velocity vector of the ice, however, exists in the tangent plane, i.e. in the flat two-dimensional tangent space to the surface of the Earth at the North Pole.

In curved four-dimensional spacetime one can define vectors of finite length at each point. They exist in the flat four-dimensional tangent spacetime at each point. We shall need these concepts in Chapter 9 where curvature is discussed. Now we go back to our simple one-dimensional space: *the curve*.

Let a curve be given parametrically by $x = x(t)$, $y = y(t)$. The tangent-vector $\vec{u}$ of the curve at a point P is an arrow with inital point at P and directed along the curve in the direction of increasing values of the parameter t.

In any dimension and in any coordinate system the components of $\vec{u}$ are defined by

$$u^i \equiv \frac{dx^i}{dt}. \tag{3.16}$$

Here i refers to the i'th component. Thus

$$\vec{u} = u^i \vec{e}_i = \frac{dx^i}{dt}\vec{e}_i. \tag{3.17}$$

In the case that a curve lies in a plane the tangent vector $\vec{u}$ may be decomposed in a two-dimensional Cartesian coordinate system. Then Eq. (3.16) represents the two component equations

$$u^x = \frac{dx}{dt} \quad \text{and} \quad u^y = \frac{dy}{dt}, \tag{3.18}$$

and Eq. (3.17) reduces to

$$\vec{u} = u^x \vec{e}_x + u^y \vec{e}_y = \frac{dx}{dt}\vec{e}_x + \frac{dy}{dt}\vec{e}_y. \tag{3.19}$$

Consider again the parabola given by equation (3.2). In this case

$$u^x = \frac{dx}{dt} = 1 \quad \text{and} \quad u^y = \frac{dy}{dt} = -2t, \tag{3.20}$$

showing that the tangent vector field of the parabola is

$$\vec{u} = \vec{e}_x - 2t\vec{e}_y. \tag{3.21}$$

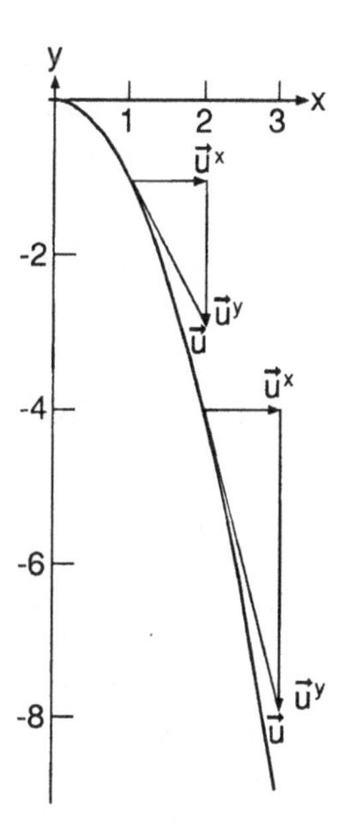

Fig. 3.4 The tangent vectors at two different points on the parabola

Since $t = x$ according to equation (3.2), we get

$$\vec{u} = \vec{e}_x - 2x\vec{e}_y. \tag{3.22}$$

Thus the x-component of this vector field is constant, and the y-component is proportional to the distance from the origin along the x-axis. This is illustrated in Fig. 3.4. The tangent vector *field* is the infinite manifold of vectors with initial points along the curve. The initial points trace completely the curve.

Here $u^x = 1$, $u^y = -1$ at $x = 1$ and $u^x = 1$, $u^y = -2$ at $x = 2$.

As a second example consider a straight line,

$$x = at + b \quad \text{and} \quad y = ct + d. \tag{3.23}$$

The line has a constant tangent vector field

$$\vec{u} = a\vec{e}_x + c\vec{e}_y. \tag{3.24}$$

The components of a tangent vector are defined, as said above, as $u^x = dx/dt$ and $u^y = dy/dt$ (when the curve is in a two-dimensional space). You can therefore use your old knowledge of differentiation when you are going to calculate the components of a tangent vector.

Tangents that we learned about in school had no definite length. We drew them suitably long on our paper, and were told that mathematically they would go on indefinitely. A decisive character of a tangent vector is its finite length, u. In Fig. 3.4 we see that the length or magnitude is that of a hypothenus of a right angled triangle.

This makes the calculation of its length an easy undertaking: $u^2 = (u^x)^2 + (u^y)^2$. Using Eq. (3.20) we get $u^2 = 1^2 + (-2t)^2 = 1 + 4t^2$ or $u = \sqrt{1+4t^2}$. For $t = 0$ we get $u = 1$. That is, having chosen 1 as the magnitude of $\vec{e}_x$, the tangent vector with root at point $(0, 0)$, goes along the x axis from 0 to 1. At the point with $t = 5$ the tangent-vector is already fairly long, $\sqrt{1^2 + 4 \cdot 5^2} = \sqrt{101} \approx 10$ units, and like the curve itself approximating the vertical slope.

3.4 Differential equations and Newton's 2. law

Curves representing paths of moving particles are of particular interest in physics. According to classical dynamics these curves are fixed if you know the positions and velocities of the particles at any moment and the forces acting upon them.

Isaac Newton (1642–1727) postulated that the acceleration of a particle is proportional to the force acting upon it. (The relativistic generalization of this will be considered in chapter 11.) This is *Newton's second law*. It is the most fundamental equation of Newtonian dynamics, and provides us with *the Newtonian equations of motion* of the particles.

Let us consider motion in one dimension, say along the x-axis, in order to simplify the mathematics. *Velocity* is defined as rate of change of position with time. In section 2.1 we noted that the derivative of a function y with respect to a variable x represents the rate of change of y with x. Now, the position x of the particle is a function of the time, t. Thus, the velocity is the derivative of x with respect to t

$$v = \frac{dx}{dt}. \tag{3.25}$$

Acceleration is defined as the rate of change of velocity with time,

$$a = \frac{dv}{dt}. \tag{3.26}$$

It follows that the acceleration is the second derivative of the position with respect to time

$$a = \frac{d^2x}{dt^2}. \tag{3.27}$$

Newton's second law says that the force F acting upon a particle with mass m is equal to its mass times its acceleration,

$$F = m\frac{d^2x}{dt^2}. \tag{3.28}$$

The corresponding vector equation is

$$\vec{F} = m\vec{a}. \tag{3.29}$$

The forces acting upon a particle may depend both on the position of the particle and upon time. We shall here consider forces that depend upon the position, but that do not change with time. Then the position of the particle will appear in the expression of the force on the left-hand side of Eq. (3.28), and derivatives of the position on the right-hand side. Equations containing both a function and derivatives of the function, are called *differential equations*. If the equation contains only the first derivative, it is called a first order differential equation. If it contains second derivatives, it is called a second order differential equation, and so forth.

If the expression of a certain force, acting upon a particle, is inserted in Newton's second law the resulting equation is a second order differential equation, which may be said to 'determine' the motion of the particle.

3.5 Integration

Having found the equation of motion of a particle, one wants to solve it in order to find the position of the particle as a function of time. Solving a differential equation one needs to perform a mathematical operation which is the 'opposite' of differentiation. One might say that one needs to 'antidifferentiate' an expression. This operation, which is called *integration*, will now be briefly approached in a highly simplified manner.

The antiderivative, $F(x)$, of a function $y = f(x)$ is defined as that function whose derivative is equal to $f(x)$,

$$\frac{dF}{dx} = f(x). \tag{3.30}$$

The antiderivative is expressed by a special sign

$$F(x) = \int y dx = \int f(x) dx \tag{3.31}$$

and is called the *indefinite integral* of the function $f(x)$.

Let us consider an example. We know from Eq. (2.42) that the derivative of x^p is equal to px^{p-1}, or increasing the exponent by 1,

$$\left(x^{p+1}\right)' = (p+1)x^p. \tag{3.32}$$

Dividing by $p+1$ we get

$$\left(\frac{1}{p+1}x^{p+1}\right)' = x^p \quad \text{with} \quad p \neq -1. \tag{3.33}$$

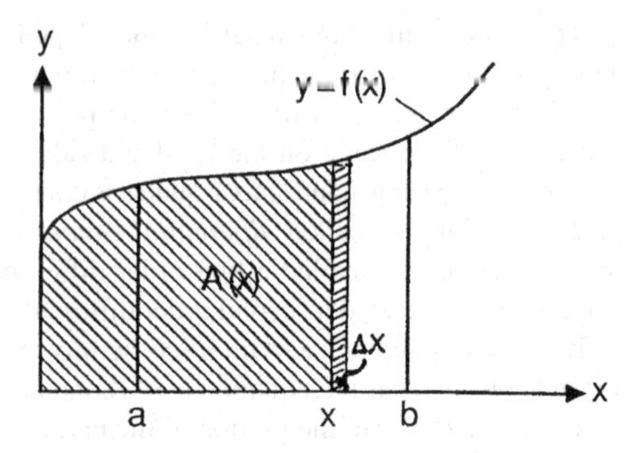

Fig. 3.5 Geometrical interpretation of integration

Hence, $x^{p+1}/(p+1)$ is that function whose derivative is x^p. Because the derivative of a constant is zero, we can add a constant C to the function. Hence, we have calculated the whole class of integrals!

$$\int x^p dx = \frac{1}{p+1} x^{p+1} + C. \tag{3.34}$$

Putting for example $p = 2$, we get

$$\int x^2 dx = \frac{1}{3} x^3 + C, \tag{3.35}$$

where C is called a 'constant of integration'.

We shall now give a geometrical interpretation of the integral. Consider the graph of a function $y = f(x)$ as shown in Fig. 3.5. Let us try to calculate the area of the region bounded by the graph, the x-axis, and the vertical lines $x = a$ and $x = b$. In general the shape of the graph is curved, so this is indeed not a trivial task.

The first known solution to this problem is due to Pierre de Fermat (1601–1665). Newton and Leibniz managed to construct a general method, independently of each other, about three hundred years ago. Their ingenious construction made clear that there is an intimate connection between the mathematical representation of an area and the concept of an integral perceived as an antiderivative.

We start by defining an area function $A(x)$, which represents the hached area in Fig. 3.5 bounded by the y and the x axis, the graph, and a vertical line with an arbitrary x-coordinate. Clearly, $A(x)$ grows when x increases. Increasing x by a small value Δx, A will increase by a small value ΔA. Consider the small column drawn from the x-axis to the graph. The left side of the column has position x and the right side has position $x + \Delta x$. Its width is Δx. The height of the left side of the column is $y = f(x)$, and of the right side $y + \Delta y$. If we cut the top of the column by a horizontal line where its left side meets the parabola, it will be shaped like a rectangle with width Δx and height $f(x)$. The area of this rectangle is $\Delta A = f(x)\Delta x$, neglecting a small triangular area on the top of the column. Dividing by Δx on both sides and taking the limit $\Delta x \to 0$, we get

$$\lim_{\Delta x \to 0} \frac{\Delta A}{\Delta x} = f(x). \tag{3.36}$$

Comparing with the definition (2.4) we see that the area function $A(x)$ has $f(x)$ as its derivative. This means that the area function is the antiderivative, or the integral, of $f(x)$, which was the great result of Newton and Leibniz. Looking at the figure it it also clear that the area function vanishes at $x = 0$, i.e. $A(0) = 0$. Thus we can define the area function formally by

$$A(x) \equiv F(x) - F(0), \tag{3.37}$$

where $F(x)$ is the antiderivative of $f(x)$ as given in Eq. (3.31).

The area, A_{ab}, between a and b is equal to the area from the y-axis to b minus the area from the y-axis to a. Hence,

$$A_{ab} = A(b) - A(a). \tag{3.38}$$

Inserting the right-hand side of the definition (3.37) leads to

$$\begin{aligned} A_{ab} &= F(b) - F(0) - [F(a) - F(0)] \\ &= F(b) - F(0) - F(a) + F(0) = F(b) - F(a). \end{aligned} \tag{3.39}$$

The area between a and b is written

$$A_{ab} = \int_a^b f(x)dx \tag{3.40}$$

and is said to be the integral of $f(x)$ from a to b. It is called a *definite integral*. We now have the following formula for calculating a definite integral

$$\int_a^b f(x)dx = F(b) - F(a) \tag{3.41}$$

which is a fundamental equation in the calculus of integration. The right-hand side is calculated by first inserting the number at the top of the integral sign, called the upper integration limit, in the antiderivative, and then subtracting the value obtained when the lower integration limit is inserted into the antiderivative.

Let us consider an example, the area of the region enclosed by the parable $y = x^2$, the x-axis and the vertical lines $x = 1$ and $x = 3$, as shown in Fig. 3.6. According to Eq. (3.40) this area is given by the definite integral

$$A_{13} = \int_1^3 x^2 dx. \tag{3.42}$$

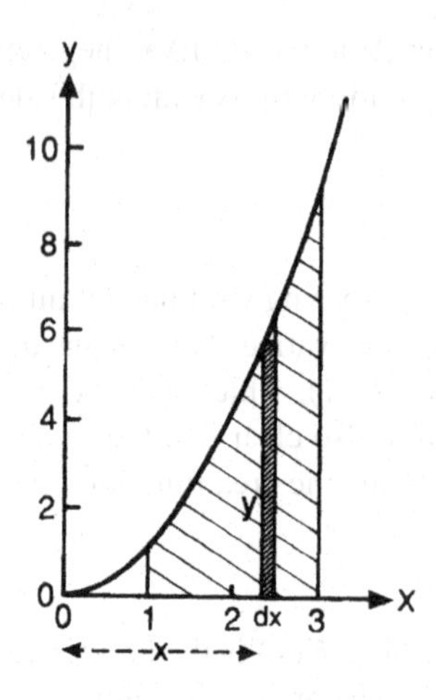

Fig. 3.6 The integral of the parable $y = x^2$ from 1 to 3

First we find the indefinite integral

$$F(x) = \int x^2 dx = \frac{1}{3}x^3 + C. \tag{3.43}$$

We may omit the constant C because it will be cancelled in the subtraction. Hence,

$$A_{13} = F(3) - F(1) = \frac{1}{3}3^3 - \frac{1}{3}1^3 = \frac{26}{3}.$$

We have developed the concept of a definite integral as an area. That is, however, not necessary. Integrals, definite as well as indefinite, are of great use in practically all fields where mathematics is applied. It may be used, for instance, to calculate the distance a planet moves along its elliptical path and with varying velocity. In such applications the integral is not something given. One has to deduce it. Then one first calculates a very small change of the quantity one is to calculate. This corresponds to calculating the area of the small column of Fig. 3.6, $dA = ydx = x^2dx$. Having obtained this the calculation proceeds by applying Eq. (3.41).

We now have a powerful mathematical tool which we can use to find the path of a particle from a knowledge of the forces acting upon it, and the position and velocity at a certain moment.

3.6 The exponential and logarithmic functions

The power function $y = x^r$ of a positive real variable x, where r is a real number, is a function satisfying the following three rules

$$x^p \cdot x^r = x^{p+r} \quad \text{and} \quad \frac{x^p}{x^r} = x^{p-r}, \tag{3.44}$$

$$(x \cdot z)^r = x^r \cdot z^r, \tag{3.45}$$

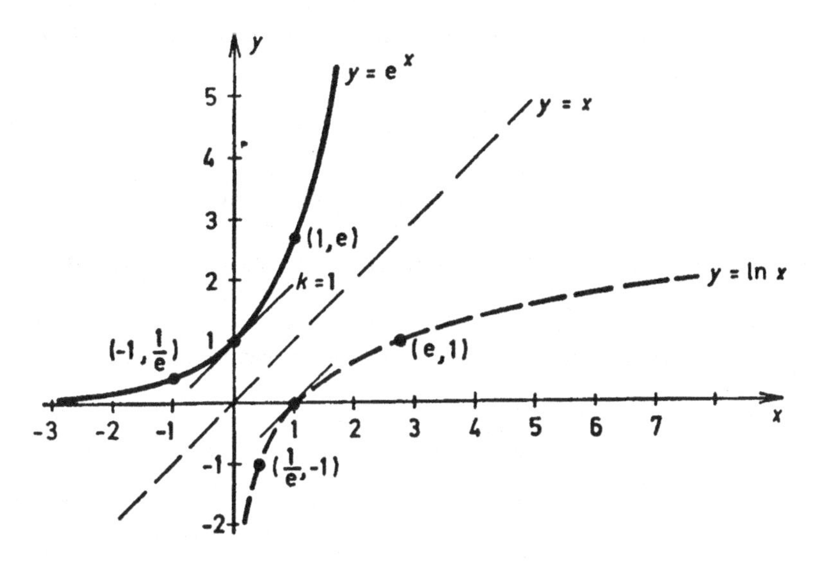

Fig. 3.7 A function and its inverse

and

$$(x^p)^r = x^{p \cdot r} \quad \text{and} \quad (x^p)^{\frac{1}{r}} = x^{\frac{p}{r}}. \tag{3.46}$$

The last equation in (3.46) is the rule for taking the r'th root of x^p. Note that the second of Eqs. (3.44) with $p = r$ gives $x^0 = 1$, and that the same equation with $p = 0$ gives $1/x^r = x^{-r}$.

The *power function* has the variable x as the base and a constant exponent. The *exponential function* on the other hand, is defined as a function with a constant base and the variable x as exponent, $y = a^x$.

The *inverse of a function* $y = f(x)$ is defined as the function obtained by exchanging x and y in the expression for $f(x)$, and then solving the resulting equation with respect to y. The graph of a function and the graph of its inverse are symmetrical about the line $y = x$, as shown in Fig. 3.7.

A new function, the so-called natural logarithm, was introduced by the Swiss mathematician Leonard Euler (1707–1783). It was a great and useful mathematical invention, based on Euler's discovery of a particular number, e, defined by

$$\mathrm{e} \equiv \lim_{n \to \infty} \left(1 + \frac{1}{n}\right)^n. \tag{3.47}$$

Inserting large values of n this number can be calculated as accurately as one wants. To fifteen decimal places, for example, the value is

$$\mathrm{e} = 2.718281828459045.$$

The natural logarithm is denoted by $\ln x$ and is defined as the inverse function of the exponential function e^x, i.e.

$$y = \ln x \Leftrightarrow x = e^y. \tag{3.48}$$

Since $\mathrm{e}^0 = 1$, it follows that $\ln 1 = 0$. Also, since $x = \mathrm{e}$ for $y = 1$, we have $\ln \mathrm{e} = 1$.

We shall now deduce the most important rules for calculating with the natural logarithm. Let $x = \ln a$ and $z = \ln b$. Then $a = \mathrm{e}^x$ and $b = \mathrm{e}^z$. Using the rule (3.44) we get $ab = \mathrm{e}^x \mathrm{e}^z = \mathrm{e}^{x+z}$. According to Eq. (3.48) this is equivalent to $\ln(ab) = x + z$. Inserting the expressions for x and z we have the rule

$$\ln(ab) = \ln a + \ln b. \tag{3.49}$$

Similarly, since $a/b = \mathrm{e}^x/\mathrm{e}^z = \mathrm{e}^{x-z}$,

$$\ln\left(\frac{a}{b}\right) = \ln a - \ln b. \tag{3.50}$$

Let now $a = \ln x^r$. This is equivalent to $\mathrm{e}^a = x^r$. Taking the r'th root of each side, using the second rule in (3.46), we obtain $x = \mathrm{e}^{a/r}$, which is equivalent to $\ln x = a/r$. Multiplying each side of this equation by r we have $a = r \ln x$. From the two expressions for a we obtain the rule

$$\ln x^r = r \ln x. \tag{3.51}$$

We shall now calculate the derivative of the natural logarithm. Let $y = \ln x$. An increment Δx in x produces an increment Δy in y, given by

$$\Delta y = \ln(x + \Delta x) - \ln x = \ln \frac{x + \Delta x}{x} = \ln\left(1 + \frac{\Delta x}{x}\right), \tag{3.52}$$

where we have used the rule (3.50). The derivative is given by the limiting value of this expression as $\Delta x \to 0$. In order to simplify the calculation, we introduce a number n by $\Delta x/x = 1/n$, so that $\Delta x \to 0$ corresponds to $n \to \infty$. By means of Eq. (3.52) we then get for the derivative of the natural logarithm

$$\begin{aligned} y' = \frac{dy}{dx} &= \lim_{\Delta x \to 0} \frac{\Delta y}{\Delta x} = \lim_{\Delta x \to 0} \frac{1}{\Delta x} \ln\left(1 + \frac{\Delta x}{x}\right) \\ &= \lim_{n \to \infty} \frac{n}{x} \ln\left(1 + \frac{1}{n}\right) = \frac{1}{x} \lim_{n \to \infty} \ln\left(1 + \frac{1}{n}\right)^n, \end{aligned} \tag{3.53}$$

where we have used the rule (3.51) in the last step. Using Eq. (3.47) we get

$$y' = \frac{1}{x} \ln \mathrm{e}.$$

Hence, since $\ln \mathrm{e} = 1$, the derivative of the natural logarithm is

$$(\ln x)' = \frac{1}{x}. \tag{3.54}$$

Let $u(x)$ be a function of x. Using Eq. (3.54) and the chain rule for differentiation, we get

$$(\ln u)' = \frac{1}{u}u' = \frac{u'}{u}. \tag{3.55}$$

Since 'integration is the opposite of differentiation', we have the integral

$$\int \frac{1}{x}dx = \ln x + C. \tag{3.56}$$

This completes Eq. (3.33) for $p = -1$. Note also, from Eq. (3.55) that

$$\int \frac{u'}{u}dx = \ln u + C, \tag{3.57}$$

for an arbitrary function u of x. In words: the integral of a fraction in which the numerator is the derivative of the denominator, is equal to the natural logarithm of the denominator.

Let us consider an example. We shall find the integral of the function $2x/(x^2+1)$. Then we write $u = x^2 + 1$ and get $u' = 2x$. Thus, we have

$$\int \frac{2x}{x^2+1}dx = \int \frac{u'}{u}dx = \ln u + C = \ln(x^2+1) + C.$$

We shall now find the derivative of the exponential function $y = a^x$. This is most easily done by first taking the (natural) logarithm of each side, and then differentiating. Applying at first the rule (3.51) we get $\ln y = x \ln a$. Differentiation by means of the rule (3.55) gives $y'/y = \ln a$. Multiplying each side by y leads to $y' = y \ln a$. We have thereby obtained the formula

$$(a^x)' = a^x \ln a. \tag{3.58}$$

Since $\ln \mathrm{e} = 1$ we also have

$$(\mathrm{e}^x)' = \mathrm{e}^x. \tag{3.59}$$

It follows immediately that

$$\int \mathrm{e}^x dx = \mathrm{e}^x + C. \tag{3.60}$$

3.7 Integrating equations of motion

A particle moving in three-dimensional space will in general have three independent components of the motion. The position, velocity and acceleration of the particle are all vectors, and so are the forces acting upon the particle. According to Newtonian dynamics the equation of motion of a particle is found by finding the forces acting upon the particle adding them vectorially and applying Newton's second law (see Eq. (3.29)) to the particle. In three-dimensional space Newton's second law is represented by three component equations,

$$m\frac{d^2x}{dt^2} = F^x, \quad m\frac{d^2y}{dt^2} = F^y, \quad \text{and} \quad m\frac{d^2z}{dt^2} = F^z. \tag{3.61}$$

We shall first consider the simple case that no force is acting upon the particle. Then the equations of motion are

$$\frac{d^2x}{dt^2} = 0, \quad \frac{d^2y}{dt^2} = 0, \quad \text{and} \quad \frac{d^2z}{dt^2} = 0. \tag{3.62}$$

(Note that if $m\frac{d^2x}{dt^2} = 0$ and $m \neq 0$, then $\frac{d^2x}{dt^2} = 0$.) The antiderivative of the second derivative is the first derivative. And the antiderivatice of zero is a constant (since the derivative of a constant is zero). Integrating Eq. (3.62) therefore gives

$$\frac{dx}{dt} = C_1, \quad \frac{dy}{dt} = C_2, \quad \text{and} \quad \frac{dz}{dt} = C_3. \tag{3.63}$$

The left-hand expressions are just the velocity components of the particle. Denoting these by v^x, v^y, and v^z we obtain

$$v^x = C_1, \quad v^y = C_2, \quad \text{and} \quad v^z = C_3. \tag{3.64}$$

This equation shows that the velocity components of the particle are constant. Since the antiderivative of dx/dt is x, and so forth, and the antiderivative of C_1 is $C_1 t$, and so forth, a new integration gives

$$x = x_0 + v^x t, \quad y = y_0 + v^y t, \quad \text{and} \quad z = z_0 + v^z t, \tag{3.65}$$

where the integration constants have been denoted by x_0, y_0, and z_0. Inserting $t = 0$ we find

$$x(0) = x_0, \quad y(0) = y_0, \quad \text{and} \quad z(0) = z_0 \tag{3.66}$$

Thus, the constants x_0, y_0, and z_0 are the coordinates of a point that the particle passes at time $t = 0$. Comparing with the parameter representation of a straight line, Eq. (3.15), we see that the path followed by a particle that is not acted upon by any forces is a straight line.

As our second example we shall calculate the path of a particle thrown out horizontally with an initial velocity v_0 in a field of weight with acceleration of gravity equal to g. During the free fall of the particle (neglecting air resistance) the only force acting upon the particle is gravity, which acts vertically. There is no horizontal force. Then it follows from our first example that the particle has constant horizontal velocity.

We choose a coordinate system with horizontal x-axis and vertical y-axis. The particle is thrown out from the origin, in the positive x-direction, at a point of time $t = 0$. In this case the equations of motion of the particle are

$$\frac{d^2x}{dt^2} = 0 \quad \text{and} \quad \frac{d^2y}{dt^2} = -g. \tag{3.67}$$

Integration with $v^y(0) = 0$ gives

$$\frac{dx}{dt} = v_0 \quad \text{and} \quad \frac{dy}{dt} = -gt. \tag{3.68}$$

Integrating once more with $x(0) = y(0) = 0$ leads to

$$x = v_0 t \quad \text{and} \quad y = -\frac{1}{2}gt^2. \tag{3.69}$$

Inserting $t = x/v_0$ into the expression for y, we find

$$y = -\left(\frac{g}{2v_0^2}\right)x^2. \tag{3.70}$$

This is the equation of a parabola of a similar shape as that of Fig. 3.1. With a suitable initial velocity the path of the particle will be just that parabola.

Chapter 4
Approaching general relativity: introducing curvilinear coordinate systems

Cartesian coordinate systems have straight coordinate lines orthogonal to each other. The coordinate basis vector fields are constant, and all the vectors of each field are unit vectors with the same direction. This is the simplest coordinate system that one can imagine. It seems strange that it can be advantageous to introduce coordinate systems with coordinate curves which are not straight lines, so-called curvilinear coordinate systems. But in fact some objects of investigation have certain symmetries, for example cylindrical or spherical symmetry that makes it advantageous to introduce curvilinear coordinates. In a flat space, however, every coordinate system can be transformed into a Cartesian system, so that in principle one could solve every problem with reference to Cartesian coordinates.

In general relativity it is necessary to introduce curvilinear coordinate systems. According to this theory spacetime[1] is curved and four-dimensional with three space dimensions and one time dimension. This curved four-dimensional spacetime seems to be the easiest way of taking into account the results from careful experimental investigations of electromagnetism and gravitation.

In fact, already about 1920, the general theory of relativity was generalized to describe spacetime with more than three spatial dimensions. More than one time dimension is problematic, however, because it leads to certain strange possibilities of backward time travel, which belongs to science fiction. These generalized versions of general relativity are called Kaluza–Klein theories, and have been thoroughly investigated recently. They lead to the possibility of giving a unified geometrical description of classical (i.e. non-quantum) electromagnetism and gravity. The problem of creating a coherent single theory for gravitation and quantum theory is still unsolved. The most recent effort is called superstring theory and involves ten-dimensional spacetime. In this book we shall be concerned with four-dimensional spacetime.

[1]See Sect. 1.7 and Ch. 5.

Ø. Grøn and A. Næss, *Einstein's Theory: A Rigorous Introduction for the Mathematically Untrained*, DOI 10.1007/978-1-4614-0706-5_4,

Consider hypothetical two-dimensional creatures living in a two-dimensional space made up of the surface of a sphere. They would have to describe their world with reference to the curved surface. No three-dimensional space exists for these creatures. And the spherical surface *cannot* be covered by a two-dimensional Cartesian coordinate system. (Look at a globus!). The situation is similar in the case of curved spaces of higher dimensions, such as our spacetime. It cannot be covered by a Cartesian coordinate system. As long as we insist that the spacetime may be curved and four-dimensional, and that the description of our universe shall be referred to a coordinate system with four dimensions only, there is no other way than introducing curvilinear coordinate systems.

The simplest curvilinear coordinate system is a system of concentric circles and straight radial lines out from the common centre of the circles. The coordinates of this system are called *plane polar coordinates*. In this chapter we shall consider the main properties of such a coordinate system. Some knowledge of trigonometric functions is then indispensable.

4.1 Trigonometric functions

We shall now introduce three so-called *trigonometric* functions, named *sinus*, *cosinus*, and *tangens*. They are widely used in physics, meteorology and astronomy—everywhere.

Consider a circle with radius r and centre at the origin of a Cartesian coordinate system, as shown in Fig. 4.1. The radial line OP makes an angle θ with the positive x-axis. The point P has coordinates (x, y).

Before we define the functions sinus, cosinus and tangens, we shall have to know how we measure angles. There are two common angular measures. In daily life we measure angles in *degrees*. For example the latitude and longitude of the position of a mountain are given in degrees. One degree may be defined in the following way: divide a circle in 360 parts by marking 360 equidistant points. Draw radial lines from the centre of the circle to all of the dividing points. One degree is the angle between two neighbouring radial lines. Thus the angle around the full circle is 360 degrees. Since there are four right angles around a circle, a right angle is equal to 90 degrees, and a half circle is 180 degrees.

The other angular measure, which is used in connection with the trigonometric functions, is called a *radian*. With reference to Fig. 4.1 we formulate the following definition: the angle, θ, i.e. the *number of radians*, is defined as the length of the circular arc between 0 and P divided by the radius. Note that this ratio is independent of the radius of the circle. The sign π is the symbol for the number of radians of a 180 degree angle. Its value is 3.141592.... It follows that the length of a circle with radius r is $2\pi r$. The angle around a circle, as measured in radians, is 2π, and a right angle is equal to $2\pi/4 = \pi/2$. Increasing an angle by 2π corresponds to going one time around a circle.

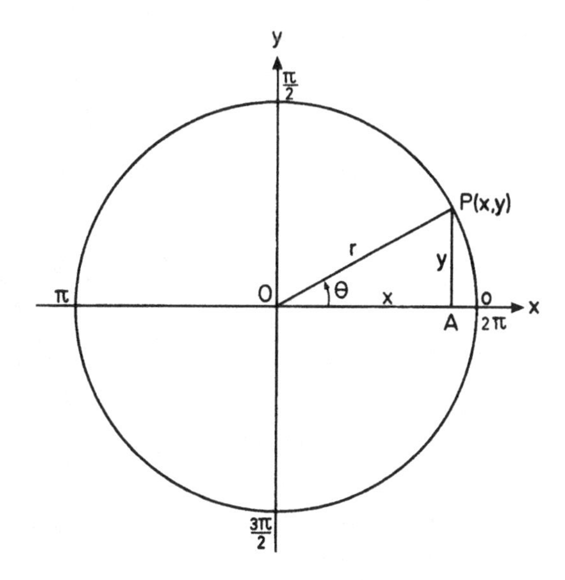

Fig. 4.1 Circle of radius r centered at the origin

The functions sinus(θ) and cosinus(θ) are denoted by $\sin\theta$ and $\cos\theta$. They are defined with reference to Fig. 4.1 by the equations

$$\sin\theta \equiv \frac{y}{r} \quad \text{and} \quad \cos\theta \equiv \frac{x}{r}, \tag{4.1a}$$

where $r = (x^2 + y^2)^{1/2}$ according to the Pythagorean theorem. Note that if $r = 1$ then $\sin\theta = y$ and $\cos\theta = x$. For θ outside the range $[0, 2\pi\rangle$ the definition implies

$$\sin(\theta + 2\pi) = \sin\theta \quad \text{and} \quad \cos(\theta + 2\pi) = \cos\theta. \tag{4.1b}$$

Equation (4.1b) shows that $\sin\theta$ and $\cos\theta$ are periodic functions with period 2π. This corresponds to the fact that if P on Fig. 4.1 is moved around the circle, so that θ passes 2π, then x and y, and thus $\sin\theta$ and $\cos\theta$ obtain just the same values as one round earlier.

The function tangens(θ) is denoted by $\tan\theta$, and is defined by

$$\tan\theta \equiv \frac{\sin\theta}{\cos\theta}. \tag{4.2}$$

Substituting for $\sin\theta$ and $\cos\theta$ from the definitions in Eqs. (4.1) leads to

$$\tan\theta = \frac{y}{x}. \tag{4.3}$$

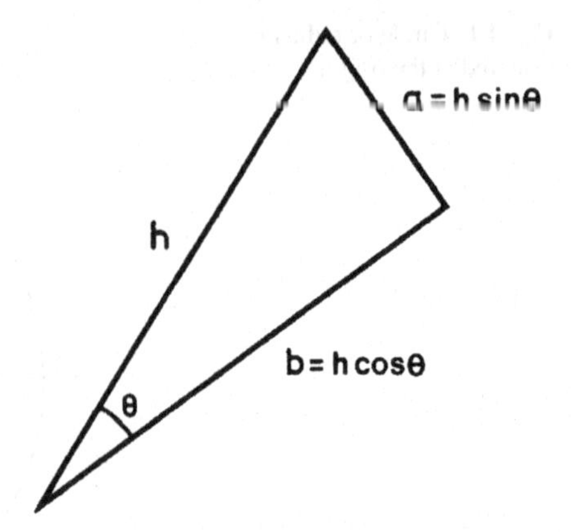

Fig. 4.2 A right-angled triangle

The function $\tan\theta$ is in principle superfluous. We can always use $\sin\theta$ and $\cos\theta$. However, the function $\tan\theta$ simplifies some calculations, as will be seen in our description of plane curves in Sect. 9.1.

It will be useful to formulate the expressions for the trigonometric functions without reference to a Cartesian coordinate system; only referring to a right angled triangle, such as OAP in Fig. 4.1. Note that in relation to the angle θ the base x is the side adjacent to θ, and the height y is the side opposite to θ. Using these terms the expressions for $\sin\theta$ and $\cos\theta$ can be formulated in the following way.

$$\sin\theta = \frac{\text{the length of the side opposite to } \theta}{\text{the length of the hypotenuse}},$$

$$\cos\theta = \frac{\text{the length of the side adjacent to } \theta}{\text{the length of the hypotenuse}},$$

and

$$\tan\theta = \frac{\text{the length of the side opposite to } \theta}{\text{the length of the side adjacent to } \theta}.$$

These formulations permit us to find expressions for the trigonometric functions by reference to a right angled triangle with arbitrary orientation, not referring to any coordinate system, see Fig. 4.2.

From Fig. 4.2 and the expressions (4.1) we get $\sin\theta = a/h$ and $\cos\theta = b/h$. Multiplying by h we can express the magnitude of the side opposite to θ and the side adjacent to θ, by θ and the hypotenuse h of the right angled triangle,

$$a = h\sin\theta \quad \text{and} \quad b = h\cos\theta. \tag{4.4}$$

Consider Fig. 4.3. The part CQD of the figure is obtained by turning the part APB an angle 90 degrees to the left. Thus the angles CQD and APB are equal. We obtain the following sentence: two angles with pairwise normal legs, are equal.

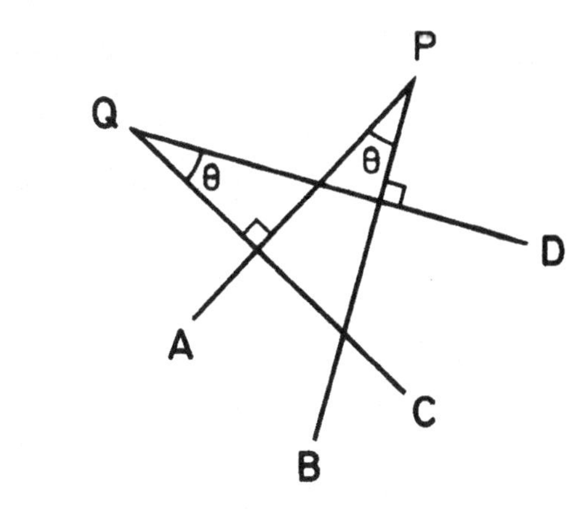

Fig. 4.3 Angles with pairwise normal legs

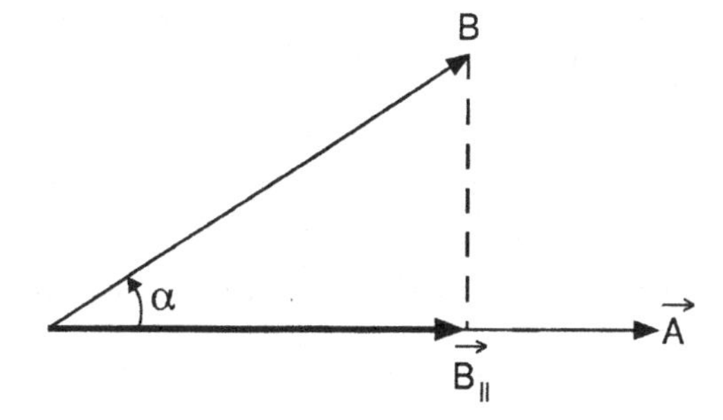

Fig. 4.4 Angle α between $\vec{A}$ and $\vec{B}$

In Eq. (1.8) we have written the dot product of two vectors $\vec{A}$ and $\vec{B}$ as 'the magnitude of $\vec{A}$ times the magnitude of $\vec{B}$'s projection, $\vec{B}_{||}$, along $\vec{A}$'. This may now be expressed in terms of the angle α between $\vec{A}$ and $\vec{B}$.

From Fig. 4.4 is seen that $|\vec{B}_{||}| = \cos\alpha$. In general

$$|\vec{B}_{||}| = \begin{cases} |\vec{B}|\cos\alpha & \text{for} \quad -\frac{\pi}{2} \le \alpha < \frac{\pi}{2} \\ -|\vec{B}|\cos\alpha & \text{for} \quad \frac{\pi}{2} \le \alpha < \frac{3\pi}{2}. \end{cases} \tag{4.5}$$

Inserting this into Eq. (1.8) gives

$$\vec{A}\cdot\vec{B} = |\vec{A}||\vec{B}|\cos\alpha \tag{4.6}$$

where the sign ambiguity in Eq. (1.8) is taken care of automatically. Since the angle that a vector makes with itself is zero, and $\cos 0 = 1$, it follows that the square of the magnitude of a vector is

$$|\vec{A}|^2 = \vec{A}\cdot\vec{A}. \tag{4.7}$$

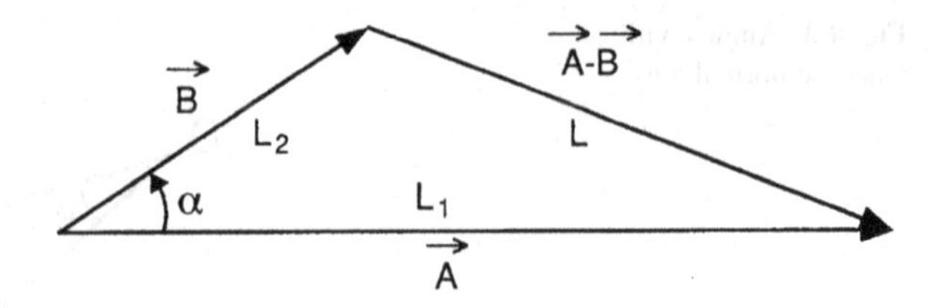

Fig. 4.5 The extended Pythagorean theorem

We shall now deduce a geometrical identity that will be used in Ch. 5. The square of $|\vec{A}-\vec{B}|$ is

$$|\vec{A}-\vec{B}|^2 = (\vec{A}-\vec{B})\cdot(\vec{A}-\vec{B}) = \vec{A}\cdot\vec{A} - 2\vec{A}\cdot\vec{B} + \vec{B}\cdot\vec{B}$$
$$= |\vec{A}|^2 + |\vec{B}|^2 - 2\vec{A}\cdot\vec{B}. \tag{4.8}$$

Substituting the expression (4.6) for the dot product $\vec{A}\cdot\vec{B}$ leads to

$$|\vec{A}-\vec{B}|^2 = |\vec{A}|^2 + |\vec{B}|^2 - 2|\vec{A}||\vec{B}|\cos\alpha. \tag{4.9}$$

Look at Fig. 1.9. Let the angle between the vectors $\vec{A}$ and $\vec{B}$ be α, the length of $\vec{A}$ be L_1, the length of $\vec{B}$ be L_2, and the length of $\vec{A}-\vec{B}$ be L. Then Eq. (4.9) may be written

$$L^2 = L_1^2 + L_2^2 - 2L_1L_2\cos\alpha. \tag{4.10}$$

This is an extension of the Pythagorean theorem to the case when the triangle has no 90 degree angle. It might have been named 'The extended Pythagorean theorem', but is usually called 'the law of cosines'. In Fig. 4.5 we have redrawn Fig. 1.9, but this time with the quantities L, L_1, L_2, and α which appear in Eq. (4.10).

The magnitudes of $\sin\theta$ and $\cos\theta$ when θ increases from 0 to 2π are seen from Fig. 4.1 by letting P start at the positive x-axis and then move around the circle. At the x-axis, $x = r$ and $y = 0$, giving and $\cos 0 = 1$. Moving on to the y-axis, x decreases towards zero and y increases towards r. When P is at the y-axis the angle θ is equal to $\pi/2$. Thus $\sin(\pi/2) = 1$ and $\cos(\pi/2) = 0$. As P comes to the negative x-axis, $x = -r$ and $y = 0$, while $\theta = \pi$, so $\sin\pi = 0$ and $\cos\pi = -1$. Letting P move further to the negative y-axis, the angle θ reaches $3\pi/2$, while $x = 0$ and $y = -r$, giving $\sin(3\pi/2) = -1$ and $\cos(3\pi/2) = 0$. We may now draw the diagrams representing $y = \sin\theta$ and $y = \cos\theta$ (in Fig. 4.6), with y along a vertical axis and θ along a horizontal axis.

An important relation between $\sin\theta$ and $\cos\theta$ should be noted. Using Eqs. (4.1) we get

$$\sin^2\theta + \cos^2\theta = \frac{x^2}{r^2} + \frac{y^2}{r^2} = \frac{x^2+y^2}{r^2} = \frac{r^2}{r^2} = 1. \tag{4.11}$$

This equation is valid for every value of the angle θ.

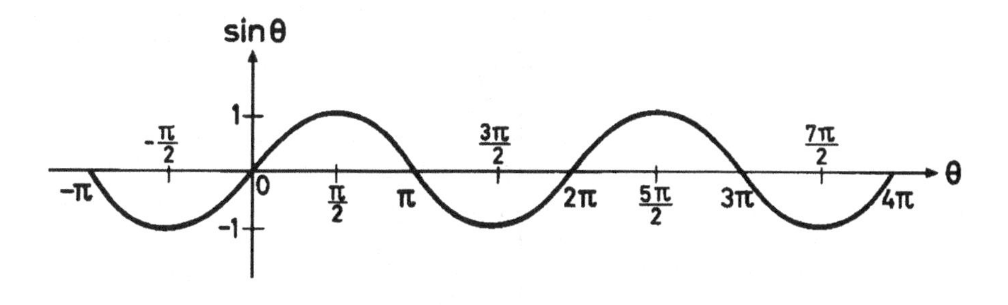

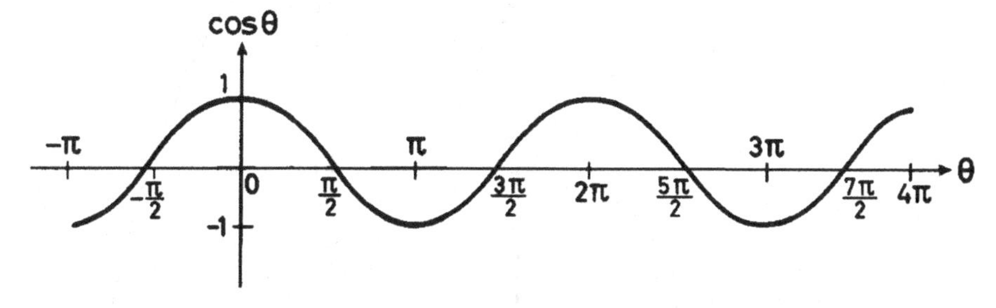

Fig. 4.6 The function $\sin\theta$ and $\cos\theta$

We shall need to calculate cosinus to the sum (or difference) of two angles. The formulae both for sinus and cosinus to a sum of two angles are found from Fig. 4.7. You will find a lot of information written on the figure. These are results that we shall now deduce. Consider first the angle DPC. PC is normal to OC and PD is normal to OA. According to Fig. 4.3 two angles with pairwise normal legs are equal. Hence, we conclude that the angle DPC is equal to the angle POC, i.e. it is equal to u.

Next we shall utilize Eq. (4.4) which is related to Fig. 4.2. Consider the triangle OCP which is similar to the triangle of Fig. 4.2. Using Eq. (4.4) we find

$$OC = \cos u \quad \text{and} \quad PC = \sin u. \tag{4.12}$$

Looking at the triangle DCP we find

$$\begin{aligned} DC &= PC \sin v = \sin u \sin v, \\ PD &= PC \cos v = \sin u \cos v. \end{aligned} \tag{4.13}$$

From the triangle OBC we get

$$\begin{aligned} OB &= OC \cos v = \cos u \cos v, \\ BC &= OC \sin v = \cos u \sin v, \end{aligned} \tag{4.14}$$

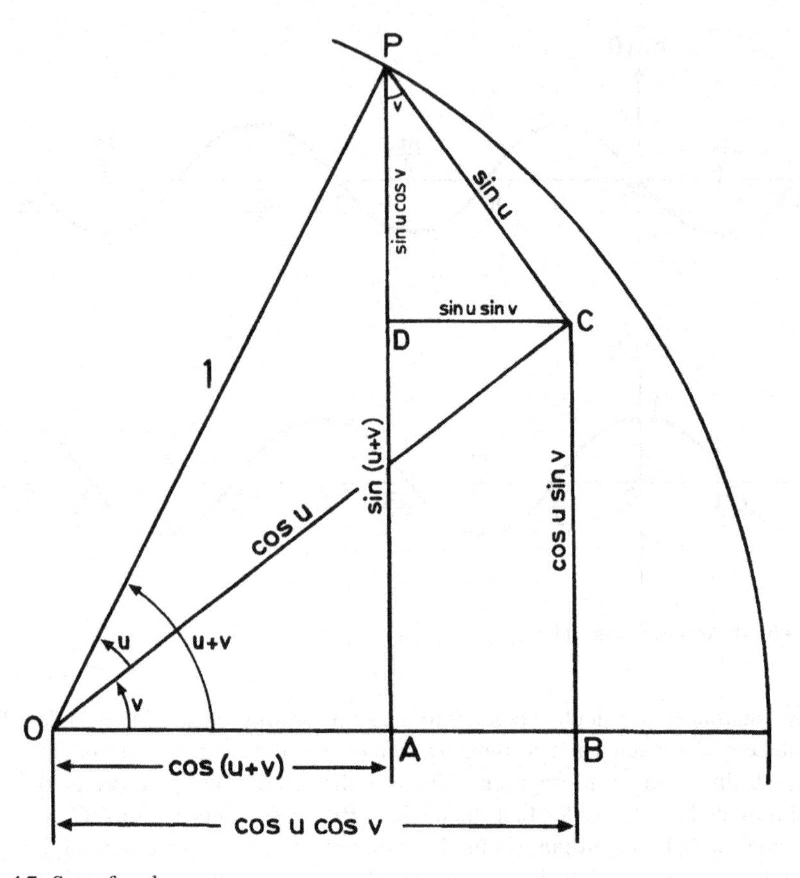

Fig. 4.7 Sum of angles

and from the triangle OAP we have

$$OA = \cos(u + v) \quad \text{and} \quad AP = \sin(u + v). \tag{4.15}$$

Looking at Fig. 4.7 we now find

$$\sin(u + v) = PD + AD = PD + BC. \tag{4.16}$$

Inserting the expressions for PD and BC from Eqs. (4.13) and (4.14), respectively, we get

$$\sin(u + v) = \sin u \, \cos v + \cos u \, \sin v \tag{4.17}$$

and

$$\cos(u + v) = OB - AB = OB - DC. \tag{4.18}$$

Inserting the expressions for OB and DC from Eqs. (4.14) and (4.13), respectively, leads to

$$\cos(u+v) = \cos u \cos v - \sin u \sin v. \tag{4.19}$$

We shall now deduce a formula which will be needed in chapter 14. Using Eq. (4.17) we get

$$\begin{aligned} \sin 2\theta &= \sin(\theta+\theta) \\ &= \sin\theta \cos\theta + \cos\theta \sin\theta = 2\sin\theta \cos\theta. \end{aligned} \tag{4.20}$$

From Eq. (4.19) we have

$$\cos 2\theta = \cos(\theta+\theta) = \cos^2\theta - \sin^2\theta.$$

Using Eq. (4.11) this may be written alternatively as

$$\cos 2\theta = 2\cos^2\theta - 1 = 1 - 2\sin^2\theta.$$

4.1.1 Differentiation of trigonometric functions

In our analysis of curved coordinate systems we shall have to differentiate the functions $\sin\theta$ and $\cos\theta$. A geometrical deduction of the expressions for the derivatives of these functions shall now be given.

Consider Fig. 4.8. A circle with radius 1 is drawn about the origin O of a Cartesian coordinate system. The radii OP and OQ make angles θ and $\theta + \Delta\theta$, respectively, with the positive x axis. The point P has coordinates (x, y) and the point Q $(x - \Delta x, y + \Delta y)$. For a small angle $\Delta\theta$, the direction of the line PQ is very close to that of the tangent at P. Then PQ is normal to the radial line OP. Thus the legs PQ and QR of the angle PQR are pairwise normal to the legs OP and OA. Since the angle POA is equal to θ, the angle PQR must also be equal to θ. From the triangle PQR we then have

$$\sin\theta = \frac{\Delta x}{\Delta\theta} \quad \text{and} \quad \cos\theta = \frac{\Delta y}{\Delta\theta}. \tag{4.21}$$

Note that the line PQ equals $\Delta\theta$ because the radius is equal to 1. Also from the triangles OAP and OBQ we get

$$\sin\theta = y \quad \text{and} \quad \cos\theta = x, \tag{4.22}$$

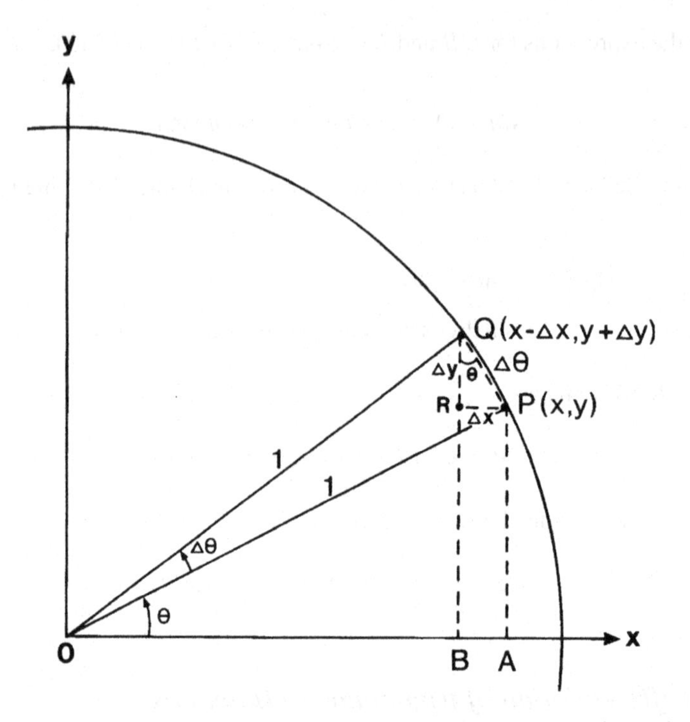

Fig. 4.8 Changes in position on the unitunit!circle circle when incrementing θ by $\Delta\theta$

and

$$\sin(\theta + \Delta\theta) = y + \Delta y \quad \text{and} \quad \cos(\theta + \Delta\theta) = x - \Delta x. \tag{4.23}$$

The expressions for the derivatives of $\sin\theta$ and $\cos\theta$ can now be deduced directly from the general definition of the derivative of a function, Eq. (2.4).

$$\begin{aligned}(\sin\theta)' &= \lim_{\Delta\theta\to 0} \frac{\sin(\theta + \Delta\theta) - \sin\theta}{\Delta\theta} = \lim_{\Delta\theta\to 0} \frac{y + \Delta y - y}{\Delta\theta} \\ &= \lim_{\Delta\theta\to 0} \frac{\Delta y}{\Delta\theta} = \cos\theta \end{aligned} \tag{4.24}$$

and

$$\begin{aligned}(\cos\theta)' &= \lim_{\Delta\theta\to 0} \frac{\cos(\theta + \Delta\theta) - \cos\theta}{\Delta\theta} = \lim_{\Delta\theta\to 0} \frac{x - \Delta x - x}{\Delta\theta} \\ &= \lim_{\Delta\theta\to 0} \left(-\frac{\Delta x}{\Delta\theta}\right) = -\sin\theta. \end{aligned} \tag{4.25}$$

where we have used Eqs. (4.21)–(4.23).

In order to calculate, in chapter 9, the curvature of a curve, we shall need the derivative of $\tan\theta$. This is most simply deduced by applying the formula (2.46) for differentiating a fraction of functions. Inserting $u = \sin\theta$ and $v = \cos\theta$ and using Eqs. (4.2),(4.24), and (4.25) we find

$$(\tan\theta)' = \left(\frac{\sin\theta}{\cos\theta}\right)' = \frac{\cos\theta\cos\theta - \sin\theta\,(-\sin\theta)}{\cos^2\theta}$$
$$= \frac{\cos^2\theta + \sin^2\theta}{\cos^2\theta} = \frac{1}{\cos^2\theta}. \tag{4.26}$$

Alternatively, from the next last expression in Eq. (4.26), we get

$$(\tan\theta)' = 1 + \frac{\sin^2\theta}{\cos^2\theta} = 1 + \tan^2\theta, \tag{4.27}$$

where we have used the definition of $\tan\theta$ in Eq. (4.2).

In Sect. 9.2 we shall need to know the MacLaurin series for $\sin\theta$. In order to collect the necessary theory of trigonometry at one place, we shall immediately deduce the form of this series.

The MacLaurin series of a function $f(\theta)$ is given in Eq. (2.84) as

$$f(\theta) = f(0) + f'(0)\,\theta + \frac{f''(0)}{2}\,\theta^2 + \frac{f'''(0)}{6}\,\theta^3 + \cdots. \tag{4.28}$$

In the case that $f(\theta) = \sin\theta$ we get

$$\begin{aligned}
f(\theta) &= \sin\theta && \Rightarrow f(0) = 0\\
f'(\theta) &= \cos\theta && \Rightarrow f'(0) = 1\\
f''(\theta) &= -\sin\theta && \Rightarrow f''(0) = 0\\
f'''(\theta) &= -\cos\theta && \Rightarrow f'''(0) = -1.
\end{aligned}$$

Substitution into Eq. (4.28) gives

$$\sin\theta = \theta - \frac{1}{6}\theta^3 + \cdots. \tag{4.29}$$

These are the first terms of the MacLaurin series for $\sin\theta$.

4.2 Plane polar coordinates

A coordinate system with plane polar coordinates, r and θ, is shown in Fig. 4.9.

Instead of vertical and horizontal coordinate lines, such as in a Cartesian coordinate system, the coordinate curves in a system with plane polar coordinates

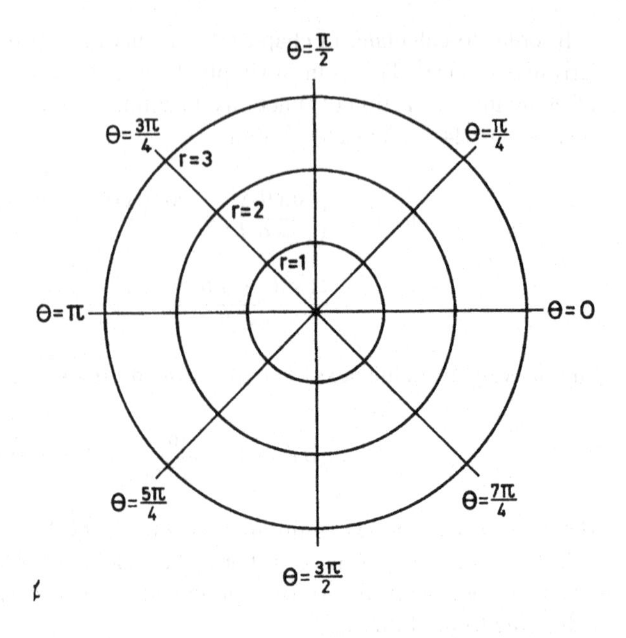

Fig. 4.9 Plane polar coordinates, r and θ

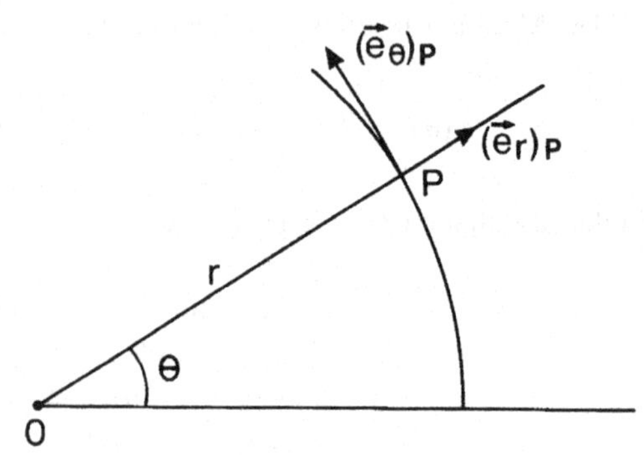

Fig. 4.10 Radial and tangential basis vectors

are concentric circles and radial lines. The Cartesian coordinates x and y are the horizontal and vertical distances from the origin. The plane polar coordinate r of an arbitrary point P, is the distance from the origin to P. The plane polar coordinate θ is the angle of the radial line from the origin to P with respect to a horizontal line.

Just as there are horizontal and a vertical basis vectors, $\vec{e}_x$ and $\vec{e}_y$, at every point in a Cartesian coordinate system, there are radial, and a tangential basis vectors, $\vec{e}_r$ and $\vec{e}_\theta$, at every point in a polar coordinate system. These basis vectors at an arbitrary point P are shown in Fig. 4.10. The point O is the origin of the coordinate system. The point P has coordinates r and θ. The basis vectors at P have been denoted by $(\vec{e}_r)_P$ and $(\vec{e}_\theta)_P$ to indicate that their directions and magnitudes depend upon the point at which they are drawn. The Cartesian basis vectors, $\vec{e}_x$ and $\vec{e}_y$, on the other hand, are the same everywhere.

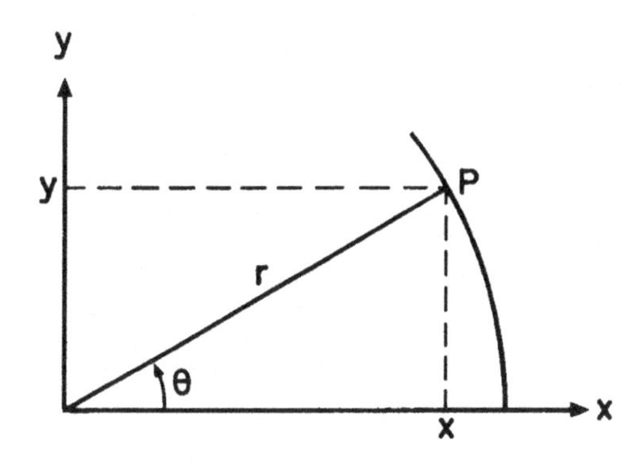

Fig. 4.11 Relation between Cartesian and polar coordinates

The simple relation between a Cartesian coordinate system and a polar coordinate system with the same origin, and with the direction $\theta = 0$ corresponding to the x-axis, is shown in Fig. 4.11.

From the figure is seen that

$$\cos\theta = \frac{x}{r} \quad \text{and} \quad \sin\theta = \frac{y}{r} \tag{4.30}$$

which gives

$$x = r\cos\theta \quad \text{and} \quad y = r\sin\theta. \tag{4.31}$$

This is our first example of a coordinate transformation equation.

Some curves have a simpler equation in polar coordinates than in Cartesian coordinates. The equation of a circle with centre at the origin and radius R is in Cartesian coordinates

$$x^2 + y^2 = R^2. \tag{4.32}$$

The equation of the same circle in polar coordinates is

$$r = R. \tag{4.33}$$

The position vector of an arbitrary point P with Cartesian coordinates (x, y) is

$$\vec{r} = x\vec{e}_x + y\vec{e}_y. \tag{4.34}$$

The Cartesian components of $\vec{r}$ are $r^x = x$ and $r^y = y$, and may be expressed by the coordinates r and θ by means of Eq. (4.31), giving

$$\vec{r} = r\cos\theta\,\vec{e}_x + r\sin\theta\,\vec{e}_y. \tag{4.35}$$

We now calculate the magnitude of the basis vectors $\vec{e}_r$ and $\vec{e}_\theta$. Using a concept introduced in section 3.3 we may say that the basis vector field $\vec{e}_r$ is a tangent vector field to the radial lines. The coordinate θ is constant along a radial line, while r

increases along it. Thus r can be used as a parameter for such a line. The parametric equation for a radial line is then

$$x = r\cos\theta \quad \text{and} \quad y = r\sin\theta$$

where θ is a constant, specific for each line. Equation (3.19) with $\vec{u} = \vec{e}_r$, $t = r$, and replacing dx/dt with $\partial x/\partial r$, and dy/dt with $\partial y/\partial r$, gives

$$\vec{e}_r = \frac{\partial x}{\partial r}\vec{e}_x + \frac{\partial y}{\partial r}\vec{e}_y, \tag{4.36}$$

which shows that

$$(\vec{e}_r)^x = \frac{\partial x}{\partial r} \quad \text{and} \quad (\vec{e}_r)^y = \frac{\partial y}{\partial r}. \tag{4.37}$$

Differentiating Eq. (4.31) with respect to r, (remember that θ is constant when we differentiate partially with respect to r), we get

$$\frac{\partial x}{\partial r} = \cos\theta \quad \text{and} \quad \frac{\partial y}{\partial r} = \sin\theta. \tag{4.38}$$

Thus

$$(\vec{e}_r)^x = \cos\theta \quad \text{and} \quad (\vec{e}_r)^y = \sin\theta \tag{4.39}$$

showing that

$$\vec{e}_r = \cos\theta\,\vec{e}_x + \sin\theta\,\vec{e}_y. \tag{4.40}$$

The magnitude of $\vec{e}_r$, as calculated by means of Eqs. (1.5) and (4.39) is

$$|\vec{e}_r| = \cos^2\theta + \sin^2\theta = 1, \tag{4.41}$$

which shows that $\vec{e}_r$ is an unit vector independent of its position. But $\vec{e}_r$ changes with θ. Since the magnitude of $\vec{e}_r$ is constant and equal to unity, the change of $\vec{e}_r$ with θ is only a change of direction.

The basis vector fields $\vec{e}_\theta$ are tangent vector fields to the circles about the origin. The coordinate r is constant along a circle, while θ increases along it. So θ can be used as a parameter for such a circle. The parametric equation for a circle about the origin is again Eq. (4.31), but this time with $r =$ constant for each circle, and with θ as variable.

Equation (3.17) with $\vec{u} = \vec{e}_\theta$, $t = \theta$, and first $i = x$, so $i = y$, then gives

$$(\vec{e}_\theta)^x = \frac{\partial x}{\partial \theta} = -r\sin\theta \quad \text{and} \quad (\vec{e}_\theta)^y = \frac{\partial y}{\partial \theta} = r\cos\theta, \tag{4.42}$$

which leads to

$$\vec{e}_\theta = -r\sin\theta\,\vec{e}_x + r\cos\theta\,\vec{e}_y. \tag{4.43}$$

The magnitude of $\vec{e}_\theta$ is found from Eqs. (1.5) and (4.43) as

$$|\vec{e}_\theta| = \left[(-r\sin\theta)^2 + (r\cos\theta)^2\right]^{1/2} = r(\sin^2\theta + \cos^2\theta)^{1/2} = r. \tag{4.44}$$

This equation shows that the magnitude of $\vec{e}_\theta$ increases proportionally to the distance from the origin. This is not a unit vector field. However, one could define a corresponding unit vector field $\vec{e}_{\hat{\theta}}$ by

$$\vec{e}_{\hat{\theta}} = \frac{\vec{e}_\theta}{|\vec{e}_\theta|} = \frac{1}{r}\vec{e}_\theta. \tag{4.45}$$

Here the hat above the index denotes that the vector has unit length.

Note that the calculation above has provided us with a method for transforming basis vectors. Using Eq. (3.19) we find the transformation of the basis vectors from a knowledge of the corresponding coordinate transformation.

We have seen that in general the coordinate basis vectors are not unit vectors. And the corresponding unit vectors, such as $\vec{e}_{\hat{\theta}}$, will in general not be coordinate basis vectors. Still they are basis vectors, in the sense that an arbitrary vector can be decomposed along the basis vectors.

Chapter 5
The metric tensor

The metric tensor is perhaps the most important mathematical quantity in the theory of relativity. From a knowledge of the metric tensor one may compute the geometry of spacetime and for example the motion of the planets in the solar system. In this chapter we shall give a thorough introduction to the metric tensor and its physical significance in the theory of relativity.

5.1 Basis vectors and the dimension of a space

In Chapter 1 basis vectors were introduced as reference vectors for directions (in flat space), and it was mentioned that at every point of an n-dimensional space there are n basis vectors. One could say that there are n independent directions in an n-dimensional space. Let us say what these terms express in a precise way. We then need to make a few definitions.

The vectors $\vec{e}_1, \vec{e}_2, \ldots, \vec{e}_n$ are said to be *linearly independent* if no set of real numbers $a_1, a_2, \ldots, a_n$ different from zero exists, such that

$$a_1\vec{e}_1 + a_2\vec{e}_2 + \cdots + a_n\vec{e}_n = 0. \tag{5.1}$$

The geometrical meaning of this definition is: The vectors $\vec{e}_1, \vec{e}_2, \ldots, \vec{e}_n$ are linearly independent if it is not possible to make a closed polygon of the vectors, even if we try our best by adjusting their lengths, that is, we cannot make a vector sum of all the vectors equal to zero, even if we scale them. Vectors that are not linearly independent are said to be *linearly dependent*. A set of three linearly dependent vectors is shown in Fig. 5.1.

In this figure we consider three vectors $\vec{v}_1 = 8\vec{e}_x$, $\vec{v}_2 = -4\vec{e}_x + 3\vec{e}_y$, $\vec{v}_3 = -4\vec{e}_x - 3\vec{e}_y$. Thus $\vec{v}_1 + \vec{v}_2 + \vec{v}_3 = 0$, which means that they are linearly dependent, and that when they are drawn as in Fig. 5.1, they make up a closed polygon.

Ø. Grøn and A. Næss, *Einstein's Theory: A Rigorous Introduction for the Mathematically Untrained*, DOI 10.1007/978-1-4614-0706-5_5,

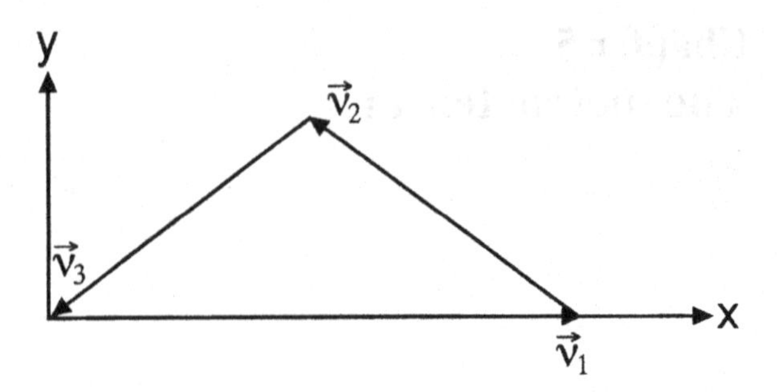

Fig. 5.1 Three linearly dependent vectors

In general two vectors that have different directions are linearly independent. Three vectors in a plane with different directions are, however, linearly dependent, since one can make a closed polygon in a plane by means of three such vectors.

A set of maximally linearly independent vectors in a space consists of the maximum number of linearly independent vectors in the space. If $\vec{e}_1, \vec{e}_2, \ldots, \vec{e}_n$ are maximally linearly independent, there exists a vector $\vec{u}$ so that the vectors $\vec{e}_1, \vec{e}_2, \ldots, \vec{e}_n, \vec{u}$ are linearly dependent, which means that there exist numbers a_1 , a_2 , ..., a_n so that

$$a_1\vec{e}_1 + a_2\vec{e}_2 + \cdots + a_n\vec{e}_n + \vec{u} = 0$$

or

$$\vec{u} = -a_1\vec{e}_1 - a_2\vec{e}_2 - \cdots - a_n\vec{e}_n. \tag{5.2}$$

By means of three vectors with different directions in a plane one can always make a closed polygon. To see this, draw the first vector from a chosen point, then add the second by drawing it from the terminal point of the first vector as in Fig. 1.8, and choose $\vec{u}$ as minus the sum of the first two vectors. Clearly there are only two vectors in any set of maximally linearly independent vectors in a plane. Correspondingly, there are three vectors in a maximally linearly independent set of vectors in any ordinary three-dimensional space, in a room for example.

The maximally linearly independent set of vectors in a space adequately represents the independent directions in the space. This motivates the following definition. A *vector basis*, or, more precisely, a *set of basis vectors* $\{\vec{e}_1, \ldots, \vec{e}_n\}$ of a space is a set of maximally linearly independent vectors in the space. Furthermore we introduce new symbols for the numbers a_i, $i = 1, \ldots, n$, namely $u^i = -a_i$. Equation (5.2) then takes the form

$$\vec{u} = u^i\vec{e}_i = u^1\vec{e}_1 + u^2\vec{e}_2 + \ldots + u^n\vec{e}_n \tag{5.3}$$

which is recognized as the component form of a vector (see Eq. (3.17)). Note that we have superscripts on the vector components and subscripts on the basis vectors (or vice versa, see Sect. 5.6). The reason for this is connected with the coordinate invariance of vectors, which will be treated later in this chapter.

Later in this chapter we shall need to use the expression 'linear combination'. We define: a *linear combination* of some quantities $q_1, q_2, \ldots, q_n$ is an expression of the form $a_1 q_1 + a_2 q_2 + \cdots + a_n q_n$, where $a_1, a_2, \ldots, a_n$ are numbers. The component form (5.3) of a vector thus implies that any vector can be written as a linear combination of the basis vectors.

The decomposition of a vector in a given space is not unique. In a two-dimensional space, for example, $\{\vec{e}_x, \vec{e}_y\}$ is the coordinate vector basis of a Cartesian coordinate system, and $\{\vec{e}_r, \vec{e}_\theta\}$ is the coordinate vector basis of a coordinate system with plane polar coordinates. This illustrates that we can have many different sets of basis vectors in a space.

The *dimension* of a space is defined as the number of vectors of a vector basis in the space. Note that this definition does not refer in any way to 'orthogonality'. We are used to thinking of different dimensions as orthogonal to each other, or independent of each other, in the sense that motion in one dimension does not involve motion in any other dimension. For example motion along the x-axis does not involve any displacement along the y-axis. However, this needs not always be so, as we shall later see in connection with rotating reference frames. The different basis vectors of a vector basis need not be orthogonal to each other, and time is not necessarily orthogonal to space.

5.2 Space and spacetime

What is space? When you say "N.N. has much space available in his home", you can usually substitute 'room' for 'space'. However, this is not the way that we shall think of 'space' in the general theory of relativity. We shall have to be familiar with very different uses. In order to avoid prejudices, let us invent a word 'wace', rather than always use the word 'space'. We start by saying: A wace is, among other things, something within which there are directions and in which things can be moved, or more technically, displaced. Forget about finite distances between points and retain rather a notion of (tiny) neighborhoods. In waces we can *feel* our way, locally.

There exist indefinitely many kinds of waces. How do we know? Because the existence is in general a purely *mathematical* one. The multitude of waces is not more mysterious than the fact that there exist infinitely many numbers. As used within physics, however, a wace has a physical aspect.

In a very special four-dimensional wace called 'spacetime', there are three dimensions called space dimensions and one called the time dimension. A tiny displacement normally affects all four dimensions.

It is useful to imagine displacement in a two-dimensional wace as a displacement 'on' a surface. We use inverted commas because the proposition 'on' suggests 'upon', that is, a third dimension, which is foreign to a two-dimensional wace. The displacements in our case are parts of the surface. In the general case a bulging surface, like a crust on a slightly boiling soup-like fluid, bubbles are appearing and disappearing, but do not burst. We must admit that we normally imagine a boiling surface within a three-dimensional room, whereas in such a wace we should try to imagine ourselves as two-dimensional beings, parts of an indefinitely thin crust.

According to the pre-relativistic conceptions, or rather, ‘models’, of our universe, we live in a three-dimensional Euclidean (flat) physical space. The properties of space, as a kind of empty container without walls, were not thought to be influenced by its contents of matter. And time was something which did not interfere with space itself.

In the *special theory of relativity* space and time are united in a four-dimensional wace, called “*spacetime*”. Space and time can no longer be separated in any absolute sense, as is the case in Newtonian physics. The relativity of simultaneity has the following strange consequence (see Sect. 5.10). What one observer perceives as a purely spatial distance between two simultaneous events, is a combination of a time interval and a spatial distance for another observer, moving relative to the first one. But the geometry of spacetime is not influenced by matter present. It is flat always and everywhere.

The situation is different in the *general theory of relativity*. Here spacetime is a part of physics. The geometrical properties of spacetime depend upon the matter and energy present in it, and spacetime in turn determines the motion of free particles. A remarkable aspect of this is that spacetime is just as dynamic as matter.

Spacetime is a four-dimensional wace with *Riemannian geometry*, and the equations of general relativity imply that *free particles move along geodesic curves* (that is, straightest possible curves) in this wace (see Ch. 12).

Since spacetime is conceived of as curved, we must develop a description that can be used with reference to curved coordinate systems. This is obtained by introducing a kind of generalized vectors called tensors. In order to appreciate those properties of tensors that make them so useful for the mathematical formulation of the general theory of relativity, we need some knowledge about how vector components change under a change of coordinate system *without any change of the vector itself.* In short: how the vector components *transform*. The next paragraph is devoted to this topic. Tensor components also transform without any change of the tensor itself. According to Einstein’s ideas, natural laws must be formulated in such a way that they retain their identity whatever reference systems, we humans, with our finite intelligence, choose to employ in our study of those laws. Tensors suit this purpose.

5.3 Transformation of vector components

What happens to the components of a vector when the vector remains identical with itself, *while the coordinate systems we place it in changes?* It turns out that the equation expressing the effect of all these changes are superbly short. Only 14 symbols are needed in Eq. (5.13) below, even less than in the case of the simple rule $(a + b)^2 = a^2 + 2ab + b^2$.

It was important for Einstein to find something like vectors which could take care of seemingly immensely complicated relations. To the essential simplicity of natural laws, there could perhaps correspond simple vector equations eliminating human fuss with coordinates.

Let x^μ be the old coordinates in an n-dimensional wace, and $x^{\mu'}$ be the new ones. The only thing we know about the latter is that they are functions of the old ones:

$$\left.\begin{aligned} x^{1'} &= f_{1'}(x^1, x^2, \ldots, x^n) \\ x^{2'} &= f_{2'}(x^1, x^2, \ldots, x^n) \\ &\vdots \\ x^{n'} &= f_{n'}(x^1, x^2, \ldots, x^n) \end{aligned}\right\}. \tag{5.4}$$

In an extreme case every new coordinate is a complicated function of all the old ones. Using the notation $x^{1'}(\ldots)$ for $f_{1'}(\ldots)$, $x^{2'}(\ldots)$ for $f_{2'}(\ldots)$, and so on, the whole system of equations (5.4) may be written simply as

$$x^{\mu'} = x^{\mu'}(x^1, \ldots, x^n) \quad \text{for} \quad \mu' = 1', 2', \ldots, n'. \tag{5.5}$$

Note that $x^{\mu'}$ should be thought of as *one* symbol, denoting the first coordinate in the new system, the second, and so forth.

Let us consider a simple example, where the 'old' coordinates are the plane polar coordinates, r and θ, of section 4.2, and the new ones are the Cartesian coordinates x and y. Then $x^1 = r$, $x^2 = \theta$, $x^{1'} = x$, and $x^{2'} = y$. The transformation equation (4.31) from the plane polar coordinates to the Cartesian coordinates, is a set of equations of the type (5.5). The first one of the equations (5.5), $x^{1'} = x^{1'}(x^1, x^2)$, is $x = r\cos\theta$, and the second one, $x^{2'} = x^{2'}(x^1, x^2)$, is $y = r\sin\theta$.

We shall find the transformation formula for vector components by thinking of an arbitrary vector as the tangent vector of a curve, $x^\mu(\tau)$. Then the components of the vector are $u^\mu = dx^\mu/d\tau$, where τ is the curve parameter. The curve is defined by specifying the coordinates as functions of the curve parameter τ. The transformation formula for the tangent vector components is found by differentiating the transformed coordinates with respect to the parameter τ. Then we need a generalization of the chain rule for differentiation, Eq. (2.31), to functions of several variables. In a similar way that Eq. (2.31) is deduced from Eq. (2.26), it follows from Eq. (2.61) that the chain rule for a function of several variables takes the form

$$\frac{dx^{\mu'}}{d\tau} = \frac{\partial x^{\mu'}}{\partial x^\mu}\frac{dx^\mu}{d\tau}. \tag{5.6}$$

Writing out the sum in the case of two dimensions Eq. (5.6) looks like this

$$\frac{dx^{\mu'}}{d\tau} = \frac{\partial x^{\mu'}}{\partial x^1}\frac{dx^1}{d\tau} + \frac{\partial x^{\mu'}}{\partial x^2}\frac{dx^2}{d\tau}. \tag{5.7}$$

Let us consider the example with plane polar coordinates and Cartesian coordinates, again. Then Eq. (5.7) represents the two equations

$$\frac{dx}{d\tau} = \frac{\partial x}{\partial r}\frac{dr}{d\tau} + \frac{\partial x}{\partial \theta}\frac{d\theta}{d\tau} = \cos\theta\frac{dr}{d\tau} - r\sin\theta\frac{d\theta}{d\tau} \tag{5.8}$$

and

$$\frac{dy}{d\tau} = \frac{\partial y}{\partial r}\frac{dr}{d\tau} + \frac{\partial y}{\partial \theta}\frac{d\theta}{d\tau} = \sin\theta\frac{dr}{d\tau} + r\cos\theta\frac{d\theta}{d\tau}. \tag{5.9}$$

Using Eq. (3.16) we can write Eq. (5.6) as

$$u^{\mu'} = \frac{\partial x^{\mu'}}{\partial x^{\mu}} u^{\mu}. \tag{5.10}$$

A transformation of some quantities is said to be *linear and homogeneous* if all new (i.e. transformed) quantities (vector components in the present case) are linear combinations of the old quantities. Equation (5.10) represents a linear and homogeneous transformation of the vector components u^{μ}.

Transforming, for example, the components of a vector in the plane polar coordinate system to its components in the Cartesian system, we get, from Eqs. (5.8)–(5.10)

$$\begin{aligned} u^x &= \cos\theta u^r - r\sin\theta u^\theta, \\ u^y &= \sin\theta u^r + r\cos\theta u^\theta. \end{aligned} \tag{5.11}$$

Written out for a vector in an n-dimensional space, the innocent-looking Eq. (5.10) represents the following set of equations

$$\left.\begin{aligned} u^{1'} &= \frac{\partial x^{1'}}{\partial x^1}u^1 + \frac{\partial x^{1'}}{\partial x^2}u^2 + \cdots + \frac{\partial x^{1'}}{\partial x^n}u^n \\ u^{2'} &= \frac{\partial x^{2'}}{\partial x^1}u^1 + \frac{\partial x^{2'}}{\partial x^2}u^2 + \cdots + \frac{\partial x^{2'}}{\partial x^n}u^n \\ &\vdots \\ u^{n'} &= \frac{\partial x^{n'}}{\partial x^1}u^1 + \frac{\partial x^{n'}}{\partial x^2}u^2 + \cdots + \frac{\partial x^{n'}}{\partial x^n}u^n \end{aligned}\right\}. \tag{5.12}$$

The linear and homogeneous character of the transformation 'law' of vector components implies that all the terms on the right-hand side of the system of equations (5.12) are proportional to an 'old' vector component, u^{μ}. This has a most important consequence. One can always orient a coordinate system (which we choose as the 'old' system), so that a vector in the old coordinate system has just one component, for example u^1, different from zero, or, as it is often expressed, one non-vanishing component. Since all the coordinates of both the old and the new system are included in every coordinate transformation, at least one of the new coordinates is a function of x^1. So, at least one of the coefficients $\partial x^{\mu'}/\partial x^1$ of u^1 is different from zero (i.e. non-vanishing). Looking at the system of equations, we then see that at least one of the new vector components are non-vanishing. The conclusion of this verbal mathematical deduction is: *No vector can be transformed away*. And neither can a new vector appear due to a coordinate transformation. The existence of any vector is coordinate invariant.

In the rest of this book, when we mention systems of equations written in a condensed form, like Eq. (5.10), we shall permit ourselves to write only 'equation' even where 'set of equations' would be more correct.

Equation (5.10) is the transformation equation of the components of a tangent vector under an arbitrary change of coordinate system. It was deduced for tangent vectors, but all vectors can in fact be regarded as tangent vectors. Therefore Eq. (5.10) holds generally for all vectors. This equation furnishes the completely general rule for transforming the components of an arbitrary vector.

By exchanging the indices we obtain the transformation equation for vector components from the new coordinate system back to the old one

$$u^{\mu} = \frac{\partial x^{\mu}}{\partial x^{\mu'}} u^{\mu'}. \tag{5.13}$$

The vector components, u^{μ}, with superscripts, transform according to Eq. (5.10) from the old coordinate system to the new one. The coefficients $\partial x^{\mu'}/\partial x^{\mu}$ have two indices, and can be arranged as a matrix, which is called the *transformation matrix* from the old to the new system. The coefficients $\partial x^{\mu'}/\partial x^{\mu}$ are called the *elements* of the transformation matrix. The vector components transform according to Eq. (5.13) back to the old system. The coefficients $\partial x^{\mu}/\partial x^{\mu'}$ of this backwards transformation are the elements of the *inverse* transformation matrix. Performing a to and fro transformation, i.e. a transformation followed by the inverse transformation, leads to the original vector components. Replacing $u^{\mu'}$ in eq (5.13) by the right-hand side of Eq. (5.10), we find the result of the to and fro transformation,

$$u^{\mu} = \frac{\partial x^{\mu}}{\partial x^{\mu'}} \frac{\partial x^{\mu'}}{\partial x^{\nu}} u^{\nu}. \tag{5.14}$$

Note that we changed the summation index μ in Eq. (5.10) to ν, since the letter μ was already used as a free (i.e. non-summation) superscript in the equation. Such changes of summation indices are always permitted. Which letter we use for a summation index, also called a *dummy index*, does not matter.

The great German mathematician Leopold Kronecker (1823–91) is known for his disrespect of other numbers than the whole ones: He coined the maxim "God created the whole numbers, the rest is human doing". A curious little symbol $\delta^{\mu}{}_{\nu}$ is called 'the Kronecker symbol'. It may be defined as follows:

$$\delta_{\mu\nu} \equiv \delta^{\mu}{}_{\nu} \equiv \begin{cases} 1 & \text{if} \quad \mu = \nu \\ 0 & \text{if} \quad \mu \neq \nu \end{cases} \tag{5.15}$$

We now utilize the Kronecker symbol to write

$$u^{\mu} = \delta^{\mu}{}_{\nu} u^{\nu}. \tag{5.16}$$

The only non-vanishing term in this sum is obtained for $\nu = \mu$. Equation (5.14) can thus be written

$$\frac{\partial x^{\mu}}{\partial x^{\mu'}} \frac{\partial x^{\mu'}}{\partial x^{\nu}} u^{\nu} = \delta^{\mu}{}_{\nu} u^{\nu},$$

which shows that the elements of a transformation matrix and of the inverse transformation matrix fulfils the relationship

$$\frac{\partial x^{\mu}}{\partial x^{\mu'}} \frac{\partial x^{\mu'}}{\partial x^{\nu}} = \delta^{\mu}{}_{\nu}. \tag{5.17}$$

5.4 The Galilean coordinate transformation

The Galilean coordinate transformation is a transformation between a system at rest and a moving system. Before we proceed to write down the mathematical form of the transformation, we need to make some conceptual preparations. The expression 'the motion of a coordinate system', has not yet been given a clear meaning. A coordinate system mapping a space is essentially a continuum of sets of numbers, where each number represents the value of a coordinate. In spacetime one such set at a point contains four numbers, the values of, say, (x, y, z, t) at the point. The coordinates are mathematical objects. In order to give a meaning to the above expression, we must give a physical interpretation of the coordinates. An intermediate step is to introduce a reference for motion. This is called a *reference frame*, and may be defined as a continuum of particles with given motion. A *comoving coordinate system* in a reference frame is a coordinate system in which the reference particles of the frame are at rest, i.e. they have constant spatial coordinates.

Let us choose the platform of a railway station as one reference frame, and a train moving past the platform with constant velocity v as another reference frame. The 'old' coordinate system, $\{x, t\}$, is comoving with the platform, i.e. it may be thought of as a system of measuring rods and clocks at rest on the platform, and the 'new' one, $\{x', t'\}$ is comoving with the train. The Galilean coordinate transformation from the platform system to the train system is

$$x' = x - vt \quad \text{and} \quad t' = t, \tag{5.18}$$

where we have omitted the y and z-coordinates. From this transformation we find, for example, that the position, in the train system, of the origin $x = 0$ of the platform system, is $x = -vt$, which shows that the platform moves with velocity v

in the negative x'-direction relative to the train. The corresponding transformation matrix is

$$\begin{bmatrix} \frac{\partial x'}{\partial x} & \frac{\partial x'}{\partial t} \\ \frac{\partial t'}{\partial x} & \frac{\partial t'}{\partial t} \end{bmatrix} = \begin{bmatrix} 1 & -v \\ 0 & 1 \end{bmatrix}. \tag{5.19}$$

The inverse transformation represents a transformation from the train system to the platform system,

$$x = x' + vt' \quad \text{and} \quad t = t'. \tag{5.20}$$

The inverse transformation matrix is

$$\begin{bmatrix} \frac{\partial x}{\partial x'} & \frac{\partial x}{\partial t'} \\ \frac{\partial t}{\partial x'} & \frac{\partial t}{\partial t'} \end{bmatrix} = \begin{bmatrix} 1 & v \\ 0 & 1 \end{bmatrix}. \tag{5.21}$$

Let us apply the Galilean transformation to the components of a velocity vector. The way this is done in elementary mechanics, is to define the velocity of a particle in the platform system by

$$\vec{u} = u\,\vec{e}_x = \frac{dx}{dt}\vec{e}_x \tag{5.22}$$

and the velocity in the train system by

$$\vec{u}' = u'\vec{e}_{x'} = \frac{dx'}{dt'}\vec{e}_{x'}. \tag{5.23}$$

Differentiating the transformation equation (5.18) and using the definitions (5.22) and (5.23), we obtain the Galilean velocity transformation

$$u' = u - v. \tag{5.24}$$

We should obtain the same result by applying Eq. (5.12). The only nonvanishing velocity component in the platform system is $u^1 = u$. Then Eq. (5.12) is reduced to

$$u' = \frac{\partial x'}{\partial x}u = u, \tag{5.25}$$

which is clearly not correct. What has gone wrong?

The reason for the failure is deeper than just a calculation error. It has to do with what we mean by a vector. It is not sufficient to think of a vector just as a quantity with magnitude and direction. From our definition of the dimension of a space follows that a vector has always the same number of components (not necessarily nonvanishing) as the number of dimensions in the space it exists in. A Galilean transformation concerns space and time. Four dimensions are involved, even if time and space are not woven together as in the theory of relativity. Therefore, in order to obtain a proper application of the Galilean transformation, Eq. (5.19), to the components of a vector, the vector must be defined not only with spatial components, but also with a time component. Still omitting the y and z dimensions, we then have

$$\vec{u} = u^x \vec{e}_x + u^t \vec{e}_t \tag{5.26}$$

with

$$u^x = \frac{dx}{dt} \quad \text{and} \quad u^t = \frac{dt}{dt} = 1. \tag{5.27}$$

The transformation of the components of the velocity vector, is now found by inserting the elements of the transformation matrix (5.19), and $u^t = 1$, into the transformation formula (5.10),

$$u^{x'} = \frac{\partial x'}{\partial x} u^x + \frac{\partial x'}{\partial t} u^t = u^x - v \tag{5.28}$$

which is the correct result.

If we consider motion in an arbitrary direction, not only along the x-axis, the Galilean transformation takes the form

$$\vec{u}' = \vec{u} - \vec{v}. \tag{5.29}$$

The fundamental dynamical law in Newtonian mechanics is Newton's second law, which we considered in section 3.4. This has a most interesting property which we shall now show.

Let us assume that a particle with mass m is acted upon by a force $\vec{f}$. In the rest frame of the platform the particle then has an acceleration $\vec{a}$ given by Newton's second law $\vec{f} = m\vec{a}$. Acceleration is the derivative of the velocity. Differentiating Eq. (5.29), and noting that $d\vec{v}/dt = 0$ since $\vec{v}$ is the constant velocity of, say, the train, we get

$$\vec{a}' = \vec{a}. \tag{5.30}$$

Hence, as measured in the rest frame of the train, the acceleration of the particle is the same as measured on the platform.

In Newtonian mechanics forces and masses are absolute quantities, i.e.

$$\vec{f}' = \vec{f} \quad \text{and} \quad m' = m. \tag{5.31}$$

I follows that $\vec{f}' = m'\vec{a}'$, showing that Newton's second law is valid in the same form in two reference frames connected by a Galilean transformation. One usually expresses this by saying that the fundamental equations of Newtonian mechanics are invariant under Galilean transformations. This is what is meant when one says that Newtonian mechanics obeys the principle of relativity.

5.5 Transformation of basis vectors

We shall deduce the transformation formula for basis vectors, and start by using that the vector sum of the component vectors is equal to the vector itself. If we transform from the old to the new coordinate system, the vector itself is not affected, but the components of the vector are in general different in the two coordinate systems,

$$\vec{u} = u^{\mu'} \vec{e}_{\mu'} = u^{\mu} \vec{e}_{\mu} \tag{5.32}$$

where $\vec{e}_{\mu'}$ are the basis vectors of the new coordinate system.

From their dishonorable slavery under the tyranny of coordinate axes, we see that basis vectors are in a sense fake vectors. They are not sovereign, immutable beings, but chained to a coordinate system. Equations (5.13) and (5.32) give

$$u^{\mu} \vec{e}_{\mu} = u^{\mu'} \vec{e}_{\mu'} = \frac{\partial x^{\mu'}}{\partial x^{\mu}} u^{\mu} \vec{e}_{\mu'} = u^{\mu} \frac{\partial x^{\mu'}}{\partial x^{\mu}} \vec{e}_{\mu'}. \tag{5.33}$$

Since this equation is valid for an arbitrary vector $\vec{u}$, which may have only one component u^{μ} different from zero, it follows that

$$\vec{e}_{\mu} = \frac{\partial x^{\mu'}}{\partial x^{\mu}} \vec{e}_{\mu'}. \tag{5.34}$$

This is the formula for transforming the basis vectors of the new system to the old one.

Unlike other vectors, a set of basis vectors is not invariant against a change of coordinate system. Like the component vectors that we talked about in Sect. 1.4, they have magnitudes and directions, but the 'identity' of the basis vectors, their magnitudes and directions, depend upon the coordinate system.

Applying Eq. (5.34) to the example after Eq. (5.5) we find the transformation from the basis vectors of the Cartesian coordinate system to those of the system with plane polar coordinates,

$$\vec{e}_r = \frac{\partial x}{\partial r} \vec{e}_x + \frac{\partial y}{\partial r} \vec{e}_y = \cos\theta \vec{e}_x + \sin\theta \vec{e}_y, \tag{5.35}$$

$$\vec{e}_\theta = \frac{\partial x}{\partial \theta} \vec{e}_x + \frac{\partial y}{\partial \theta} \vec{e}_y = -r\sin\theta \vec{e}_x + r\cos\theta \vec{e}_y, \tag{5.36}$$

in accordance with Eqs. (4.40) and (4.43).

By exchanging the indices μ' and μ we obtain an equation for transforming the basis vectors of the old coordinate system to those of the new system.

$$\vec{e}_{\mu'} = \frac{\partial x^{\mu}}{\partial x^{\mu'}} \vec{e}_{\mu}. \tag{5.37}$$

The basis vectors transform by means of the elements of the inverse transformation matrix, i.e. they transform inversely, i.e. in the opposite 'direction', relative to the transformation (5.13) of the vector components.

5.6 Covariant and contravariant vector components

Vectors continue to be the same when described in different coordinate systems, that is vectors are *invariant*. This means that the product sums of vector components and basis vectors, $v^\mu \vec{e}_\mu = v^1\vec{e}_1 + v^2\vec{e}_2 + \cdots$, must be invariant (see Eq. (5.32)). This invariance may be expressed in two different ways. Either by using vector components, u^μ, that transform by means of the elements of the transformation matrix, and basis vectors, $\vec{e}_\mu$, that transform by means of the elements of the inverse matrix. Or one could introduce vector components, u_μ, that transform by means of the inverse matrix, and basis vectors, $\vec{e}^\mu$, that transform by means of the ordinary transformation matrix. The first way is called a *contravariant* decomposition of a vector, 'contra' because the basis vectors transform by means of the inverse transformation matrix in this case. This is the usual decomposition which we have used all the time, with superscripts for the contravariant vector components and subscripts for the ordinary basis vectors, $\vec{u} = u^\mu \vec{e}_\mu$. The second way of expressing the invariance of a vector, is called the *covariant* decomposition of a vector, with subscripts for the covariant vector components and superscripts for the basis vectors, $\vec{u} = u_\mu \vec{e}^\mu$. The basis vectors $\vec{e}^\mu$ are called *covectors*.

The covectors, $\vec{e}^\mu$, are defined implicitely by

$$\vec{e}^\mu \cdot \vec{e}_\nu = \delta^\mu{}_\nu, \tag{5.38}$$

where $\vec{e}_\nu$ are the usual basis vectors, and $\delta^\mu{}_\nu$ is the Kronecker symbol. The transformation formula for the covectors may be deduced as follows. From Eqs. (5.37) and (5.38) we get

$$\frac{\partial x^\nu}{\partial x^{\nu'}}\vec{e}^{\mu'} \cdot \vec{e}_\nu = \vec{e}^{\mu'} \cdot \frac{\partial x^\nu}{\partial x^{\nu'}}\vec{e}_\nu = \vec{e}^{\mu'} \cdot \vec{e}_{\nu'} = \delta^{\mu'}{}_{\nu'}. \tag{5.39}$$

Multiplying by the elements $\partial x^{\nu'}/\partial x^\alpha$ of the transformation matrix, leads to

$$\frac{\partial x^{\nu'}}{\partial x^\alpha}\frac{\partial x^\nu}{\partial x^{\nu'}}\vec{e}^{\mu'} \cdot \vec{e}_\nu = \frac{\partial x^{\nu'}}{\partial x^\alpha}\delta^{\mu'}{}_{\nu'}. \tag{5.40}$$

Applying Eqs. (5.38) and (5.15) we obtain

$$\delta^\nu{}_\alpha \vec{e}^{\mu'} \cdot \vec{e}_\nu = \vec{e}^{\mu'} \cdot \vec{e}_\alpha = \frac{\partial x^{\mu'}}{\partial x^\alpha} = \frac{\partial x^{\mu'}}{\partial x^\mu}\delta^\mu{}_\alpha. \tag{5.41}$$

Using the definition (5.38) we get

$$\vec{e}^{\mu'} \cdot \vec{e}_\alpha = \frac{\partial x^{\mu'}}{\partial x^\mu} \vec{e}^\mu \cdot \vec{e}_\alpha.$$

Hence

$$\vec{e}^{\mu'} = \frac{\partial x^{\mu'}}{\partial x^\mu} \vec{e}^\mu. \tag{5.42}$$

Comparing with Eq. (5.13) we see that the covectors transform in the same way as the contravariant vector components. The transformation from the marked to the unmarked system is

$$\vec{e}^\mu = \frac{\partial x^\mu}{\partial x^{\mu'}} \vec{e}^{\mu'}. \tag{5.43}$$

A vector may be decomposed along the covectors as follows

$$\vec{u} = u_\mu \vec{e}^\mu. \tag{5.44}$$

The transformation formula for the covariant vector components are found by using Eq. (5.43)

$$\vec{u} = u_\mu \vec{e}^\mu = u_\mu \frac{\partial x^\mu}{\partial x^{\mu'}} \vec{e}^{\mu'} = \frac{\partial x^\mu}{\partial x^{\mu'}} u_\mu \vec{e}^{\mu'} = u_{\mu'} \vec{e}^{\mu'},$$

giving

$$u_{\mu'} = \frac{\partial x^\mu}{\partial x^{\mu'}} u_\mu, \tag{5.45}$$

which shows that the covariant vector components transform in the same way as the ordinary basis vectors.

The relationship between the ordinary basis vectors and the covectors, and between the contravariant and covariant components of a vector $\vec{A}$, is illustrated in Fig. 5.2.

5.7 Tensors

Until now we have considered scalar functions and vectors. The value of a scalar function at a point is specified by one number. Such functions are adequate to describe simple fields, such as the temperature in a region. However, in order to represent mathematically the wind at the surface of the Earth, it does not suffice to specify one number at each point. One needs to specify two numbers, for example the velocity of the wind in the East-West direction and in the North-South direction. A still more complicated problem is to represent mathematically the stresses in a body. Then one has to specify both the direction of a force and the direction of the

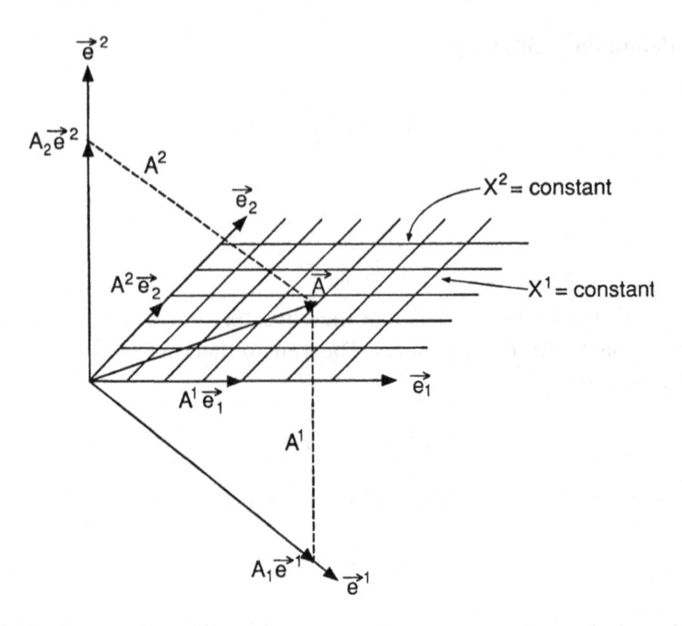

Fig. 5.2 Basis vectors and covectors

surface that the force acts upon. One needs a quantity consisting of two vectors. Such quantities of a certain type shall be defined below. They are called tensors of rank two. Vectors are called tensors of rank one, and scalars are called tensors of rank zero. Tensors have components, and the rank is just the number of indices of their components.

When we talk of 'vector fields' we think of vectors that are functions of the coordinates. However, a vector can also be thought of as a linear function acting on other vectors and giving out a real number. When one vector acts upon another in this way, the real number it gives out is the scalar product between the other vector and itself. There is an additional convention here. A vector, thought of as a tensor of rank one, is only allowed to act upon a vector of opposite type than itself. A basis vector is called a contravariant tensor of rank one, and a covector is called a covariant tensor of rank one. A contravariant tensor of rank one can only act upon a covariant basis vector of rank one, and vice versa,

$$\vec{e}_\mu(\vec{e}^\nu) = \vec{e}_\mu \cdot \vec{e}^\nu \quad \text{and} \quad \vec{e}^\mu(\vec{e}_\nu) = \vec{e}^\mu \cdot \vec{e}_\nu.$$

The scalar product is defined to be symmetrical. Using Eq. (5.38) we find that the real numbers obtained when the basis vectors act upon each other, are the values of the Kronecker symbols,

$$\vec{e}_\mu(\vec{e}^\nu) = \vec{e}^\mu(\vec{e}_\nu) = \delta^\mu{}_\nu. \tag{5.46}$$

In order to be able to write a tensor of rank two or higher in component form we must introduce the tensor product, which is denoted by $\otimes$ and defined by

$$\vec{e}_\mu \otimes \vec{e}_\nu(\vec{e}^\alpha, \vec{e}^\beta) = \vec{e}_\mu(\vec{e}^\alpha)\vec{e}_\nu(\vec{e}^\beta). \tag{5.47}$$

There are three types of basis tensors of rank two or greater:

1. A *contravariant* basis tensor acts only upon covectors.
2. A *covariant* basis tensor acts only upon ordinary vectors.
3. A *mixed* basis tensor acts upon both ordinary vectors and covectors.

Tensors of rank two can be written in component form as linear combinations of basis tensors of the form $\vec{e}_\mu \otimes \vec{e}_\nu$, $\vec{e}^\mu \otimes \vec{e}^\nu$, $\vec{e}^\mu \otimes \vec{e}_\nu$, or $\vec{e}_\mu \otimes \vec{e}^\nu$:

$$E = E^{\mu\nu}\vec{e}_\mu \otimes \vec{e}_\nu, \quad F = F_{\mu\nu}\vec{e}^\mu \otimes \vec{e}^\nu,$$
$$G = G_\mu{}^\nu\vec{e}^\mu \otimes \vec{e}_\nu, \quad H = H^\mu{}_\nu\vec{e}_\mu \otimes \vec{e}^\nu. \tag{5.48}$$

Tensors are a sort of generalized vectors. Like a vector, a tensor has a coordinate independent existence. Under a coordinate transformation the tensor components change, but not the tensor itself. The transformation formulae for the contravariant components of a tensor of rank two can be deduced by using the transformation formula Eq. (5.34) for the basis vectors,

$$\begin{aligned} S &= S^{\mu'\nu'}\vec{e}_{\mu'} \otimes \vec{e}_{\nu'} = S^{\mu\nu}\vec{e}_\mu \otimes \vec{e}_\nu \\ &= S^{\mu\nu}\frac{\partial x^{\mu'}}{\partial x^\mu}\vec{e}_{\mu'} \otimes \frac{\partial x^{\nu'}}{\partial x^\nu}\vec{e}_{\nu'} = \frac{\partial x^{\mu'}}{\partial x^\mu}\frac{\partial x^{\nu'}}{\partial x^\nu}S^{\mu\nu}\vec{e}_{\mu'} \otimes \vec{e}_{\nu'}. \end{aligned} \tag{5.49}$$

Hence

$$S^{\mu'\nu'} = \frac{\partial x^{\mu'}}{\partial x^\mu}\frac{\partial x^{\nu'}}{\partial x^\nu}S^{\mu\nu}. \tag{5.50}$$

In the same way one finds the transformation formulae for the covariant and the mixed components

$$S_{\mu'\nu'} = \frac{\partial x^\mu}{\partial x^{\mu'}}\frac{\partial x^\nu}{\partial x^{\nu'}}S_{\mu\nu}, \tag{5.51a}$$

$$S^{\mu'}{}_{\nu'} = \frac{\partial x^{\mu'}}{\partial x^\mu}\frac{\partial x^\nu}{\partial x^{\nu'}}S^\mu{}_\nu, \tag{5.51b}$$

and

$$S_{\mu'}{}^{\nu'} = \frac{\partial x^\mu}{\partial x^{\mu'}}\frac{\partial x^{\nu'}}{\partial x^\nu}S_\mu{}^\nu. \tag{5.51c}$$

5.8 The metric tensor

As we have mentioned above, the geometrical properties of coordinate systems reflect the geometrical properties, such as symmetries and curvature, of the space they cover. This information about the geometry of a space is imprinted in the way that the coordinate basis vector fields (see Sect. 1.3 on vector fields) vary with position. This, again, is encoded in the position and time dependence of the scalar products, $\vec{e}_\mu \cdot \vec{e}_\nu$, of all the basis vectors in a coordinate system.

In systems of curved coordinates both the angle between basis vectors and their length may change from neighbourhood to neighbourhood. The situation is so complicated that we need an exceedingly clever 'notation', that is, mathematical terminology, in order to be able to easily survey the wilderness of changes. When we deal with n dimensions, we get n^2 scalar products between the basis vectors, all of which may change from neighbourhood to neighbourhood,

$$\vec{e}_\mu \cdot \vec{e}_\nu, \quad \mu = 1, 2, \ldots, n, \quad \nu = 1, 2, \ldots, n. \tag{5.52}$$

Spelled out this is a short-hand for a collection of scalar products

$$\begin{bmatrix} \vec{e}_1 \cdot \vec{e}_1 & \vec{e}_1 \cdot \vec{e}_2 & \cdots & \vec{e}_1 \cdot \vec{e}_n \\ \vec{e}_2 \cdot \vec{e}_1 & \vec{e}_2 \cdot \vec{e}_2 & \cdots & \vec{e}_2 \cdot \vec{e}_n \\ \vdots & \vdots & & \vdots \\ \vec{e}_n \cdot \vec{e}_1 & \vec{e}_n \cdot \vec{e}_2 & \cdots & \vec{e}_n \cdot \vec{e}_n \end{bmatrix}. \tag{5.53}$$

The $\vec{e}$'s are basis vectors along coordinate curves that are, in the general case, not straight. In what follows we shall explain why and how the n^2 quantities in (5.53) tell us everything we need about the geometrical properties of our coordinate system.

Figure 5.3 shows two basis vectors $\vec{e}_\mu$ and $\vec{e}_\nu$ with an angle α between them. Equation (4.5) applied to the present case shows that the projection of $\vec{e}_\nu$ onto $\vec{e}_\mu$, written $\vec{e}_{\nu\parallel}$, has a magnitude $|\vec{e}_{\nu\parallel}| = |\vec{e}_\nu| \cos\alpha$. From Eq. (1.8) for the scalar product of two vectors then follows

$$\vec{e}_\mu \cdot \vec{e}_\nu = |\vec{e}_\mu||\vec{e}_\nu| \cos\alpha. \tag{5.54}$$

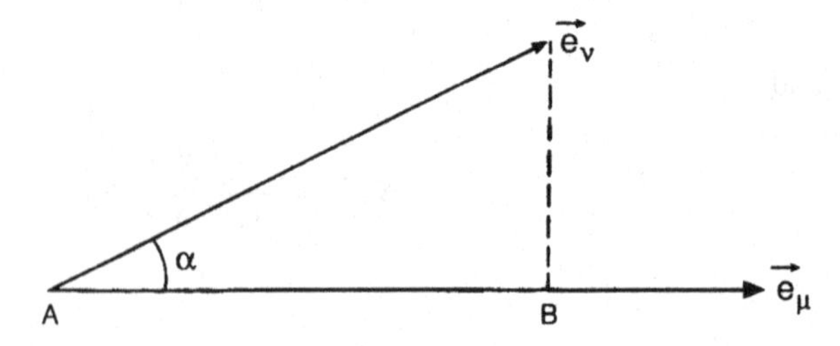

Fig. 5.3 Basis vectors

Note from Eq. (5.54) that the order of the vectors means nothing for the magnitude or the sign of the scalar product. This is expressed by saying that the scalar product is *symmetrical* under exchange of the vectors:

$$\vec{e}_\mu \cdot \vec{e}_\nu = \vec{e}_\nu \cdot \vec{e}_\mu. \tag{5.55}$$

The mathematicians' response to the demand for a clever terminology, making it possible to treat all the scalar products in a systematic way, was to invent a new concept: the metric tensor. In Ch. 1 we introduced, just as an anticipation of things to come, the famous expression of the *components of the metric tensor*

$$g_{\mu\nu} \equiv \vec{e}_\mu \cdot \vec{e}_\nu \tag{5.56}$$

where each $\vec{e}_\mu$ is a basis vector field, like the vector field in Fig. 1.4, i.e. the magnitudes and directions of the vectors depend upon the position. This implies that the components, $g_{\mu\nu}$, of the metric tensor are in general functions of the coordinates.

The three lines indicate that (5.56) is not an equation. It is a regulative definition announcing that in what follows the expression $g_{\mu\nu}$ will be used as a short-hand for $\vec{e}_\mu \cdot \vec{e}_\nu$.

According to the definition (5.56) and Eq. (5.55) *the metric tensor is symmetrical*,

$$g_{\nu\mu} = g_{\mu\nu}. \tag{5.57}$$

The number of independent components, $g_{\mu\nu}$, of the metric tensor is thereby reduced from n^2 to $n(n+1)/2$ in an n-dimensional space. In spaces of one, two, three and four dimensions, for example, the metric tensor has respectively one, three, six and ten independent components. These are usually written in matrix form. For the mentioned number of dimensions they are, respectively

$$g_{\mu\nu} = \begin{bmatrix} g_{11} \end{bmatrix},$$

$$g_{\mu\nu} = \begin{bmatrix} g_{11} & \\ g_{21} & g_{22} \end{bmatrix},$$

$$g_{\mu\nu} = \begin{bmatrix} g_{11} & & \\ g_{21} & g_{22} & \\ g_{31} & g_{32} & g_{33} \end{bmatrix},$$

and

$$g_{\mu\nu} = \begin{bmatrix} g_{11} & & & \\ g_{21} & g_{22} & & \\ g_{31} & g_{32} & g_{33} & \\ g_{41} & g_{42} & g_{43} & g_{44} \end{bmatrix},$$

where we have written only the independent components. The unwritten components on the upper right half of the matrices are equal to those on the lower left part.

In a coordinate system where the basis vectors are orthogonal to each other, the scalar products of different basis vectors are zero, according to Eq. (5.54) since $\alpha = \pi/2$ for orthogonal vectors, and $\cos(\pi/2) = 0$. Only the scalar products of each vector by itself are different from zero. This means that in a coordinate system with orthogonal basis vectors, only the components of the metric tensor with equal indices are different from zero. Since they are found along the diagonal when the components are written as a matrix, such a metric tensor is called *diagonal*. Thus, in the case of a right angled coordinate system in four-dimensional spacetime, the metric tensor has the form

$$g_{\mu\nu} = \begin{bmatrix} g_{11} & & & \\ & g_{22} & & \\ & & g_{33} & \\ & & & g_{44} \end{bmatrix}. \tag{5.58}$$

To save space this is usually written as

$$g_{\mu\nu} = \mathrm{diag}[g_{11}, g_{22}, g_{33}, g_{44}]. \tag{5.59}$$

Example 5.1. The simplest two-dimensional case is the metric of a Cartesian coordinate system in a plane. Then the only scalar products of basis vectors different from zero are $\vec{e}_x \cdot \vec{e}_x = 1$ and $\vec{e}_y \cdot \vec{e}_y = 1$, and the only non-vanishing components of the metric tensor are,

$$g_{11} = g_{22} = 1. \tag{5.60}$$

Example 5.2. Let us now, as an illustration, calculate the components of the metric tensor in the coordinate system with plane polar coordinates. The basis vectors are given in Eqs. (5.35) and (5.36). The scalar products are

$$\begin{aligned} \vec{e}_r \cdot \vec{e}_r &= \left(\cos\theta\, \vec{e}_x + \sin\theta\, \vec{e}_y\right)\left(\cos\theta\, \vec{e}_x + \sin\theta\, \vec{e}_y\right) \\ &= \cos^2\theta\, \vec{e}_x \cdot \vec{e}_x + \cos\theta\, \sin\theta\, \vec{e}_x \cdot \vec{e}_y \\ &\quad + \sin\theta\, \cos\theta\, \vec{e}_y \cdot \vec{e}_x + \sin^2\theta\, \vec{e}_y \cdot \vec{e}_y \\ &= \left(\cos^2\theta\right) 1 + (\cos\theta\, \sin\theta)\, 0 \\ &\quad + (\sin\theta\, \cos\theta)\, 0 + \left(\sin^2\theta\right) 1 \\ &= \cos^2\theta + \sin^2\theta = 1, \end{aligned} \tag{5.61}$$

$$\vec{e}_\theta \cdot \vec{e}_\theta = \left(-r \sin\theta\, \vec{e}_x + r\cos\theta\, \vec{e}_y\right)\left(-r \sin\theta\, \vec{e}_x + r\cos\theta\, \vec{e}_y\right) = r^2 \sin^2\theta + r^2\cos^2\theta = r^2\left(\sin^2\theta + \cos^2\theta\right) = r^2, \tag{5.62}$$

and

$$\vec{e}_r \cdot \vec{e}_\theta = \left(\cos\theta\, \vec{e}_x + \sin\theta\, \vec{e}_y\right)\left(-r \sin\theta\, \vec{e}_x + r\cos\theta\, \vec{e}_y\right) = -r\cos\theta\,\sin\theta + r\sin\theta\,\cos\theta = 0. \tag{5.63}$$

Equation (5.63) shows that the basis vectors of the coordinate system with plane polar coordinates are orthogonal, as shown in Fig. 4.10. Thus the metric tensor is diagonal, and—concluding this example—the only nonvanishing components are

$$g_{11} = 1 \quad \text{and} \quad g_{22} = r^2. \tag{5.64}$$

The transformation of the scalar products of the basis vectors under a change of coordinate system follows from Eq. (5.37)

$$\vec{e}_{\mu'} \cdot \vec{e}_{\nu'} = \frac{\partial x^\mu}{\partial x^{\mu'}}\vec{e}_\mu \cdot \frac{\partial x^\nu}{\partial x^{\nu'}}\vec{e}_\nu = \frac{\partial x^\mu}{\partial x^{\mu'}}\frac{\partial x^\nu}{\partial x^{\nu'}}\vec{e}_\mu \cdot \vec{e}_\nu. \tag{5.65}$$

From the definition (5.56) and Eq. (5.65) follows that the components of the metric tensor transform in the same way as the scalar products of the basis vectors

$$g_{\mu'\nu'} = \frac{\partial x^\mu}{\partial x^{\mu'}}\frac{\partial x^\nu}{\partial x^{\nu'}}\, g_{\mu\nu}. \tag{5.66}$$

Comparing with the first of Eqs. (5.49) we see that $g_{\mu\nu}$ transform as covariant components of a tensor of rank 2. The new components are linear combinations of the old ones.

Note that expressions such as $\frac{\partial x^\mu}{\partial x^{\mu'}}\frac{\partial x^\nu}{\partial x^{\nu'}}$ are products of two factors, and not a double differentiation. In the case of a two-dimensional space, with one coordinate system having coordinates $x^1 = x$ and $x^2 = y$ and another one $x^{1'} = x'$ and $x^{2'} = y'$, Eq. (5.66) represents the following system of equations

$$g_{1'1'} = \frac{\partial x^1}{\partial x^{1'}}\frac{\partial x^1}{\partial x^{1'}}g_{11} + \frac{\partial x^1}{\partial x^{1'}}\frac{\partial x^2}{\partial x^{1'}}g_{12} + \frac{\partial x^2}{\partial x^{1'}}\frac{\partial x^1}{\partial x^{1'}}g_{21} + \frac{\partial x^2}{\partial x^{1'}}\frac{\partial x^2}{\partial x^{1'}}g_{22}$$

$$
\begin{aligned}
g_{1'2'} &= \frac{\partial x^1}{\partial x^{1'}} \frac{\partial x^1}{\partial x^{2'}} g_{11} + \frac{\partial x^1}{\partial x^{1'}} \frac{\partial x^2}{\partial x^{2'}} g_{12} \\
&\quad + \frac{\partial x^2}{\partial x^{1'}} \frac{\partial x^1}{\partial x^{2'}} g_{21} + \frac{\partial x^2}{\partial x^{1'}} \frac{\partial x^2}{\partial x^{2'}} g_{22} \\
g_{2'1'} &= \frac{\partial x^1}{\partial x^{2'}} \frac{\partial x^1}{\partial x^{1'}} g_{11} + \frac{\partial x^1}{\partial x^{2'}} \frac{\partial x^2}{\partial x^{1'}} g_{12} \\
&\quad + \frac{\partial x^2}{\partial x^{2'}} \frac{\partial x^1}{\partial x^{1'}} g_{21} + \frac{\partial x^2}{\partial x^{2'}} \frac{\partial x^2}{\partial x^{1'}} g_{22} \\
g_{2'2'} &= \frac{\partial x^1}{\partial x^{2'}} \frac{\partial x^1}{\partial x^{2'}} g_{11} + \frac{\partial x^1}{\partial x^{2'}} \frac{\partial x^2}{\partial x^{2'}} g_{12} \\
&\quad + \frac{\partial x^2}{\partial x^{2'}} \frac{\partial x^1}{\partial x^{2'}} g_{21} + \frac{\partial x^2}{\partial x^{2'}} \frac{\partial x^2}{\partial x^{2'}} g_{22}
\end{aligned}
\tag{5.67}
$$

Example 5.3. Let us, as an illustrating application of this equation, transform the components of the metric tensor in a plane from a Cartesian coordinate system to a system with plane polar coordinates. Here we choose $x^1 = x$, $x^2 = y$, $x^{1'} = r$, and $x^{2'} = \theta$. Equations (5.67), (4.25) and (5.60) give the non-vanishing components

$$
\begin{aligned}
g_{1'1'} &= \frac{\partial x}{\partial r} \frac{\partial x}{\partial r} g_{11} + \frac{\partial y}{\partial r} \frac{\partial y}{\partial r} g_{22} \\
&= (\cos\theta)(\cos\theta)\, 1 + (\sin\theta)(\sin\theta)\, 1 \\
&= \cos^2\theta + \sin^2\theta = 1
\end{aligned}
\tag{5.68}
$$

and

$$
\begin{aligned}
g_{2'2'} &= \frac{\partial x}{\partial \theta} \frac{\partial x}{\partial \theta} g_{11} + \frac{\partial y}{\partial \theta} \frac{\partial y}{\partial \theta} g_{22} \\
&= (-r\sin\theta)(-r\sin\theta)\, 1 \\
&\quad + (r\cos\theta)(r\cos\theta)\, 1 \\
&= r^2\sin^2\theta + r^2\cos^2\theta \\
&= r^2\left(\sin^2\theta + \cos^2\theta\right) = r^2,
\end{aligned}
\tag{5.69}
$$

in accordance with Eq. (5.64). Note that we need to mark the indices only in a transformation equation, where we relate the components in two different coordinate systems, not in an equation such as Eq. (5.64) where only the components in one coordinate system appear.

The components, in a certain coordinate system, of the metric tensor of a space is often called just *the metric*. From the definition (5.56) is seen that the metric is directly related to a coordinate system, it is a 'personal signature', its 'fingerprint'.

However, the general theory of relativity is not a theory of coordinate systems, but of the space we supposedly live in: physics, not mathematics. The metric is a fundamental element of this theory. And it does indeed contain information not only of coordinate systems. In fact, from the metric we can extract everything we want to know about the geometry of the mathematically defined space that the coordinate system fills. Mathematically defined spaces are used as 'models' of the physical space. How all this comes about you are invited to learn in the following chapters. It requires particular attention to the contents of Ch. 6—the famous Christoffel symbol—and Ch. 9 which introduces the Riemann curvature tensor.

5.9 Covariant and contravariant tensor components

In Sects. 5.6 and 5.7 we met covariant, contravariant and mixed tensor components. Even if one can perform all calculations in general relativity in terms of one sort of components only, some expressions are simplified when different types of tensor components are allowed. Therefore it has become common practice to use both covariant, contravariant and mixed tensor components in relativistic calculations.

We shall here introduce the contravariant and mixed components of the metric tensor, and we shall show how the different sorts of tensor components can be calculated from each other by means of the covariant and contravariant components of the metric tensor.

The contravariant components of the metric tensor are defined by

$$g^{\mu\nu} = \vec{e}^{\mu} \cdot \vec{e}^{\nu}. \tag{5.70}$$

From Eqs. (5.70), (5.15) and (5.39) we have

$$\vec{e}^{\mu} \cdot \vec{e}^{\nu} = g^{\mu\nu} = g^{\nu\alpha} g^{\mu}{}_{\alpha} = g^{\nu\alpha} \vec{e}^{\mu} \cdot \vec{e}_{\alpha} = \vec{e}^{\mu} \cdot g^{\nu\alpha} \vec{e}_{\alpha}. \tag{5.71}$$

Hence

$$\vec{e}^{\nu} = g^{\nu\alpha} \vec{e}_{\alpha}. \tag{5.72}$$

From this, together with Eq. (5.56), we get

$$g^{\mu\nu} g_{\nu\alpha} = g^{\mu\nu} \vec{e}_{\nu} \cdot \vec{e}_{\alpha} = \vec{e}^{\mu} \cdot \vec{e}_{\alpha}. \tag{5.73}$$

Applying Eq. (5.38) we have

$$g^{\mu\nu} g_{\nu\alpha} = \delta^{\mu}{}_{\alpha}. \tag{5.74}$$

Using this equation one can calculate the contravariant components of the metric tensor from the covariant components. If the metric tensor is diagonal, so that $g_{\nu\alpha} \neq 0$ only for $\alpha = \nu$, Eq. (5.74) implies $g^{\mu\mu} g_{\nu\nu} = 1$ for $\nu = \mu$, which gives

$$g^{\mu\mu} = 1/g_{\mu\mu}. \tag{5.75}$$

We shall now find how the covariant vector components can be calculated from the contravariant components and the metric tensor. Multiplying the equation

$$u_\alpha \vec{e}^\alpha = u^\alpha \vec{e}_\alpha$$

by $\vec{e}_\mu$ we get

$$u_\alpha \vec{e}^\alpha \cdot \vec{e}_\mu = u^\alpha \vec{e}_\alpha \cdot \vec{e}_\mu.$$

Using Eq. (5.38) on the left-hand-side and Eqs. (5.56) and (5.57) on the right-hand side give

$$u_\alpha \delta^\alpha{}_\mu = u^\alpha g_{\alpha\mu} = g_{\mu\alpha} u^\alpha$$

or

$$u_\mu = g_{\mu\alpha} u^\alpha. \tag{5.76}$$

Calculating u_μ from Eq. (5.76) is called *lowering a tensor index.*

A corresponding formula for *raising a tensor index* is found by using, successively, Eqs. (5.15), (5.74) and (5.76),

$$u^\mu = \delta^\mu{}_\alpha u^\alpha = g^{\mu\nu} g_{\nu\alpha} u^\alpha = g^{\mu\nu} u_\nu. \tag{5.77}$$

Hence, by means of the metric tensor the contravariant components are mapped upon the covariant ones, and vice versa.

The general expression (1.26) for the scalar product of two vectors is

$$\vec{u} \cdot \vec{v} = g_{\mu\nu} u^\mu v^\nu. \tag{5.78}$$

In three dimensions, for example, this is a short hand notation for the nine terms in Eq. (1.27). Inserting first $u_\nu = g_{\nu\mu} u^\mu = g_{\mu\nu} u^\mu$ from Eq. (5.76), and then $v_\mu = g_{\mu\nu} v^\nu$, Eq. (5.78) may be written in two alternative ways

$$\vec{u} \cdot \vec{v} = u_\nu v^\nu = u^\mu v_\mu.$$

The mixed components of the metric tensor, $g^\mu{}_\nu$, are defined by

$$g^\mu{}_\nu = \vec{e}^\mu \cdot \vec{e}_\nu.$$

From Eq. (5.38) then follows

$$g^\mu{}_\nu = \delta^\mu{}_\nu. \tag{5.79}$$

Equation (5.79) shows that the mixed components of the metric tensor are the values of the Kronecker symbol.

As mentioned above, if the components of a tensor have one index, the tensor is said to be of *rank* 1. Thus, vectors are, in fact, tensors of rank 1. The components of the metric tensor are a set of quantities with two indices. It is said to be of *rank* 2.

This holds whatever the dimensions of the space. The transformation formulae for tensors of rank 2 are given in Eqs. (5.50) and (5.51). There are sets of quantities which also transform similarly, but have more than two indices, let us say three, μ, ν, and α. If three, the tensor is said to be of *rank* 3. They may be of any rank. The transformation formulae of the tensor components are deduced in the same way as in Eq. (5.50). The contravariant components, for example, transform according to the following rules

$$\left.\begin{aligned} T^{\mu'} &= \frac{\partial x^{\mu'}}{\partial x^{\mu}} T^{\mu} \\ T^{\mu'\nu'} &= \frac{\partial x^{\mu'}}{\partial x^{\mu}} \frac{\partial x^{\nu'}}{\partial x^{\nu}} T^{\mu\nu} \\ T^{\mu'\nu'\alpha'} &= \frac{\partial x^{\mu'}}{\partial x^{\mu}} \frac{\partial x^{\nu'}}{\partial x^{\nu}} \frac{\partial x^{\alpha'}}{\partial x^{\alpha}} T^{\mu\nu\alpha} \\ \vdots \qquad & \qquad \vdots \qquad\qquad \vdots \\ T^{\mu_1'\mu_2'\ldots\mu_n'} &= \frac{\partial x^{\mu_1'}}{\partial x^{\mu_1}} \frac{\partial x^{\mu_2'}}{\partial x^{\mu_2}} \cdots \frac{\partial x^{\mu_n'}}{\partial x_n^{\mu}} T^{\mu_1\mu_2\ldots\mu_n} \end{aligned}\right\} . \tag{5.80}$$

For all these quantities it is sufficient that they transform similarly to vectors. This secures the decisive coordinate independence, which is used in mathematical formulations of the basic natural laws whatever the complicated circumstances, say the steadiness and invariance of laws of oceanic waves in hurricanes. The variation of wind direction results in infinitely complicated forms of interfering wave-ridges, but presumably not in new basic 'laws'.

The tensor-notation is superbly economical: it makes sets of equations surveyable and understandable, which otherwise would be completely impossible to grasp. A formula in Eddington's elegant *The Mathematical Theory of Relativity* (p. 108) is worth looking at

$$\begin{aligned} (4!) J^2 g' = {} & \epsilon_{\alpha\beta\gamma\delta} \epsilon_{\iota\zeta\eta\theta} g'_{\alpha\iota} g'_{\beta\zeta} g'_{\gamma\eta} g'_{\delta\theta} \\ & \epsilon_{\iota\kappa\lambda\mu} \epsilon_{\nu\xi o\varpi} \frac{\partial x'_\nu}{\partial x_\iota} \frac{\partial x'_\zeta}{\partial x_\kappa} \frac{\partial x'_\eta}{\partial x_\lambda} \frac{\partial x'_\theta}{\partial x_\mu} \\ & \epsilon_{\rho\sigma\tau\upsilon} \epsilon_{\phi\chi\psi\omega} \frac{\partial x'_\phi}{\partial x_\rho} \frac{\partial x'_\chi}{\partial x_\sigma} \frac{\partial x'_\psi}{\partial x_\tau} \frac{\partial x'_\omega}{\partial x_\upsilon} \cdots \end{aligned}$$

His comment is laconic: "There are about 280 billion $= 280 \times 10^{12}$ terms on the right, and we proceed to rearrange those which do not vanish." He ends up with a very short formula—thanks to the ingenious tensor-notation. Without that instrument it might have been several million kilometres long.

5.10 The Lorentz transformation

In the general theory of relativity gravitation is described geometrically in terms of curved four-dimensional spacetime. This spacetime has a strange property which is apparent even in the flat spacetime of the special theory of relativity. Although flat, its geometry is not Euclidean. The distance between two points in spacetime is not given by the usual form of the Pythagorean theorem. The new form of the Pythagorean theorem, valid for four-dimensional spacetime, will be deduced in the next section, from the postulates of the special theory of relativity; that the velocity of light does not depend upon the motion of the light source and is equal in all directions, and that it is impossible, by performing experiments inside a laboratory, to measure any velocity of the laboratory. Thus the laws of nature should be formulated so that they are valid, unchanged, whatever the movement of the laboratory. This means that if we write down an equation representing a physical law, with reference to a certain laboratory frame, and then translate to another frame moving relatively to the first, the equation should not change. Thus, each term of the equation must change in the same way under the transformation between the two frames. In the special theory of relativity such a change of frame is done mathematically by applying the so-called Lorentz transformation that we shall now introduce.

Imagine a gardener sitting on a bench, getting more and more excited as he is observing what looks like an odd hummingbird. The hummingbird flies with constant velocity along a straight line. She has a watch, a yardstick and an organ that emits an extremely short flash of light just as she passes the gardener. The gardener, who is a hobby mathematician, decides to imagine a Cartesian coordinate system $\{x, y, z\}$ with x-axis along the travelling line of the hummingbird and with himself sitting at the origin, $x = 0$. Also he adjusts his clocks to show zero when the hummingbird passes him. In the gardener's coordinate system the position of the hummingbird at a point of time t is

$$x = vt, \quad \text{and} \quad y = z = 0. \tag{5.81}$$

When the hummingbird flies straight away with velocity v along the gardener's x-axis, she considers herself as being at rest at point zero of her x'-axis. Both the gardener and the odd hummingbird knows the Galilean transformation equations

$$x' = x - vt, \quad y' = y, \quad z' = z, \quad \text{and} \quad t' = t. \tag{5.82}$$

Equations (5.81) and (5.82) imply that the position of the hummingbird on her own axis is

$$x' = x - vt = 0 \quad \text{and} \quad y' = z' = 0, \tag{5.83}$$

showing that the hummingbird is indeed at rest at the origin of the marked coordinate system.

Neither mechanical nor optical experiments, nor any other sorts of experiments can decide whether the gardener or the hummingbird is at rest. They may both consider themselves at as rest and the other one as moving. This is the gist of Einstein's special principle of relativity expressed in layman terms.

Observing the flash of light emitted by the hummingbird they both find (surprisingly) that light travels isotropically with the speed c and expands into a spherical surface with radius ct. This is essentially what Einstein's second postulate says.

Because the hummingbird inevitably flies along a radius of the sphere away from the centre $x = 0$, the gardener says he occupies a privileged position and remains at the centre, whereas she, at $x' = 0$, gets further and further away from the centre. "Not so", says the hummingbird, "I am at the centre all the time". The gardener replies: "As you like it, honey, but see to it that I need to change my way of thinking as little as possible."

What is the least possible change of the Galilean transformation equations (5.18) such that the gardener and the hummingbird both can maintain that they are permanently at rest at the centre of the light wave, although they move relative to each other? The hummingbird promises to find a way out. But there are a few preliminary steps, she says. We must acknowledge that the equations of the two spherical surfaces are

$$x^2 + y^2 + z^2 = (ct)^2 \tag{5.84}$$

$$x'^2 + y'^2 + z'^2 = (ct')^2. \tag{5.85}$$

The radii ct and ct' expand proportionally to time. By subtracting the first equation from the second:

$$\left(x'^2 - x^2\right) + \left(y'^2 - y^2\right) + \left(z'^2 - z^2\right) = c^2\left(t'^2 - t^2\right). \tag{5.86}$$

We postulate that motion along the x axis does not influence the results of measuring distances in the (transverse) y and z directions. (Remember the promise of least possible change.) Thus

$$y' = y \quad \text{and} \quad z' = z$$

are still valid in our new transformation. Equation (5.86) then is reduced to

$$x'^2 - x^2 = c^2\left(t'^2 - t^2\right)$$

or

$$x'^2 - c^2t'^2 = x^2 - c^2t^2. \tag{5.87}$$

These were the preliminary steps. To proceed further we demand that the new transformation shall deviate as little as possible from the Galilean transformation. So we assume that the new transformation has the form $x' = \gamma(x - vt)$ and $t' = t$, where γ is a positive constant to be determined by inserting the expressions for x' and t' in Eq. (5.87). However, inserting $t' = t$ into Eq. (5.87) gives $x' = x$. This is only an 'identity transformation', so the proposed form of the transformation does not work.

A more radical solution is necessary in order to obtain a consistent kinematics, so that both the gardener and the hummingbird can regard themselves as permanently at the centre of the spherical light wave emitted by the hummingbird as it passed the gardner. We try a time transformation of the form

$$t = ax + bt, \tag{5.88}$$

where a and b are constants. Here t' is time measured with clocks at rest relative to the hummingbird, and t time measured with clocks at rest relative to the gardener. The physical consequence of a time transformation of the form (5.88) is unexpected. It implies that simultaneous events in the gardner's system are not simultaneous in the hummingbird's system, and vice versa. This can be seen as follows.

The proposed coordinate transformation has the form

$$x' = \gamma (x - vt) \quad \text{and} \quad t' = at + bx. \tag{5.89}$$

Consider two events with coordinates x_1 and x_2 that the gardner says are simultaneous. They happen for example at $t_1 = t_2 = 0$. The hummingbird then says that the events happen at the points of time $t_1' = bx_1$ and $t_2 = bx_2$. Events that are simultaneous according to the gardener, happens with a time difference $t_2' - t_1' = b(x_2 - x_1)$ according to the hummingbird.

This so-called 'relativity of simultaneity' is what makes the seemingly contradictory observations of the gardener and the hummingbird possible. What the hummingbird says is the light wave at a certain instant, the gardner will say is a succession of pictures of different parts of the wave at different points of time. The hummingbird will say the same about the gardener's wave.

We shall now go on and calculate the constants γ, a, and b in terms of the relative velocity v between the gardener and the hummingbird and the velocity of light, c. Inserting the expressions for x' and t' into Eq. (5.87) results in

$$[\gamma (x - vt)]^2 - c^2 (at + bx)^2 = x^2 - c^2 t^2.$$

Multplying out the left-hand side we get

$$\gamma^2 x^2 - 2\gamma^2 vxt + \gamma^2 v^2 t^2 - c^2 a^2 t^2 - 2c^2 abxt - c^2 b^2 x^2 = x^2 - c^2 t^2.$$

Collecting terms with the same powers of x and t gives

$$\left(\gamma^2 - c^2 b^2\right) x^2 - 2\left(\gamma^2 v + c^2 ab\right) xt + \left(\gamma^2 v^2 - c^2 a^2\right) t^2 = x^2 - c^2 t^2.$$

In order that the left-hand and right-hand side of this equation shall be equal for arbitrary values of x and t, the coefficients in front of x^2, of xt and of t^2, must be the same at each side of the equality sign. This leads to

$$\gamma^2 - c^2 b^2 = 1, \tag{5.90}$$

$$\gamma^2 v + c^2 ab = 0, \tag{5.91}$$

and

$$\gamma^2 v^2 - c^2 a^2 = -c^2. \tag{5.92}$$

Equation (5.91) gives

$$b = -\frac{\gamma^2 v}{c^2 a}. \tag{5.93}$$

Inserting this into Eq. (5.90), we get

$$\gamma^2 - c^2 \frac{\gamma^4 v^2}{c^4 a^2} = 1,$$

which may be written

$$\frac{\gamma^4 v^2}{c^2 a^2} = \gamma^2 - 1.$$

Thus

$$c^2 a^2 = \frac{\gamma^4 v^2}{\gamma^2 - 1}. \tag{5.94}$$

Inserting this into Eq. (5.92) gives

$$\gamma^2 v^2 - \frac{\gamma^4 v^2}{\gamma^2 - 1} = -c^2.$$

Multiplying each term by $\gamma^2 - 1$ leads to

$$\gamma^4 v^2 - \gamma^2 v^2 - \gamma^4 v^2 = -\gamma^2 c^2 + c^2.$$

The first and third term on the left-hand-side cancel each other. Collecting the terms with γ^2 on the left-hand-side gives

$$\left(c^2 - v^2\right) \gamma^2 = c^2$$

from which we get

$$\gamma^2 = \frac{c^2}{c^2 - v^2} = \frac{1}{1 - v^2/c^2}.$$

Taking the positive square root leads to

$$\gamma = \frac{1}{\sqrt{1 - v^2/c^2}}. \tag{5.95}$$

It follows that

$$\gamma^2 - 1 = \frac{c^2}{c^2 - v^2} - 1 = \frac{c^2 - c^2 + v^2}{c^2 - v^2} = \frac{v^2}{c^2 - v^2}$$
$$= \frac{v^2}{c^2} \frac{c^2}{c^2 - v^2} = \frac{v^2}{c^2} \gamma^2.$$

We insert this into Eq. (5.94),

$$c^2a^2 = \frac{\gamma^4 v^2}{(v^2/c^2)\gamma^2} = c^2\gamma^2,$$

from which follows

$$a = \gamma. \tag{5.96}$$

Subsituting this into Eq. (5.93), we obtain

$$b = -\frac{\gamma^2 v}{c^2\gamma} = \frac{v}{c^2}\gamma. \tag{5.97}$$

Inserting the expressions for γ, a, and b from Eqs. (5.95), (5.96) and (5.97) into Eq. (5.89) finally gives us the simplest transformation that is in accordance with Eq. (5.86)

$$x' = \gamma(x - vt), \quad y' = y, \quad z' = z, \quad \text{and}$$
$$t' = \gamma\left(t - \frac{v}{c^2}x\right), \quad \text{where} \quad \gamma = \frac{1}{\sqrt{1 - \frac{v^2}{c^2}}}. \tag{5.98}$$

These famous transformation equations are called the "Lorentz transformation". Lorentz was the contemporary phycisist Einstein admired most. "Everything that emanated from his surpremely great mind was as clear and beautiful as a good work of art."

Suppose we are in a strange civilization where we are invited to travel just behind a beam of the light of a distant star. (When he was 16 Einstein dreamed about such a travel.) If we follow the light, it would be nearly at rest relative to us, we might think. It would not! Experience suggests to us that light travels ahead of every observer with the same velocity c. It forces us to concede that the universe may be radically different from what we might expect. Our conceptions about the properties of the universe on the whole are based upon experiences involving very small velocities compared to that of light. We have had no good reason to expect anything as dramatic as such a property of light. It is therefore justifiable to call it a near miracle that our simple changes of the Galilei transformation take wonderful care of this strange kind of observation. How could one expect such a strange miracle? It is understandable that Einstein and others played with the idea of God as a mathematician fond of simple, beautiful and 'deep' formulae.

It is sometimes said: After all, the special theory of relativity, in which the Lorentz transformation plays a significant role, means only a slight correction of the Galilean and Newtonian kinematics. Perhaps it may be looked upon only as a generalization or widening of the old theory. Nothing could be more misleading!

Newton's theory includes Newton's conceptual framework and his general understanding of what he considered the physical world to be like. To this theory belongs Universal Time, Absolute Space and a wealth of more or less abstract general theorems. The Newtonian theory is not 'approximately relativistic', even if its predictions are nearly the same as those of the theory of relativity for moderate velocities, because the theoretical structure is deeply different. Newton asserts something completely incompatible with the special theory of relativity. From observational reports alone you can never *deduce* any theory. A set of reported mesurements can be 'explained' through indefinitely many theorems. These may even be inconsistent with each other.

Through observation we *discover* things. Abstract theories are *inventions*. The fabulous success of some of these inventions surprises us. There is, perhaps, some kind of hidden harmony between structures of thought and structures of physical reality? Something established through hundred of millions of years of evolution?

The clash with habitual thinking is formidable compared to the small change in mathematical formulae. Suppose the hummingbird emits a flash while flying 9/10 the speed of light. The gardener thinks: she flies just behind the surface of the expanding sphere of light created by the flash, observing a velocity equal to 30,000 km/sec for the light. But she knows otherwise: a measurement would reveal that the light she emitted travels not with a velocity 30,000 km/sec, but with the velocity 300,000 km/sec, away from her, as if she had been sitting quietly with the gardener. Whether you fly toward it or away from it, you measure the same speed of light.

Such measurements were actually performed by Michelson and Morley in 1887. They knew that the Earth moves in an elliptical orbit around the Sun with a velocity of about 30 km/s, and wanted to measure this velocity from its effect upon the propagation of light. The velocity of light relative to the Earth should be 60 km/s less in the forward direction than in the backwards direction. Michelson and Morley found a method by which they could measure such a difference in the velocity of light, even if it were only a tenth of the magnitude they sought for due to the motion of the Earth. The shocking result of the experiment was that they measured no difference at all for the velocity of light in the forward and the backward directions.

Whatever the intensity of democratic feelings, the notion that we all remain at the centre, whatever our speed relative to the light source was, and still is, felt as an affront to common sense. Einstein, when he learned about the result of the Michelson–Morley experiment, was not shocked. He had thought for many years about space and time, and gained an openness for what would be a deepening of common sense. He saw that measurements involving speed, for instance by the gardener and the hummingbird, rests on postulates about what simultaneity is and how it is measured by means of light beams. He arrived at new notions of time and space which satisfied reason. These notions made it possible not to consider the Michelson–Morley results as a brute, strange fact, but to place them in an orderly understandable whole.

One can make a correspondence between (mathematical) space and physical objects by associating the points of a space with the position of particles (without spatial extension). This is the space of our daily life, and also that of the pre-relativistic physics. Understood in this way, space is part of our material world.

There is a similar correspondence between points in spacetime and *events* (i.e. idealized physical events without extension in space and time).

We shall briefly return to the relativity of simultaneity. Let a train, moving past a station with velocity v, be our reference frame Σ' with comoving coordinates x' and t'. The station is a frame Σ with coordinates x and t. A passenger in Σ' observes that the front and back doors of his wagon suddenly open (by a deplorable technical accident), at a point of time $t' = 0$. The co-moving coordinates of the doors are $x' = 0$ and $x' = L_0$, where L_0 is the length of the wagon.

The position and point of time of the events that the doors were opened, as observed from the station, are given by the Lorentz transformation, Eq. (5.98). The inverse transformation, from Σ' to Σ, is obtained from Eq. (5.98) just by replacing v by $-v$, giving

$$x = \gamma(x' + vt') \quad \text{and} \quad t = \gamma\left(t' + \frac{v}{c^2}x'\right).$$

Inserting $x' = t' = 0$ for the opening of the back door, gives $x = t = 0$. Inserting $x' = L_0$ and $t' = 0$ for the opening of the front door, gives $x = \gamma L_0$ and $t = \gamma(v/c^2)L_0$. We see that even if the doors open simultaneously as observed from the wagon, the front door opens later than the back door as observed from the station.

In the theory of relativity the Newtonian absolute simultaneity does not exist, i.e. two events that are simultaneous to one observer, are not simultaneous to another one, moving relative to the first. Hence, space and time cannot be regarded as independent. This motivated Minkowski to open his famous address to the Congress of Scientists, Cologne, September 21, 1908 with the words, "Henceforth space by itself, and time by itself, are doomed to fade away in mere shadows, and only a kind of union of the two will preserve an independent reality."

5.11 The relativistic time dilation

The invariance of the velocity of light has an interesting consequence which concerns the rate of time as measured with a moving so-called 'standard clock'. By definition the effect we shall deduce does not depend upon the nature of the clock, so we can choose to analyze a particularly simple clock, consisting essentially of a light signal being reflected between a floor and a ceiling. Such a clock is sometimes called a 'photon clock'.

Imagine that our clock is at rest in the 'train frame' Σ'. The height between the floor and the ceiling is L_0. Each reflection is a 'tick'. The light signals are assumed to move vertically as observed in the train system. Then the time interval between two ticks is

$$\Delta t' = \frac{L_0}{c}. \tag{5.99}$$

This is how fast the clock ticks in the frame where it is at rest.

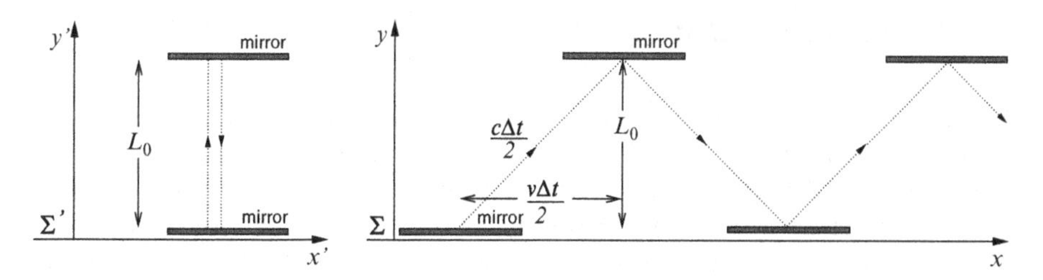

Fig. 5.4 Light signal path in two reference frames

The train with the clock moves past a station with velocity v. Within the theory of relativity the letter v always represents a relative velocity, here the velocity of the train relative to the station. In the 'station frame' Σ the path of the light signal is zig zag shaped as shown in Fig. 5.4.

Since the velocity of light is c in every direction as observed on the station as well as in the train, the time intervals between two ticks is gives by the Pythagorean theorem as follows

$$c^2\Delta t^2 = v^2\Delta t^2 + {L_0}^2$$

or

$$\left(c^2 - v^2\right)\Delta t^2 = {L_0}^2$$

which gives

$$\Delta t^2 = \frac{{L_0}^2}{c^2 - v^2} = \frac{{L_0}^2/c^2}{1 - v^2/c^2}.$$

Thus

$$\Delta t = \frac{L_0/c}{\sqrt{1 - v^2/c^2}}.$$

Inserting from Eq. (5.99) in the numerator we finally arrive at

$$\Delta t = \frac{\Delta t'}{\sqrt{1 - v^2/c^2}}. \tag{5.100}$$

Here Δt is the time interval between the ticks of the train clock as observed in the station frame. In this frame the clock moves with velocity v. Equation (5.100) implies that $\Delta t > \Delta t'$, which means that the time intervals between the ticks of the train clock are greater as observed from the station than observed from the train. Thus the clock is observed to go slower in the frame in which it moves than as observed in the frame in which it is at rest. In short: *a moving clock goes slower than a clock at rest*. This is *the special relativistic velocity dependent time dilation*.

At the limit that the velocity of the clock approaches the velocity of light, the time intervals between the ticks of the clock become indefinitely large. A clock that moves with the velocity of light goes infinitely slowly.

5.12 The line element

As mentioned above, Einstein decided to describe gravity geometrically in terms of curved spacetime. In order to understand how the metric tensor enters the theoretical structure invented to describe a curved space, we should note that it is not possible to 'fill' such a space with a Cartesian coordinate system, where all coordinate curves are orthogonal and straight. This is easily realized if you think of mapping the Earth. In order that a geometer who is feeling his way in an arbitrary wace shall be able to extract information concerning the geometry of the wace from his local measurements, he must be able to describe the measuring results in a coordinate independent way.

Remember that any wace has geometrical properties of a local character. Think of the surface of the Earth, for example. You would, perhaps, like to define a distance-vector between your home and the North Pole. But such a vector does not exist *on the surface of the Earth.* Vectors may be imagined as arrows, straight in the usual, Euclidean sense. A distance vector would have to pass through a tunnel *inside* the Earth, and this is not part of our wace, which is only the *surface* of the Earth. However, indefinitely short distance vectors between points very near each other remain arbitrarily close to the surface of the Earth. They are reckoned as part of the wace.

You may compare a distance vector with the tangent vector introduced in Chapter 3. There we considered a one-dimensional wace, namely the parabola of Fig. 3.4. It is clearly seen from the figure that any finite tangent vector $\vec{u}$ departs from the parabola. It does not exist in the curved wace represented by the parabola.

However, it is possible to measure finite distances on the surface of the Earth. This is done by 'adding indefinitely small distances' (integrating) along a curve on the surface of the Earth. The small 'distance differentials' are defined in the same way for curved waces in general, as for a flat wace.

We need not introduce any coordinate system in order to define the distance between two (indefinitely close) points in a wace. Mathematically the distance can be expressed as a coordinate independent quantity, namely the indefinitely small distance vector between the points, i.e. the differential of the distance vector $\vec{r}$, which we denote by $d\vec{r}$.

If we introduce a coordinate system, the components of $\vec{dr}$ along the coordinate curves are called *coordinate differentials* and are written dx^{μ}, i.e.

$$d\vec{r} = dx^{\mu}\vec{e}_{\mu}. \tag{5.101}$$

We now introduce a coordinate independent expression for the square of the distance, $d\ell^2$, between the points, by taking the dot product of the indefinitely small distance vector by itself,

$$d\ell^2 \equiv d\vec{r} \cdot d\vec{r}. \tag{5.102}$$

In order to obtain some feeling for what we talk about, let us consider a simple example. A geometer is going to map a flat, two-dimensional plane. He may introduce a Cartesian coordinate system with coordinates x and y. He is at a point P with coordinates (x, y), moves a distance dx along the x axis, and then a distance dy along the y axis, and arrives at a point Q with coordinates $(x + dx, y + dy)$. Then the distance vector from P to Q is

$$\vec{dr} = dx\vec{e}_x + dy\vec{e}_y.$$

From the definition (5.102) and Eq. (1.9) follows

$$d\ell^2 = dx^2 + dy^2. \tag{5.103}$$

This corresponds to calculating the distance by means of the Pythagorean theorem.

If we introduce plane polar coordinates r and θ (see Fig. 4.9) the infinitesimal distance vector, Eq. (5.101), takes the form

$$\vec{dr} = dr\vec{e}_r + d\theta\vec{e}_\theta.$$

Taking the dot product of this vector by itself we get

$$\begin{aligned} d\ell^2 &= \left(dr\,\vec{e}_r + d\theta\,\vec{e}_\theta\right) \cdot \left(dr\,\vec{e}_r + d\theta\,\vec{e}_\theta\right) \\ &= dr^2\vec{e}_r \cdot \vec{e}_r + dr\,d\theta\,\vec{e}_r \cdot \vec{e}_\theta \\ &\quad + d\theta\,dr\,\vec{e}_\theta \cdot \vec{e}_r + d\theta^2\,\vec{e}_\theta \cdot \vec{e}_\theta. \end{aligned} \tag{5.104}$$

Since the basis vectors $\vec{e}_r$ and $\vec{e}_\theta$ are normal to each other, $\vec{e}_r \cdot \vec{e}_\theta = \vec{e}_\theta \cdot \vec{e}_r = 0$. According to Eqs. (4.41) and (4.44) $\vec{e}_r \cdot \vec{e}_r = 1$ and $\vec{e}_\theta \cdot \vec{e}_\theta = r^2$. Using these results, Eq. (5.104) is reduced to

$$d\ell^2 = dr^2 + r^2 d\theta^2. \tag{5.105}$$

Even if the expressions (5.103) and (5.105) are different, they lead to the same number of metres for the distance between the points P and Q. The definition (1.31) of 'distance' is such that it doesn't matter what sort of coordinate system one uses; the calculated distance is the same whatever system one uses. This is more technically expressed by saying that the distance between two points is a coordinate invariant quantity.

Let us show this explicitly, for our example. According to Eq. (4.31) the transformation between the Cartesian coordinates (x, y) and the plane polar coordinates (r, θ) is

$$x = r\cos\theta \quad \text{and} \quad y = r\sin\theta.$$

By differentiation, using Eq. (5.4) with $x^{1'} = x$, $x^{2'} = y$, $x^1 = r$, and $x^2 = \theta$, we get

$$dx = \cos\theta\,dr - r\sin\theta\,d\theta \quad \text{and} \quad dy = \sin\theta\,dr + r\cos\theta\,d\theta.$$

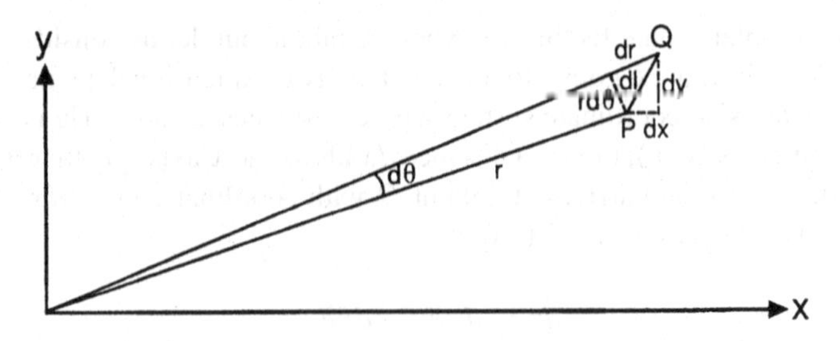

Fig. 5.5 The length $d\ell$ expressed in terms of Cartesian and polar coordinates

Inserting this into Eq. (5.103) yields

$$d\ell^2 = (\cos\theta\, dr - r\sin\theta\, d\theta)^2 + (\sin\theta\, dr + r\cos\theta\, d\theta)^2.$$

Squaring leads to

$$\begin{aligned} d\ell^2 &= \cos^2\theta\, dr^2 - 2r\cos\theta\, \sin\theta\, drd\theta + r^2\sin^2\theta\, d\theta^2 \\ &\quad + \sin^2\theta\, dr^2 + 2r\sin\theta\, \cos\theta\, drd\theta + r^2\cos^2\theta\, d\theta^2 \\ &= \left(\cos^2\theta + \sin^2\theta\right) dr^2 + r^2\left(\sin^2\theta + \cos^2\theta\right) d\theta^2. \end{aligned}$$

Using the identity $\sin^2\theta + \cos^2\theta = 1$ we obtain

$$d\ell^2 = dr^2 + r^2 d\theta^2.$$

This is the same equation as Eq. (5.105). It is the Pythagorean theorem for a (two-dimensional) Euclidean plane as expressed in polar coordinates. The equality of the expressions (5.103) and (5.105), i.e.

$$d\ell^2 = dx^2 + dy^2 = dr^2 + r^2\, d\theta^2$$

may be seen most directly geometrically, as shown in Fig. 5.5.

Let us now direct the attention towards four-dimensional spacetime. Again we shall consider the simplest case, *flat* spacetime, as in special relativity. Imagine there are two laboratories, with observers and measuring apparatus, one moving in the x direction with velocity v relative to the other. Their description of physical phenomena and geometrical relationships are related by the Lorentz transformation, Eq. (5.98). Since $y' = y$ and $z' = z$, we may discard the y and z dimensions without losing anything of interest. We therefore consider flat two-dimensional spacetime with one space dimension, that one along the x axis, and one time dimension.

What is the mathematical expression for the distance between two points O and P near each other in this two-dimensional spacetime? Let us choose two coordinate systems, $\{x, t\}$ and $\{x', t'\}$ with O as common origin. The other point P have coordinates (dx, dt) and (dx', dt'), respectively in the two coordinate systems. It would perhaps seem natural to introduce a quantity 'remoteness' in spacetime analogous to distance in space, namely

$$d\zeta = \left(dx^2 + c^2 dt^2\right)^{1/2}. \tag{5.106}$$

(The reason that we have multiplied dt^2 with c^2 is that mathematically it is only possible to add commensurable quantities, i.e. quantities that may be counted by the same types of units. You cannot add a number of metres and a number of seconds. However, a fraction of a 'light second', $c\,dt$, where dt is a fraction of a second, is a certain number of metres, and can be added to a distance dx. And the square of a number of light seconds can be added to the square of a distance, which is what we have done above.)

The above expression implies that an event which is not remote from us must be close to us both in space and time. An event can be remote because it happens far away or because it happens at a time very different from our now, or both.

In order that remoteness shall be a useful quantity for the description of physical phenomena, the remoteness between two events should be the same number of metres whether it is measured by an observer in a moving laboratory with coordinates $\{x', t'\}$, or in a stationary laboratory with coordinates $\{x, t\}$, i.e. the remoteness should be Lorentz invariant. From the Lorentz transformation, Eq. (5.98), we have

$$dx' = \frac{dx - vdt}{\sqrt{1 - \frac{v^2}{c^2}}} \quad \text{and} \quad dt' = \frac{dt - \frac{v}{c^2}dx}{\sqrt{1 - \frac{v^2}{c^2}}}. \tag{5.107}$$

Inserting this into the expression (5.106) for remoteness, squaring and simplifying, leads to

$$d\zeta'^2 = \frac{1 + \frac{v^2}{c^2}}{1 - \frac{v^2}{c^2}} d\zeta^2 - \frac{4v}{1 - \frac{v^2}{c^2}} dx\,dt.$$

This shows that remoteness is not a Lorentz invariant quantity. If, for example the remoteness between two events, as measured in the unprimed system, is $d\zeta = 3$, then the remoteness between the same two events, as measured in the primed system is $d\zeta' \neq 3$. Because of this dependence upon the velocity of the observer, the quantity remoteness has not been introduced in the standard version of the theory of relativity.

However, the experiences of the hummingbird and the gardener suggest the existence of a Lorentz invariant sort of 'spacetime interval'. The velocity of light is equal to c both as measured by the hummingbird and the gardener. Hence

$$\frac{dx'}{dt'} = c \quad \text{and} \quad \frac{dx}{dt} = c,$$

which gives

$$dx'^2 = c^2 dt'^2 \quad \text{and} \quad dx^2 = c^2 dt^2$$

or

$$dx'^2 - c^2 dt'^2 = 0 = dx^2 - c^2 dt^2. \tag{5.108}$$

This equation shows that the quantity $dx^2 - c^2 dt^2$ is Lorentz invariant and it is equal to zero for coordinate differentials associated with the propagation of a light signal. The value zero is specific for light signals, but the Lorentz invariance of the expression may be valid in general. Let us investigate if this is really the case.

We start by inserting the coordinate differentials (5.107) into $ds^2 \equiv dx'^2 - c^2 dt'^2$ where we have introduced ds^2 as a short-hand notation for the new spacetime interval. This gives

$$ds^2 = dx'^2 - c^2 dt'^2 = \left(\frac{dx - v dt}{\sqrt{1 - v^2/c^2}} \right)^2 - c^2 \left(\frac{dt - \frac{v}{c^2} dx}{\sqrt{1 - v^2/c^2}} \right)^2 .$$

Then we calculate the squares and simplify step by step

$$\begin{aligned}
ds^2 &= dx'^2 - c^2 dt'^2 \\
&= \frac{dx^2 - 2v dx\, dt + v^2 dt^2}{1 - v^2/c^2} \\
&\quad - \frac{c^2 \left(dt^2 - 2\frac{v}{c^2}\, dt\, dx + \frac{v^2}{c^4} dx^2 \right)}{1 - v^2/c^2} \\
&= \frac{dx^2 - 2v\, dx\, dt + v^2 dt^2}{1 - v^2/c^2} \\
&\quad - \frac{c^2 dt^2 - 2v\, dt\, dx + \frac{v^2}{c^2} dx^2}{1 - v^2/c^2} \\
&= \frac{\left(1 - v^2/c^2\right) dx^2 + \left(v^2/c^2 - 1\right) c^2 dt^2}{1 - v^2/c^2} \\
&= \frac{\left(1 - v^2/c^2\right) dx^2 - \left(1 - v^2/c^2\right) c^2 dt^2}{1 - v^2/c^2} \\
&= dx^2 - c^2 dt^2.
\end{aligned} \tag{5.109}$$

Note that the Lorentz invariance of $dx^2 - c^2\, dt^2$ should be expected, since the expression is essentially the same as Eq. (5.87), and the differentials dx and dt transform as x and t.

The first and last short expression of the series of expressions in (5.109) is the so-called line element of our two-dimensional flat spacetime, as expressed in Cartesian coordinate systems. Equation (5.109) shows that the line element is Lorentz invariant. In general, the line element is denoted by ds^2. It is the square of an infinitesimal 'distance' in spacetime. If we add the y and z dimensions, the line element of four-dimensional flat spacetime, as expressed in a Cartesian coordinate system, takes the form

$$ds^2 = dx^2 + dy^2 + dz^2 - c^2dt^2. \tag{5.110}$$

This expression may be thought of as an extension of the Pythagorean theorem from ordinary space with a Cartesian coordinate system, to a corresponding coordinate system in four-dimensional flat spacetime.

Before we proceed to express the line element, ds^2, in arbitrary coordinate-systems, we shall interpret it physically. The physical interpretation of the line element of spacetime does not depend upon the geometrical properties of spacetime, or what sort of coordinates we use. So we can discuss the interpretation with reference to the special form of the line element in Eq. (5.110) with confidence that the interpretation we arrive at will be valid in general.

5.13 Minkowski diagrams and light cones

In order to exhibit the causal structure of spacetime we shall in this section consider flat spacetime. This structure does not depend upon the number of spatial dimensions, so we may, for the present purpose, consider a flat two-dimensional spacetime. We introduce a two-dimensional Cartesian coordinate system $\{x, ct\}$. A spacetime diagram with a vertical ct axis and a horizontal x axis is called a *Minkowski-diagram.*

Let $x^4 = ct$. The path of a particle of light, i.e. a *photon*, passing through the origin at the point of time $t = 0$, and moving in the positive x direction, is given by $x = ct = x^4$. This is called the *worldline* of the photon. It is a straight line through the origin of the Minkowski-diagram, making an angle at 45 degrees with the axes, which means that the component of the spacetime velocity of light in the time direction is equal to its component in the space direction. Thus, the velocity of light is isotropic in spacetime as well as in ordinary three-dimensional space.

Photons are said to have vanishing rest mass. Particles with non-vanishing rest mass move slower than light. They cover a smaller distance in the x direction per second than light. Accordingly the worldline of a massive particle passing $x = 0$ at $t = 0$ is a line closer to the ct axis in the Minkowski-diagram than the worldline of a photon (see Fig. 5.6). The worldline of a photon moving in the negative x direction is also shown on the figure.

Imagine that there is a person at $x = 0$ emitting a flash of light a the point of time $t = 0$. Using material particles, or reflecting electromagnetic signals in a suitable way, he may influence events taking place in the region of spacetime between the worldlines of the two photons. This region may be causally connected to an event at

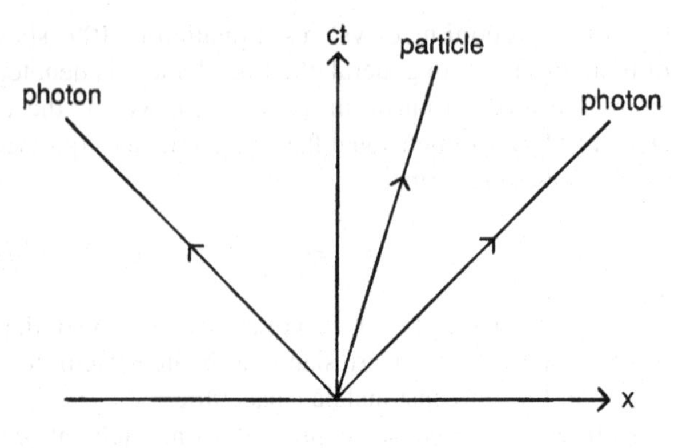

Fig. 5.6 Worldlines

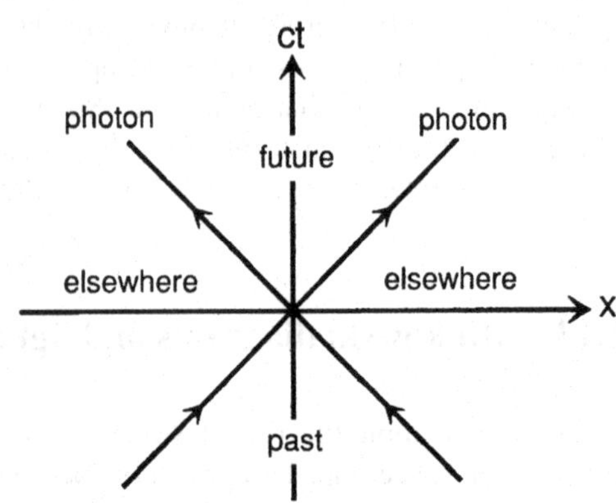

Fig. 5.7 The absolute future and past

the origin. It represents the *absolute future* of this event, in the sense that no event exists in this region that is contemporary or past relative to the event at the origin.

Imagine two photons approaching a person at the origin. In Fig. 5.7 we have included their worldlines below the x axis. Events at the region of spacetime below the x axis and between the photon worldlines, may influence the origin-event. This region is the *absolute past* of this event. What happens at the origin may be influenced by events happening in the absolute past of this point in spacetime. The region of spacetime to the left and right of the absolute future and absolute past is the *elsewhere* of the origin-event.

Let us then include one more space dimension, say the y direction. If the light signal moves a distance r in an arbitrary direction with x component x and y component y, then, from the Pythagorean theorem (see Fig. 5.8),

$$ct = r = \sqrt{x^2 + y^2}.$$

This is the equation of a cone in the three-dimensional Minkowski diagram. It is called the *light cone*. It may be obtained graphically by rotating the worldlines of the photons in Fig. 5.6 about the ct axis. From what was said above it is clear that

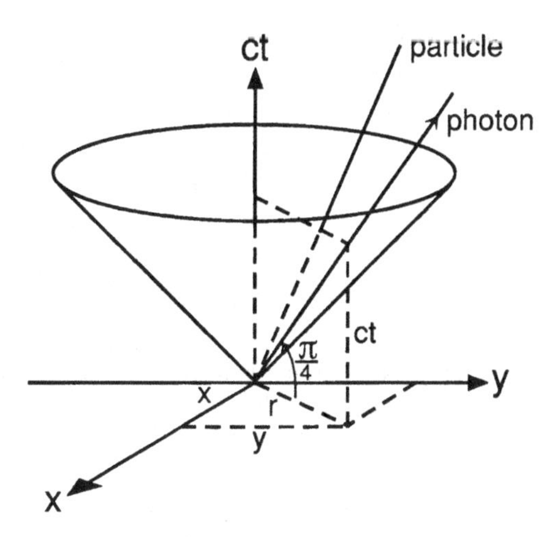

Fig. 5.8 A light cone

it is not possible for any particle (of the type that has been observed so far) to move so that its worldline passes through the vertex of the cone and then proceeds outside it. The worldlines are inside the light cone (see Fig. 5.8).

5.14 The spacetime interval

In order to understand the physical interpretation of the line element we shall proceed step by step starting with the most simple case. We shall first deal with two (physically idealized) events (without extension in space and time) that are *simultaneous* in our reference frame. Then $dt = 0$, and the line element (5.110) reduces to

$$ds^2 = dx^2 + dy^2 + dz^2.$$

As we know from the Pythagorean theorem, this is the square of the distance between the events. In this case $\sqrt{ds^2}$ is just the distance in space between the simultaneous events.

As our next case we shall consider two events happening, in our reference frame at the same place, but not at the same time. Then $dx = dy = dz = 0$. Consequently Eq. (5.110) is reduced to

$$ds^2 = -c^2 dt^2. \tag{5.111}$$

This equation looks like a contradiction, since our symbols represent quantities for which we insert numbers when the equations are applied to physical phenomena. According to high-school mathematics, dealing with so-called 'real numbers', the square of a number can never be negative. However, mathematicians have invented a new sort of numbers, so-called imaginary numbers, with the property that the square

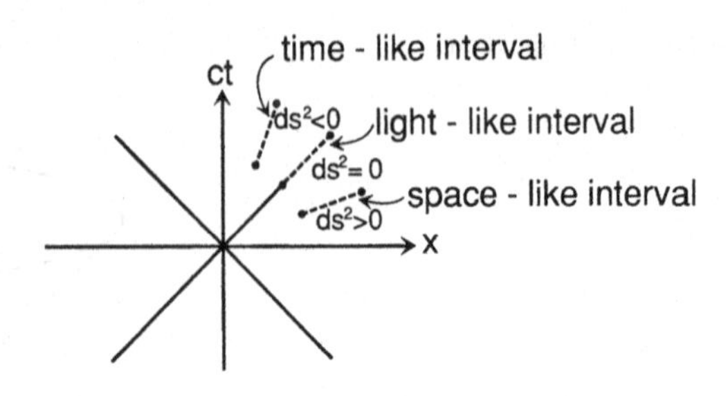

Fig. 5.9 Three classes of spacetime intervals

of an imaginary number is negative. The imaginary number whose square is equal to -1 is denoted by i. Thus

$$i^2 = -1 \quad \text{or} \quad i = \sqrt{-1}.$$

An arbitrary imaginary number is a real number times i.

The value of any measured quantity is a real number. In the situation with two events happening at the same place, the quantity ds is imaginary. In this situation ds does not represent a measurable quantity. However, dt in Eq. (5.111) represents a measurable time interval. Its value is a real number. This number is the time difference between two events, as measured on clocks at rest in our reference frame. Equation (5.111) shows that $-ds^2 = c^2\,dt^2$. Taking the square root,

$$\sqrt{-ds^2} = c\,dt, \tag{5.112}$$

which is the distance that light can travel during the time dt. This distance is termed the *spacetime interval* between the events, but is usually simply called the *interval* between the events.

We have now interpreted physically the quantity $\sqrt{-ds^2}$ for the case that the events happen at the same place, and the quantity $\sqrt{ds^2}$ for the case that the events happen simultaneously. In both cases $\sqrt{|ds^2|}$ is a distance, in the first case the distance that light can travel in a time-interval dt, and in the second case a distance in space. It has become customary in the theory of relativity to use the word 'interval' as a common name covering both cases. This suggests the following definition: The *interval* in spacetime between two events is $\sqrt{|ds^2|}$, i.e. the square root of the absolute value of the line element.

Traditionally we distinguish between three types of intervals with the fancy names space-like, light-like, and time-like. In the case that the line element is positive, $ds^2 > 0$, the interval is said to be *space-like*. In the case that the line element is zero, $ds^2 = 0$, it is called *light-like*. And in the case that the line element is negative, $ds^2 < 0$, we call it *time-like* (see Fig. 5.9).

Let us look at the physical meaning of this classification. Imagine two events P_1 and P_2 very close to each other. The square of the spatial distance between the positions of the events is

$$d\ell^2 = dx^2 + dy^2 + dz^2.$$

The line element (5.110) can now be written

$$ds^2 = d\ell^2 - c^2 dt^2. \tag{5.113}$$

In general the line element is expressed in terms of coordinates in a given reference frame, and the time interval dt is assumed to be measured by clocks at rest in the frame.

Consider first the case that the interval between the events is light-like. Then, according to the definition above, $ds^2 = 0$, which in tern implies

$$\frac{d\ell}{dt} = c.$$

A consequence of this equation is that a signal which is emitted at P_1 and absorbed at P_2 moves with the velocity of light. For example, the spacetime interval between a supernova explosion and our observation of it, is light-like.

If the interval between P_1 P_2 is a positive number, i.e. $ds^2 > 0$, which implies that $d\ell/dt > c$, then the interval according to our definition is space-like. In this case one must move faster than light in order to be present both at P_1 and P_2, which is impossible for ordinary matter moving slower than light.

As stated above two events are said to happen simultaneously if $dt = 0$. Then, according to Eq. (5.113), $ds^2 = d\ell^2$ and $d\ell^2 > 0$, showing that the interval between two simultaneous events is space-like. In order to be present at these two events we have to move infinitely fast—squarely impossible!

The interpretation when the line element is negative, and the interval is time-like, is quite different. The line element (5.110) can be written

$$\begin{aligned} ds^2 = d\ell^2 - c^2 dt^2 &= \left[\left(\frac{d\ell}{dt} \right)^2 - c^2 \right] dt^2 \\ &= \left(v^2 - c^2 \right) dt^2, \end{aligned} \tag{5.114}$$

where $v = d\ell/dt$ is the velocity of a person, moving (or staying at rest in the space dimensions if P_1 and P_2 happen at the same place in space) in such a way that she is present both at P_1 and P_2. For a time-like interval $ds^2 < 0$ which gives $v < c$. In this case it is possible to have been present at both P_1 and P_2, travelling in ordinary three-dimensional space with a velocity v less than the velocity of light c. This case includes ordinary physical travelling. Note that the person will necessarily move in the time direction on her way through spacetime from the event P_1 to P_2. While it is possible to stay at rest in ordinary three-dimensional space, it seems impossible to prevent moving in the time direction, as we incessantly travel from event to event in spacetime.

We want a Lorentz invariant measure of time telling 'truly' how fast an object (or subject) gets older. From Eq. (5.111) we see that $\sqrt{-ds^2}/c$ represents the time measured on a clock at rest in a reference frame. Remember that the line element is Lorentz invariant. From these two points it follows that if you make a Lorentz transformation from a reference frame where a clock is at rest to a frame where it moves, then the value of ds^2 between two events at which the clock is present, is the same in both reference frames. It is the time interval between the events as shown by the clock.

Imagine that the clock is comoving with a certain body. In order to make clear that we talk about time as measured on a clock carried by a given body, it is called the *proper time* of the body. A proper time interval $d\tau$ is defined mathematically by

$$d\tau \equiv \sqrt{-ds^2}/c \tag{5.115}$$

for time-like intervals, which shows that the proper time interval is Lorentz invariant. It is often useful to write Eq. (5.115) in the form

$$ds^2 = -c^2 d\tau^2. \tag{5.116}$$

Equation (5.115) looks very similar to Eq. (5.112). They have, however, different contents. Equation (5.115) is a definition telling what we mean by the proper time interval of a clock between two events. These events need not happen at the same place. The proper time interval between two events is the time interval between the events as shown by a clock moving so that it is present at both events. Equation (5.112), on the other hand, represents the special case that dt is the time interval between two events happening at the same place.

Let us denote a proper time interval as measured by a clock at rest by $d\tau_0$. From Eq. (5.114) with $v = 0$ and Eq. (5.116) then follows

$$-c^2 d\tau_0^2 = -c^2 dt^2$$

or

$$d\tau_0 = dt.$$

As was shown in Sect. 5.10 by considering a photon clock, a moving clock goes slower. This can also be deduced directly from Eq. (5.114) and the definition of proper time, in the form of Eq. (5.116), i.e.

$$-c^2 d\tau^2 = \left(v^2 - c^2\right) dt^2$$

or

$$c^2 d\tau^2 = \left(c^2 - v^2\right) dt^2 = \left(1 - \frac{v^2}{c^2}\right) c^2 dt^2.$$

Dividing by c^2 and taking the square root,

$$d\tau = \sqrt{1 - v^2/c^2}\, dt. \tag{5.117}$$

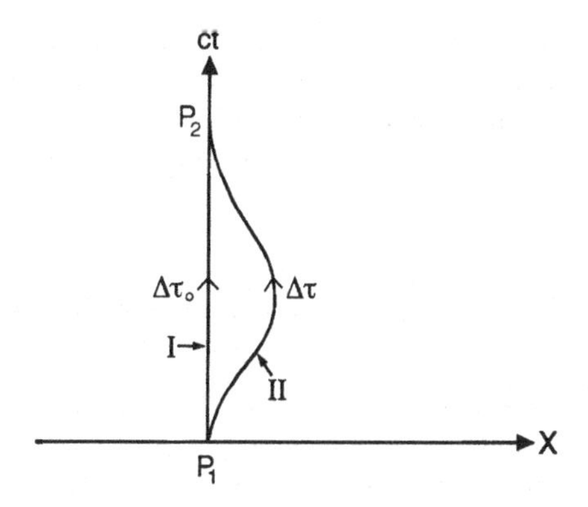

Fig. 5.10 Worldlines of a stationary and a moving observer both present at the events P_1 ad P_2

The strange minus-sign in the expression (5.110) for the line element has interesting consequences, and shows that spacetime is something very different from the usual 3-space. In Euclidean space the *shortest* distance between two points is measured along the straight line between them.

Consider two points P_1 and P_2 in spacetime connected by a time-like interval. We can choose the coordinate system so that the events happen at the same place, $d\ell = 0$. Then the the proper time interval between them, measured on a clock at rest in the coordinate-system, is

$$\Delta\tau_0 = \Delta t. \tag{5.118}$$

The path, I, of this clock is a straight line (see Fig. 5.10). On the other hand, the interval between the same two events, as measured on a clock that moves away and then comes back again, so that it can be present at both events, is found by summarizing (by integration) proper time intervals $d\tau = \sqrt{1 - v^2/c^2}dt$. This is clearly less than summarizing just the dt's, as was done in Eq. (5.118). The worldline, II, of this clock cannot be straight. Thus the proper time interval between P_1 and P_2 along any curved path is less than the proper time interval between the same two events measured along a straight curve,

$$\Delta\tau < \Delta\tau_0.$$

We have arrived at the following strange (but generally true) result: The *greatest* time-like interval in spacetime between two points, is the interval, $\Delta\tau_0$, measured along the straightest possible curve between them. There is, however, no paradox here. It is all a consequence of our definitions. A time-like interval is in general not equal to 'distance', but to proper time. The slower we travel in space the faster we travel in time. If our worldline between two events (two points in spacetime) is a straight line, then the spatial distance we travel between the events is as short as it can be, and we travel as slowly as possible in space from the first event to the second one. The proper time between the events, measured along this route, is accordingly maximal. This is what is meant by saying that a time-like interval between two points in spacetime is maximal along the straightest possible curve between the points. (The quantity 'remoteness', which we introduced above, reminds more of distance, and is *minimal* along the straightest possible curves in spacetime.)

5.15 The general formula for the line element

Knowing the meaning of the spacetime line element we may now proceed mathematically to the general (coordinate independent) expression for the line element. This is of vital importance because the special form (5.110) of the line element is valid only in Cartesian systems, and the general theory of relativity is concerned with describing the properties of curved spacetime from *arbitrary* coordinate-systems.

As was proved above the form (5.110) of the line element is invariant against a Lorentz transformation. Now we seek a generalization of this expression that is invariant against arbitrary coordinate transformations. This will secure for example that we can calculate the proper-time of a clock with arbitrary motion (accelerated or not) by means of the expression ($\sqrt{-ds^2}/c$, not only of a clock with constant velocity. In order to find an invariant expression for ds^2 we first note that the dot product between two vectors is an *invariant* quantity. So if ds^2 can be expressed as the dot product of two vectors we shall have obtained our goal.

The general component-expression for the scalar product of two vectors, $\vec{u} = u^\mu \vec{e}_\mu$ and $\vec{v} = v^\nu \vec{e}_\nu$, is

$$\vec{u} \cdot \vec{v} = u^\mu \vec{e}_\mu \cdot v^\nu \vec{e}_\nu = \vec{e}_\mu \cdot \vec{e}_\nu u^\mu v^\nu = g_{\mu\nu} u^\mu v^\nu, \tag{5.119}$$

where we have used the definition (5.56) of the components of the metric tensor. The dot product of a vector by itself is the square of the length, or magnitude, of the vector.

In Sect. 5.12 the line element was introduced as "the square of an infinitesimal 'distance' in spacetime". Such a 'distance' is an interval between two events in spacetime. Let

$$d\vec{r} = dx^\mu \vec{e}_\mu$$

be the distance vector between the events. The line element, ds^2, is given by the dot product of $d\vec{r}$ by itself,

$$ds^2 = d\vec{r} \cdot d\vec{r}.$$

Using Eq. (5.119) with $\vec{u} = \vec{v} = d\vec{r}$ we get

$$ds^2 = g_{\mu\nu} dx^\mu dx^\nu. \tag{5.120}$$

This formula is the starting point for most investigations of the consequences of the general theory of relativity. In the rest of this book it will be studied extensively.

Let us for a moment go back and consider the four-dimensional flat space with the strange time-dimension, the Minkowski spacetime. What is its metric? Writing Eq. (5.120) for the case of a so-called diagonal metric, i.e. a metric where $g_{\mu\nu}$ is different from zero only if $\mu = \nu$, we get

$$ds^2 = g_{xx}dx^2 + g_{yy}dy^2 + g_{zz}dz^2 + g_{tt}dt^2.$$

Comparing with Eq. (5.110) for the line element of Minkowski spacetime as described in a Cartesian coordinate system, we find

$$g_{xx} = 1, \quad g_{yy} = 1, \quad g_{zz} = 1, \quad \text{and} \quad g_{tt} = -c^2. \tag{5.121}$$

From the definition (5.56) of the components of the metric tensor follows that the scalar products of the basis vectors are

$$\vec{e}_x \cdot \vec{e}_x = 1, \quad \vec{e}_y \cdot \vec{e}_y = 1, \quad \vec{e}_z \cdot \vec{e}_z = 1, \quad \text{and} \quad \vec{e}_t \cdot \vec{e}_t = -c^2.$$

The fact that the dot product of the time-like basis vectors is negative shows the strange non-Euclidean character of Minkowski spacetime. In fact the magnitude of $\vec{e}_t$ must be an imaginary number; a number whose square is negative. This tells us that this vector is only part of the mathematical machinery. It does not represent a physical quantity, such as for example proper time. The result of measuring physical observables is reported in terms of real numbers.

We should distinguish between the terms 'Minkowski metric' and 'Minkowski spacetime'. Minkowski spacetime is flat four-dimensional spacetime. We need not introduce any coordinate system in order to be able to talk about the Minkowski spacetime. However, the term 'Minkowski metric' refers necessarily to the basis vectors of a coordinate system, since the 'metric' is defined as the scalar products of the basis vectors. We could have other metrics than the Minkowski metric in the Minkowski spacetime. Only the (unit) basis vectors of a Cartesian coordinate system with spatial basis vectors orthogonal to the time-direction gives the Minkowski metric.

One more point concerning the meaning of our symbols: the coordinate differentials, dx^μ, of an arbitrary coordinate system do not in general correspond to distances measured with measuring rods. Consider, for example, the plane polar coordinate system of a two-dimensional Euclidean plane. The coordinate differential $d\theta$ is not a distance. It is an angular interval. The corresponding distance along a circle with radius r is $d\ell_\theta = rd\theta$, according to the definition of radians (see Sect. 4.1). A displacement along the circle is a displacement in the $\vec{e}_\theta$ direction, and according to Eq. (4.44) the basis vector $\vec{e}_\theta$ has a magnitude r. We may therefore write $d\ell_\theta = |\vec{e}_\theta| d\theta$. In general the distance corresponding to a particular coordinate differential dx^μ is

$$dl_\mu = |\vec{e}_\mu| dx^\mu \quad \text{(no summation)}.$$

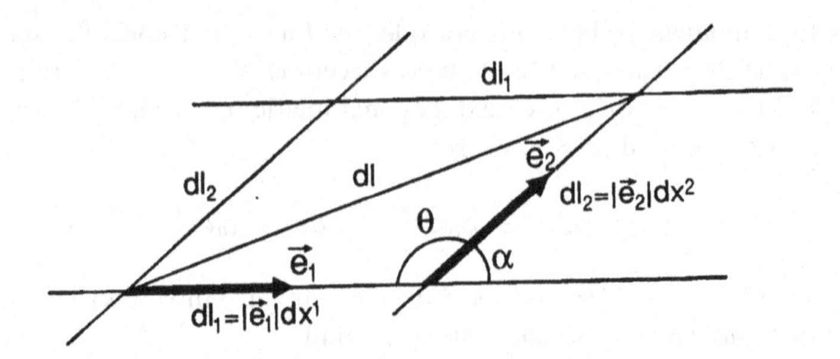

Fig. 5.11 Skew-angled coordinates

(By convention we write 'no summation' when we refer to a particular value of μ, rather that performing an Einstein summation.)

From Eq. (5.54) with $\vec{e}_\nu = \vec{e}_\mu$ and $\alpha = 0$, giving $\cos\alpha = 1$, and Eq. (5.56), we get for the square of the length of a coordinate basis vector

$$\left|\vec{e}_\mu\right|^2 = \left|\vec{e}_\mu \cdot \vec{e}_\mu\right| = \left|g_{\mu\mu}\right|$$

and

$$\left|\vec{e}_\mu\right| = \sqrt{\left|g_{\mu\mu}\right|}.$$

The distance corresponding to the coordinate differential dx^μ can therefore be written

$$d\ell_\mu = \sqrt{\left|g_{\mu\mu}\right|}dx^\mu \quad \text{(no summation).}$$

Sometimes the line element contains terms with products of different coordinate differentials. The significance of such terms may be clearly understood by considering a skew-angled coordinate system, as shown in Fig. 5.11. The figure shows a small coordinate parallelogram with sides

$$d\ell_1 = \left|\vec{e}_1\right| dx^1 = \sqrt{g_{11}}dx^1$$

and

$$d\ell_2 = \left|\vec{e}_2\right| dx^2 = \sqrt{g_{22}}dx^2,$$

respectively. Here $\vec{e}_1$ and $\vec{e}_2$ are the corresponding coordinate basis vectors, with an angle α between them. From the law of cosines, Eq. (4.10), follows

$$d\ell^2 = d\ell_1{}^2 + d\ell_2{}^2 - 2d\ell_1 d\ell_2 \cos\theta.$$

From Fig. 5.11 is seen that $\theta = \pi - \alpha$. Using Eq. (4.19), and the numerical values $\cos\pi = -1$ and $\sin\pi = 0$, we obtain

$$\begin{aligned}\cos\theta &= \cos(\pi - \alpha) = \cos\pi\,\cos(-\alpha) - \sin\pi\,\sin(-\alpha)\\ &= -\cos(-\alpha) = -\cos\alpha.\end{aligned}$$

Hence

$$d\ell^2 = d\ell_1{}^2 + d\ell_2{}^2 + 2d\ell_1 d\ell_2 \cos\alpha.$$

Inserting the expressions for $d\ell_1$ and $d\ell_2$ we get

$$\begin{aligned} d\ell^2 &= g_{11}(dx^1)^2 + g_{22}(dx^2)^2 \\ &\quad + 2|\vec{e}_1|\,|\vec{e}_2|\,\cos\alpha\, dx^1\, dx^2. \end{aligned} \tag{5.122}$$

Using Eqs. (5.54) and (5.56) together with the symmetry of the metric tensor, the last term can be written

$$\begin{aligned} 2\left|\vec{e}_1\right| \cdot \left|\vec{e}_2\right| \cos\alpha\, dx^1\, dx^2 &= 2\left(\vec{e}_1 \cdot \vec{e}_2\right) dx^1\, dx^2 = 2\, g_{12}\, dx^1\, dx^2 \\ &= g_{12} dx^1\, dx^2 + g_{21} dx^2\, dx^1. \end{aligned}$$

Inserting this into Eq. (5.122) leads to

$$d\ell^2 = g_{11}\,(dx^1)^2 + g_{12}\, dx^1\, dx^2 + g_{21}\, dx^2\, dx^1 + g_{22}(dx^2)^2.$$

Using Einstein's summation convention this may be written

$$d\ell^2 = g_{ij}\, dx^i dx^j, \quad \text{where} \quad i = 1,2 \quad \text{and} \quad j = 1,2.$$

This is Eq. (5.120) applied to the present two-dimensional case.

From this calculation is seen that the non-diagonal components of the metric tensor, i.e. $g_{ij} \neq 0$ for $i \neq j$, will be present in skew-angled coordinate systems. This corresponds to terms with products between different coordinate differentials in the line element. There are no such components or terms in coordinate systems where the basis vectors are orthogonal.

5.16 Epistemological comment

The fundamental kinematical concepts are position, direction and motion. Note, by the way, that motion (kinesis in Greek) implies position and direction. To each of these physical concepts, there corresponds an independent type of mathematical entities by means of definitions of correspondence. If you are going to arrange a meeting with a person (an event) you have to tell where and when you shall meet. In order to tell where, you can in principle specify three numbers: the longitude, the latitude and the height above the sea level. (The fact that only longitude and latitude is necessary in actual life has only to do with practical circumstances—that we usually arrange to meet persons on the surface of the Earth.) You must also specify a fourth number; the point of time for the meeting. Thus, to each event corresponds a set of four numbers telling the position of the event in spacetime. A continuum of such sets of numbers make up a mathematically defined set of coordinates (a coordinate system). The direction of a 'rod' is related to a (mathematical) basis vector field. The motion of a particle is relative to a set of particles with given motion, called a (physical) reference frame. The mathematical relevance of a reference frame is, maybe, more vividly grasped if one represents the reference points by their worldlines. Then one can define a reference frame as a continuum of worldlines in spacetime.

In this way a large set of physical phenomena are related by definitions of correspondence to mathematical entities. It is therefore natural to use the name 'mathematical physics', but one must keep in mind that physics is not supposed to be mathematics, and mathematics not supposed to be physics. It is worth a desperate fight of theoretical *physics* to keep in touch with *the world of experience.*

The above-mentioned types of references can be introduced in a physical description independently of each other. And there are several sorts of each type. What type, and which sort of reference one introduces, is a matter of convenience. The choice is determined by the properties of the system as a whole that is to be described.

Everyone knows that the Earth is round, and most people have a clear mental picture of what this means, thinking for example of a globe. Also we think that the universe expands. This is a consequence of the general relativistic interpretation of Hubble's remarkable observations in the 1920s. What could be the proper mental picture of *that*? Psychologically it is possible, and even natural, to think of the expansion of the universe as an expansion of the galaxies *through* space to more remote regions. However this conception cannot be consistently carried through in the context of relativistic cosmology (see Ch. 14). According to the relativistic models of the universe *it is space itself that expands*. But what is meant by 'space itself' here? We have to be careful not to stick narrowly to terms as used in our daily life. It is a more abstract meaning we have to depend upon in our text.

In physics the meaning of a word, such as 'space', is theory dependent. Its most important meaning is found by making clear what function the word has in what is considered the best physical theory available. We do not want to describe, here, the mathematician's abstract space, but the physical spacetime that Einstein's theory pretends that we live in.

Live in spacetime? A brief discussion of such use of words is in order. Note first that the Newtonian universe is free of certain paradoxical features deducable from Einstein's theory. One example: when we sit quietly and read this page, we move with the velocity of light in the time direction (see Sect. 10.8). The crucial, very special invariance of the speed of light makes some of us feel it is natural to stress the function of the general theory of relativity as an abstract *model*. The physical world is something very different from the concepts of the models, even if there is some sort of correspondence between the two. A theoretical model does not describe qualitatively what it models, and we do not live in models. The two authors of this book do enjoy slightly disagreeing about the 'realism' of certain concepts used in General Relativity.

Space is a concept used in models of something physical. So we have to establish correspondences, co-relations, or parallels between physical relations and mathematical relations.

It is a decisive character of the abstract mathematical 'models' and 'formalism' that they somehow are connected with observations. But the lifeline between theory and observation is thin and debateable.

5.17 Kant or Einstein: are space and time human inventions?

Hundreds of articles and many books written by philosophers and physicists have been devoted to questions of how, why and with what right Einstein seems somehow to bring time and space together, invading their fundamental difference.

What Immanuel Kant wrote in his *Kritik der reinen Vernuft* in 1781 about space and time as, completely different, fundamentally different *Anschauungsformen*, has been looked upon as presenting one of the peaks of ingenuity in Western philosophy. Views heavily inspired and in important ways similar to that of Kant—Kantian views, Kantianism, have been, and are, and will presumably and unavoidably be compared to what general relativity has to say.

One of the smallest problems is linguistic: For the German term *Anschauung* there is no single English term available. It is bound up with German culture. But one can adequately render what is meant through explanations.

What are space and time—are they real beings, asks Kant. If real, in what sense are they real? Perhaps the terms express real relations between things, relations which are there independent of our ways of perceiving them? Or are they only basic forms, ways, through which we preceive things and events: human Anschauungsformen. Are they forms which reflect the nature of our minds, that is, something subjective without which it would be perfectly impossible to experience anything in here and out there. Kant looks into his own mind, and out of the window, and votes for the latter. We *always* perceive things in space, and we always conceive our selves in time. We do not *learn* through experience the existence of space and time. We need not generalize what we learn. Out of inner necessity we perceive in certain ways.

As a consequence any sentence expressing a property of space that *must* be there in order to be a space, is a valid or true sentence, and its certainty is *apodictic*. It is *inconceivable* that it could be otherwise. If it follows from the very concept of space, the insight need not, and cannot, be confirmed by experience: Experience can never supply us with absolute certainty. Because of the ways of perceiving, we can find apodictic truths in geometry as long as it deals with space.

In the same way, our inner life is such that we cannot avoid perceiving in terms of before and after, and there is an *irreversibility*: it makes no sense to *reverse* the order, time has an arrow. But the difference between the outer perception and the inner perception is such that *space and time do not mix in any way*.

As may easily be understood there seems to be a polarity between Kantian and Einsteinian way of thought. And it is tempting to draw the conclusion that a Kantian way of thinking is obsolete. Does it not absolutize Euclid? Or prohibit a spacetime? It is a thinking that was well adapted to that of Newton, not of Einstein. This is the conclusion of Hans Reichenbach, an outstanding researcher both in physics and philosophy. He is one of the contributors to the great work edited by Arthur Schilpp, *Albert Einstein: Philosopher–Scientists*, published in 1949. In all 25 philosophers and scientists discuss Einstein's theories in this volume. Kant is mentioned in more than 30 pages and by authors representing different philosophical currents.

The "relativity" of simultaneity elicited strong negative, in part angry, reactions among both philosophers and laymen. But what Einstein says is that physical measurements and theoretical requirements, that is the results of the Michelson–Morley experiment combined with the requirement of the validity of the relativity principle for electromagnetic as well as mechanical phenomena, forces upon us a definition of simultaneity with a *relational* character. But a philosopher and a layman are completely free to stick to a 'intuitive' notion. There is nothing wrong with that from a scientific point of view as long as its supporters accept a physical concept indispensible to the science of physics. Generalizing from immediate perception we may at this very movement *imagine* events at other planets—extremely far away—happening at the very same moment. The notion—let me not call it concept—we have is in a certain sense untouched by Einstein's relationism. It is only when we bring in a requirement of an encompassing theoretical structure modelling physical reality, that the formidable relatedness to velocity occurs. Kant believed Newton had said the last word: there is an 'absolute' space and time independent of us, but a Kantian may say: *not* absolutely. A deep change of the human mind may result in different *Anschauungsformen*. A post-human being may not have an inner sense such that it necessarily perceive things in a series, involving before, after and at-the-same-time. With a modified mind there may also be no necessity of an outer sense which perceives these things spatially. But—here Kant may be completely wrong—not much can be *derived* if absoluteness is accepted. Kant believed very much could be derived from his hypothesis about absoluteness of time.

Similarly as regards spatial relations. We do not need the arrogance to think at our more or less spontaneous experience of spatial relations, such as the one expressed by 'A is between B and C' furnishes us with a concept guaranteeing *testability* which is required by the science of physics. We may agree with Kant that the present human mind is constituted in such a way that what is not perceived to be in ourselves necessarily is perceived to be *outside*—in space, or better in a spatial arrangement. Of course, we would have to explain what we mean by the "inner" and "outer" sense, and what we mean by the supposedly very different outer and inner *world*, but that goes beyond our discussion of relations to Einstein.

When it is said that the relational character of s imultaneity goes straight, not only against Kant, but against common sense, it may be objected that "common sense" scarcely has thought about what might happen when bodies move with enormous speed in relation to each other, and we try to find out what is happening simultaneously. To say "what reason did I have to think that something strange would be observed in such cases?" The depth or definiteness of intention is always limited and largely determined by requirements of action, not reflection.

But how did the many very competent philosophers and physicists conclude about 'Kant versus Einstein'? So far, I have only mentioned Hans Reichenbach's thoroughly negative conclusions. The philosopher F. S. C. Northrop is fairly negative but very specific: "[...] in the epistemology of Albert Einstein, the structure of spacetime is the structure of the scientific object of knowledge"; its real basis is not "solely in the character of the scientist as knower." This "follows from the tensor equation of gravitation [...] Thus spacetime has all the contingent character that the field strengthens, determined by the contingent distribution of matter throughout nature, possess."

Einstein is aware of many "pure deductions" used to arrive at his equation. He even declares that Henri Poincaré is right: one may suppose that there are at t_1 "any number of possible systems of theoretical physics all with an equal amount to be said for them; and this opinion [of Poincaré] is no doubt correct, theoretically. But evolution has shown that at any given moment, out of all conceivable constructions, a single one has always proved itself absolute superior to all the rest." [4, p 22] The important point for us when discussing Kant is that nothing seems to be *a priori and certain* in physics. Einstein seems, however, to have an uncritical belief that all possibilities can be compared, and one found to be superior—physically, mathematically, or in Einstein's special terminology, *simpler* (!).

After all this criticism of Kantian conceptions, let us look at more positive evaluations. There are many but highly overlapping: Theories are constructions, free creations of the intellect, in Einstein's own terminology. In this they are of course part of the human mind. They are made *prior* to certain experiences in the sense of observations, and even prior to generalizations from observations. But they have no *certainty*, and cannot possibly have "apodictic certainty". And they should, according to Einstein, at least at some points, be experimentally testable. Never verifiable, never falsifiable in any strict sense.

Ernst Cassirer is one of the great names in the philosophy of this century. He has developed a Kantian sort of philosophy, a "critical theory" which he thinks is completely compatible with Kant's [1]. The fundamental traits of this theory is completely compatible with general relativity. They are not dependent upon the invariances assumed by Euclid's formalism. The Kantian views about geometry must be modified taking into account the appearance of non-Euclidian kinds. A book by the Dutch philosopher Alfred Elsbach ends essentially with the same positive conclusions. General relativity does not imply that time and space are similar in some ways. Centimetres and seconds are related in some formulas, but through the multiplication by the imaginary number $\sqrt{-1}$. If it is reported from another planet that an athlete there has jumped 9 metres multiplied by $\sqrt{-1}$, we cannot accept it, not even understand what was going on on the planet. Elsbach says that in the equation of general relativity a centimetre plays "the same role" as $\sqrt{-1}/c$ seconds—and why not? It does not affect the updated theory of the two *Anschauungsformen*. Between $1/c$ and $\sqrt{-1}/c$ there is a fundamental difference.

After Bergson's brilliant analysis of the abyss between the intuitively given *dureé* and the measureable *temps*, any theory that seems to deflate the difference, or any theory which seems to equate time intervals with spatial distances, must expect strong, firmly based protests. The "t" in Newtonian physics is measureable and the same of course, holds in general relativity. But the latter does not in any way support a concept of time that reduces the difference between time and space as we experience it in our daily life. The loss of absolute simultaneity is due to a new kind of *relation* between certain measurementss of distance of time and measurements of velocity of motion. It does not in any way bring what we feel to be a time interval nearer to what we perceive as a difference in velocity. The "common sense" is not negated or undermined in this matter. There is still room for Kantian and other philosophies which assume a "categorical" difference between velocity intervals and time intervals. But if intersubjective, intercultural testability is required, questions are relevant which are completely new and unrelated to those pertinent to our common practical problems. The conclusion is simple: We are all predisposed to understand in terms of spatial and time relations. (But of course not only in terms of space and time, we, for instance, are predisposed to ask for causes of events.) It is difficult to imagine a decisive change in this, for example, that concepts of information could take over. Kant undermines conceptions of science and of physics which underestimate the influence of the working of the mind—and of society and culture—upon what we accept as scientific knowledge. Einstein contributes heavily to our appreciation of the role of intuition, imagination and also logical deductions—not only in physics, but in science in general. We learn from experience mainly because of the role of what we have not learned from experience.

Chapter 6
The Christoffel symbols

Choreography is the art of composing dances and the recording of movements on paper by means of convenient signs and symbols. Consider for a moment a most disturbing and uncanny experience suffered by a well established choreographer. He was supposed to record on paper the movements of certain fairly simple dances, but in a faraway, strange place.

He found himself in a big beautiful room where he could easily observe the movements of the dancers. But what horror! He felt to his abysmal dismay that the floor itself, and the walls, even the ceiling were changing shape all the time, like reflections on the surface of waves. Also elongations and contractions occurred in an utterly confusing, erratic way. To his consternation, looking at himself he saw that he himself and his paper took part in all this. How could he possibly describe the dance when deprived of a Cartesian coordinate system? Deprived of a familiar straight unmoving flat floor, stable walls and ceiling! To his surprise he managed, listening to a German named Elwin B. Christoffel (1829–1900) and to others who dropped down from the ceiling at the right moment.

In what follows we are not ambitious enough to describe a dance of a particle, or some other thing of point-like simplicity, inside weird rooms. We are postponing that. We shall limit ourselves to find a way that will be used to describe the dance of the floor in such a room. Our efforts include the introduction of the famous Christoffel symbols $\Gamma^{\nu}{}_{\mu\alpha}$, symbols that Einstein initially had found difficult to understand, but soon mastered magnificently; he simply had to.

We shall now start on a level surpassed long ago, the chapter 2 level. The curvature of a curve $y = f(x)$ is described by the second derivative $y'' = d^2 f/dx^2$. The first derivative $y' = df/dx$ furnishes information of the steepness of a curve at a given point. Fortunately, we shall lose nothing on our way to the Christoffel symbol by considering only the first derivatives. But instead of talking directly about curves, we talk about basis vectors and their changes. When these are thought of as being placed tightly together along a curve, their initial points trace that curve.

Enough introductory talk! We start in earnest.

Ø. Grøn and A. Næss, *Einstein's Theory: A Rigorous Introduction for the Mathematically Untrained*, DOI 10.1007/978-1-4614-0706-5_6,

6.1 Geometrical calculation of the Christoffel symbols of plane polar coordinates

In Fig. 6.1 a plane polar coordinate system (see fig.4.7) is illustrated with particular vectors of the basis vector fields $\vec{e}_r$ and $\vec{e}_\theta$ drawn at four different points P, Q, R, and S. The $\vec{e}_r$ vector field fills the total two-dimensional plane pointing outwards from the origin O in every direction. The magnitude is constant. In Fig. 6.1 we have only drawn four radial basis vectors, those with roots at the points P, Q, R, and S. Similarly, the $\vec{e}_\theta$ vector field fills the two-dimensional plane, each being a tangential vector to every circle with centre at O. Their directions are perpendicular to the radial basis vectors, pointing in the direction of increasing angle θ. The magnitudes of these vectors are proportional to their distance from the origin.

We shall be concerned with changes of the vectors of the basis vector fields with position. In order to prepare for a general formula representing such changes, we shall first consider a particular case: the change of the vector $\vec{e}_\theta$ by a small displacement in the θ direction.

In Fig. 6.2 we have drawn two vectors $(\vec{e}_\theta)_P$ and $(\vec{e}_\theta)_S$ of the basis vector field $\vec{e}_\theta$ at the points P and S, respectively. The points P and S are separated by a finite angle $\Delta\theta$. The vector $(\vec{e}_\theta)_{\parallel P}$ is the vector $(\vec{e}_\theta)_P$ parallel transported from P to S.

The vector $\Delta_\theta\, \vec{e}_\theta$ is the change of the vector field $\vec{e}_\theta$ from P to S, that is, a change by a displacement in the $\vec{e}_\theta$ direction. A 'change' is generally defined to mean 'final value minus initial value'. Finite changes are denoted by Δ and indefinitely small ones by d.

We shall now examine the vector triangle defined by the vectors $(\vec{e}_\theta)_{\parallel P}$, $(\vec{e}_\theta)_S$, and $\Delta_\theta\, \vec{e}_\theta$. This is shown in greater detail in Fig. 6.3.

From Fig. 6.3 and the definition of an angle as measured in radians we see that

$$L = |\vec{e}_\theta|\Delta\theta. \tag{6.1}$$

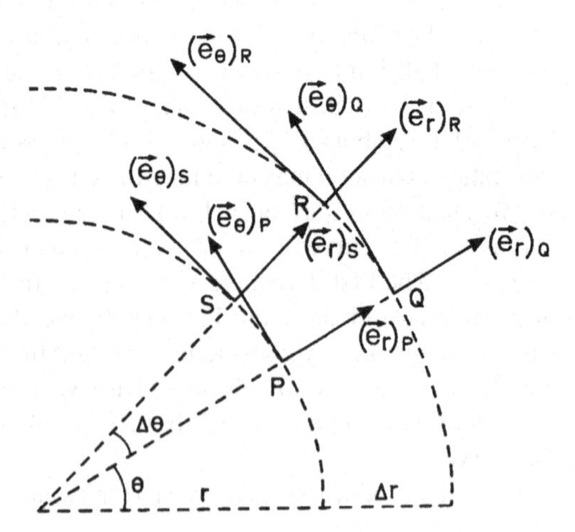

Fig. 6.1 Basis vectors of a plane polar coordinate system

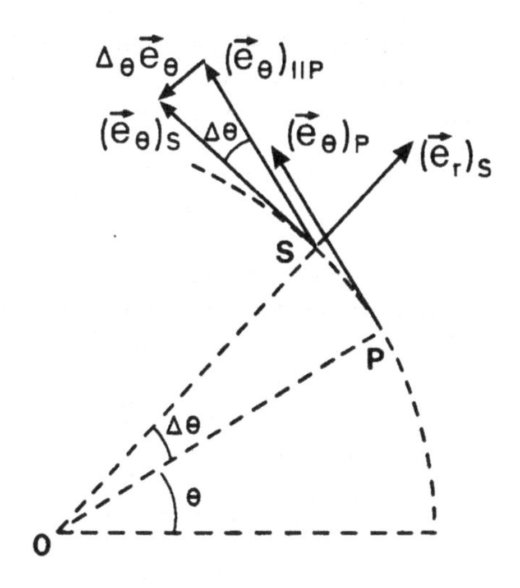

Fig. 6.2 Change of $\vec{e}_\theta$ due to an angular displacement

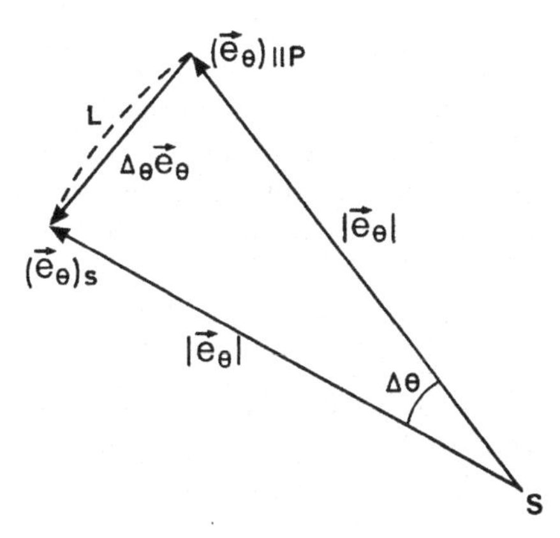

Fig. 6.3 Difference between the parallel transported vector $(\vec{e}_\theta)_{\parallel P}$ and $(\vec{e}_\theta)_S$

Since $|\vec{e}_\theta|$ has magnitude r (see Eq. (4.44)),

$$L = r\Delta\theta.$$

The difference between L and $|\Delta_\theta \, \vec{e}_\theta|$ approaches zero as $\Delta\theta \to 0$, so we obtain

$$|d_\theta \, \vec{e}_\theta| = r d\theta. \tag{6.2}$$

From Fig. 6.2 we see that $d_\theta \, \vec{e}_\theta$ points in the opposite direction of $\vec{e}_r$. Equation (6.2) then gives

$$d_\theta \, \vec{e}_\theta = -r \, d\theta \, \vec{e}_{\hat{r}}$$

where $\vec{e}_{\hat{r}}$ (pronounced 'e vector r hat') is a radial unit vector. Since the coordinate basis vector $\vec{e}_r$ is itself a unit vector field, we may drop the 'hat' in $\vec{e}_{\hat{r}}$, and write

$$d_\theta \, \vec{e}_\theta = -r \, d\theta \, \vec{e}_r. \tag{6.3}$$

We shall now consider differentiation of vectors. Sometimes the vectors $\vec{e}$ of a vector field are functions of only one coordinate, say x. Then the derivative of the vector field $\vec{e}$ is defined by

$$\frac{d\vec{e}}{dx} \equiv \lim_{\Delta x \to 0} \frac{\vec{e}(x + \Delta x) - \vec{e}(x)}{\Delta x}.$$

It may also happen that the vectors are functions of several coordinates, say $\vec{e}(x, y)$. The partial derivative of the vector function with respect to a coordinate x is defined by (compare with Eq. (2.52) for the partial derivative of a scalar function).

$$\frac{\partial \vec{e}}{\partial x} = \lim_{\Delta x \to 0} \frac{\vec{e}(x + \Delta x, y) - \vec{e}(x, y)}{\Delta x} = \lim_{\Delta x \to 0} \frac{\Delta_x \vec{e}}{\Delta x} = \frac{d_x \vec{e}}{dx}.$$

Applying this definition to the case considered in Fig. 6.3 we obtain

$$\frac{d_\theta \vec{e}_\theta}{d\theta} = \frac{\partial \vec{e}_\theta}{\partial \theta} = \left(\frac{\partial \vec{e}_\theta}{\partial \theta}\right)^r \vec{e}_r + \left(\frac{\partial \vec{e}_\theta}{\partial \theta}\right)^\theta \vec{e}_\theta. \tag{6.4}$$

The vector $(\vec{e}_\theta)_S$ is connected to the vector $(\vec{e}_\theta)_P$ by means of the 'difference vector' or 'connection vector' $d_\theta \vec{e}_\theta = (\vec{e}_\theta)_S - (\vec{e}_\theta)_P$. Therefore the components of this vector per unit coordinate distance are called *connection coefficients*, and are in the present case denoted by $\Gamma^r{}_{\theta\theta}$ and $\Gamma^\theta{}_{\theta\theta}$, i.e.

$$\frac{d_\theta \vec{e}_\theta}{d\theta} = \Gamma^r{}_{\theta\theta}\, \vec{e}_r + \Gamma^\theta{}_{\theta\theta}\, \vec{e}_\theta. \tag{6.5}$$

If the basis vectors represent coordinate basis vector fields, which will be the case in the whole of our text, the connection coefficients are called Christoffel symbols. From Eqs. (6.4) and (6.5) follow

$$\Gamma^r{}_{\theta\theta} = \left(\frac{\partial \vec{e}_\theta}{\partial \theta}\right)^r \quad \text{and} \quad \Gamma^\theta{}_{\theta\theta} = \left(\frac{\partial \vec{e}_\theta}{\partial \theta}\right)^\theta.$$

Using Eq. (6.3) we find

$$\Gamma^r{}_{\theta\theta} = -r \quad \text{and} \quad \Gamma^\theta{}_{\theta\theta} = 0.$$

We have thus calculated our first Christoffel symbols. And they have been calculated in an elementary and purely geometrical way.

It is natural, now, to define an arbitrary Christoffel symbol in the following way: The Christoffel symbol $\Gamma^\nu{}_{\mu\alpha}$ is the ν component of the change of the coordinate basis vector $\vec{e}_\mu$ by an infinitesimal coordinate displacement dx^α per unit coordinate distance. Hence

$$\Gamma^\nu{}_{\mu\alpha} \equiv \left(\frac{\partial \vec{e}_\mu}{\partial x^\alpha}\right)^\nu.$$

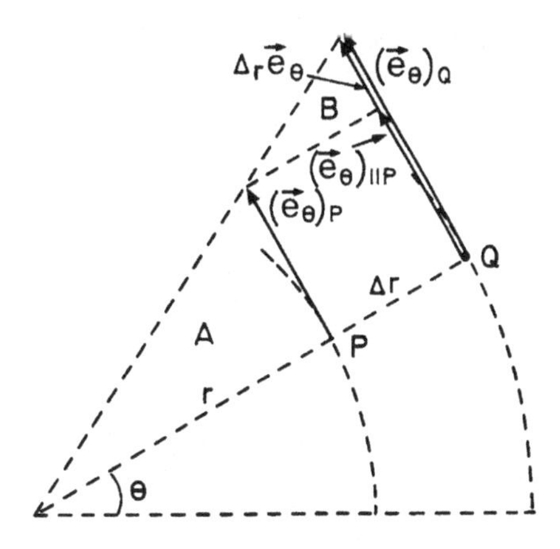

Fig. 6.4 Change of $\vec{e}_\theta$ due to a radial displacement

In order to see more clearly the geometrical significance of the Christoffel symbols, we shall proceed to calculate the remaining Christoffel symbols of the polar coordinate system in similar geometrical ways as above.

We now consider the change of the basis vector field $\vec{e}_\theta$ due to a radial displacement. This is illustrated in Fig. 6.4. The vector $\Delta_r\vec{e}_\theta$ is the change of the vector field $\vec{e}_\theta$ from P to Q. From the triangles A and B is seen that

$$\frac{|\Delta_r\vec{e}_\theta|}{\Delta r} = \frac{|\vec{e}_\theta|}{r}.$$

Hence

$$|\Delta_r\vec{e}_\theta| = |\vec{e}_\theta|(1/r)\Delta r.$$

In the limit that $\Delta r \to 0$ we get

$$|d_r\vec{e}_\theta| = |\vec{e}_\theta|(1/r)dr.$$

Since $d_r\vec{e}_\theta$ is directed along $\vec{e}_\theta$ this leads to

$$d_r\vec{e}_\theta = (1/r)\,dr\,\vec{e}_\theta$$

and therefore

$$\Gamma^\theta{}_{\theta r} = \left(\frac{\partial\vec{e}_\theta}{\partial r}\right)^\theta = \frac{1}{r} \quad \text{and} \quad \Gamma^r{}_{\theta r} = \left(\frac{\partial\vec{e}_\theta}{\partial r}\right)^r = 0.$$

Next we consider the basis vector field $\vec{e}_r$. From Fig. 6.1 is seen that this vector field does not change by radial displacements. Thus

$$d_r\vec{e}_r = 0.$$

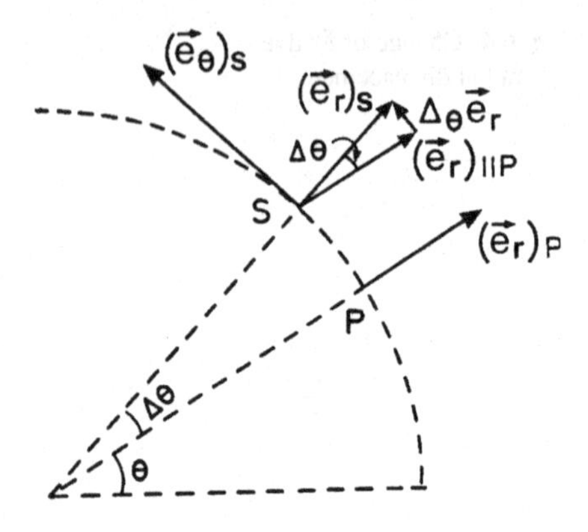

Fig. 6.5 Change of $\vec{e}_r$ due to an angular displacement

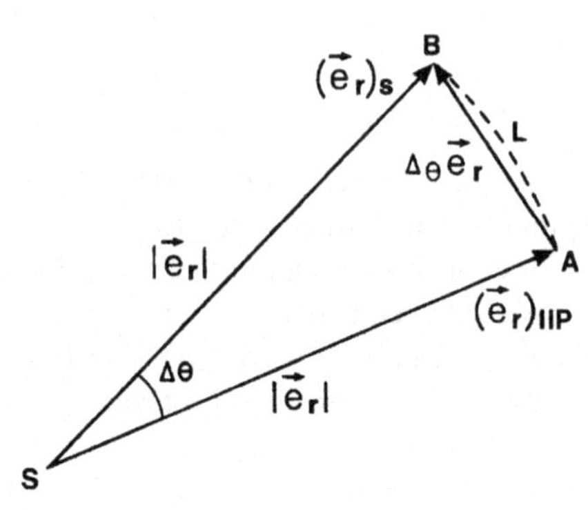

Fig. 6.6 Difference between the parallel transported vector $(\vec{e}_r)_{\|P}$ and $(\vec{e}_r)_S$

If a vector is zero, then each of its components must be zero. This implies that

$$\Gamma^r{}_{rr} = \Gamma^\theta{}_{rr} = 0.$$

We shall now calculate the change of the vector field $\vec{e}_r$ by an angular displacement $d\theta$. Then we need to consider Figs. 6.5 and 6.6. Let us examine the vector triangle defined by the vectors $(\vec{e}_r)_{\|P}$, $(\vec{e}_r)_S$, and $\Delta_\theta\vec{e}_r$. The triangle is shown in greater detail in Fig. 6.6. The arc-length between point A and B in Fig. 6.6 has been denoted by L. The drawn radius of the circle about S is $|\vec{e}_r|$. Again (see Eq. (6.1)), from the definition of an angle as measured in radians, we get

$$L = |\vec{e}_r|\Delta\theta. \tag{6.6}$$

Since $\mathbf{\vec{e}}_r$ has constant magnitude equal to one, see Eq. (4.41), Eq. (6.6) is simplified to

$$L = \Delta\theta.$$

If $\Delta\theta \to 0$, the difference between L and $|\Delta_\theta\vec{e}_r|$ approaches zero. Thus we obtain

$$|d_\theta\vec{e}_r| = d\theta. \tag{6.7}$$

From Fig. 6.5 is seen that $d_\theta \vec{e}_r$ has the same direction as $\vec{e}_\theta$. By means of Eq. (6.7) we get

$$d_\theta \vec{e}_r = d\theta \, \vec{e}_{\hat{\theta}},$$

where $\vec{e}_{\hat{\theta}} = (1/r)\, \vec{e}_\theta$ is the unit vector introduced in Eq. (4.45). Hence

$$d_\theta \vec{e}_r = (1/r)\, d\theta \, \vec{e}_\theta. \tag{6.8}$$

This equation gives the change of the basis vector $\vec{e}_r$ by an infinitesimal displacement (of the point of observation) in the $\vec{e}_\theta$ -direction. As seen from Fig. 6.5 and as stated in Eq. (6.8) this change of $\vec{e}_r$ has only a θ component. From this we obtain the two last Christoffel symbols of the plane polar coordinate system

$$\Gamma^\theta{}_{r\theta} = \left(\frac{\partial \vec{e}_r}{\partial \theta}\right)^\theta = \frac{1}{r} \quad \text{and} \quad \Gamma^r{}_{r\theta} = \left(\frac{\partial \vec{e}_r}{\partial \theta}\right)^r = 0.$$

This completes our geometrical calculation of the Christoffel symbols of the coordinate system with plane polar coordinates. We have seen in detail how the Christoffel symbols describe the change of basis vector field with position. This is the essential significance of the Christoffel symbols. We would like to emphasize that they describe coordinate systems (or more generally basis vector fields) and not space itself. In particular we note that the Christoffel symbols themselves do not give any information about the curvature of space, although they give full information about the geometrical properties of the coordinate system that we have chosen as reference of our description of that space.

The above shows that the geometrical meaning of the Christoffel symbols can be understood more or less intuitively. As a young man (before 1912) Einstein would probably disapprove of the complicated-looking Christoffel symbols with three indices, because he then had the suspicion, according to his friend Philipp Frank, that the sophisticated 'higher' mathematics used by some physicists was not intended to clarify, but rather to dumbfound the reader. He radically but reluctantly changed his opinion.

The many zeros are due to our choice of coordinate system: only one of the coordinate-curves are curved. If both were curved we would in the general case have no zeros. If the two curved coordinates are called x^1 and x^2 we get eight Christoffel symbols

$$\Gamma^1{}_{11}, \quad \Gamma^1{}_{12}, \quad \Gamma^2{}_{11}, \quad \Gamma^2{}_{12},$$

$$\Gamma^1{}_{21}, \quad \Gamma^1{}_{22}, \quad \Gamma^2{}_{21}, \quad \text{and} \quad \Gamma^2{}_{22}.$$

In the case of n dimensions the three indices can all have the values $1, 2, \ldots, n$. Thus, in an n-dimensional coordinate system there will be n^3 Christoffel symbols with different indices. In Sect. 6.4 we shall see that some of them are always equal, so the number of different Christoffel symbols is not as great as this, due to a symmetry that the Christoffel symbols fulfil. In the case of two, three, and four dimensions there are 8, 27 and 64 Christoffel symbols, respectively, with different sets of indices.

6.2 Algebraic calculation of Christoffel symbols

We shall now show how the Christoffel symbols can most efficiently be calculated by means of differentiation, instead of by geometric considerations.

Consider the basis vector field $\vec{e}_r = \vec{e}_r(r,\theta)$ and $\vec{e}_\theta = \vec{e}_\theta(r,\theta)$. According to Eq. (6.4) the change of $\vec{e}_\theta$ due to a displacement $d\theta$ is

$$d_\theta \, \vec{e}_\theta = \left(\frac{\partial \vec{e}_\theta}{\partial \theta}\right)^r d\theta \, \vec{e}_r + \left(\frac{\partial \vec{e}_\theta}{\partial \theta}\right)^\theta d\theta \, \vec{e}_\theta,$$

and from Eq. (6.5) this can be written

$$d_\theta \, \vec{e}_\theta = \Gamma^r{}_{\theta\theta} \, d\theta \, \vec{e}_r + \Gamma^\theta{}_{\theta\theta} \, d\theta \, \vec{e}_\theta. \tag{6.9}$$

Similarly the change of $\vec{e}_\theta$ under a displacement in the r direction, has the form

$$d_r \, \vec{e}_\theta = \Gamma^r{}_{\theta r} \, dr \, \vec{e}_r + \Gamma^\theta{}_{\theta r} \, dr \, \vec{e}_\theta. \tag{6.10}$$

The total change of $\vec{e}_\theta$ due to a displacement in an arbitrary direction, i.e. with both a component dr and $d\theta$ is

$$d\vec{e}_\theta = d_r \, \vec{e}_\theta + d_\theta \, \vec{e}_\theta.$$

Inserting the right-hand sides of Eqs. (6.9) and (6.10) leads to

$$d\vec{e}_\theta = \Gamma^r{}_{\theta r} \, dr \, \vec{e}_r + \Gamma^\theta{}_{\theta r} dr \, \vec{e}_\theta + \Gamma^r{}_{\theta\theta} \, d\theta \, \vec{e}_r + \Gamma^\theta{}_{\theta\theta} \, d\theta \, \vec{e}_\theta. \tag{6.11}$$

In the same way the total change of $\vec{e}_r$ is

$$d\vec{e}_r = \Gamma^r{}_{rr} \, dr \, \vec{e}_r + \Gamma^\theta{}_{rr} \, dr \, \vec{e}_\theta + \Gamma^r{}_{r\theta} \, d\theta \, \vec{e}_r + \Gamma^\theta{}_{r\theta} \, d\theta \, \vec{e}_\theta. \tag{6.12}$$

In the present case Einstein's summation convention results in a considerable simplification. Equations (6.11) and (6.12) can be written in the remarkably elegant form

$$d\vec{e}_\mu = \Gamma^\nu{}_{\mu\alpha} \, dx^\alpha \, \vec{e}_\nu, \tag{6.13}$$

where μ, ν, and α can all take the 'values' r and θ.

The algebraic method works according to the following procedure. First the basis vectors are decomposed in a (locally) Cartesian coordinate system with constant basis vectors. Then one performs the differentiation, and the resulting expressions are decomposed in the original coordinate system. Finally the Christoffel symbols are identified as the coefficients in front of $dx^\alpha \vec{e}_\nu$.

This method will now be applied to the basis vectors $\vec{e}_r$ and $\vec{e}_\theta$ as given in Eqs. (4.40) and (4.43),

$$\vec{e}_r = \cos\theta \, \vec{e}_x + \sin\theta \, \vec{e}_y \quad \text{and} \quad \vec{e}_\theta = -r \, \sin\theta \, \vec{e}_x + r \, \cos\theta \, \vec{e}_y.$$

The total differentials of these vectors are calculated from

$$d\vec{e}_r = \left(\frac{\partial \vec{e}_r}{\partial r}\right)^x dr\,\vec{e}_x + \left(\frac{\partial \vec{e}_r}{\partial r}\right)^y dr\,\vec{e}_y + \left(\frac{\partial \vec{e}_r}{\partial \theta}\right)^x d\theta\,\vec{e}_x + \left(\frac{\partial \vec{e}_r}{\partial \theta}\right)^y d\theta\,\vec{e}_y. \tag{6.14}$$

and

$$d\vec{e}_\theta = \left(\frac{\partial \vec{e}_\theta}{\partial r}\right)^x dr\,\vec{e}_x + \left(\frac{\partial \vec{e}_\theta}{\partial r}\right)^y dr\,\vec{e}_y + \left(\frac{\partial \vec{e}_\theta}{\partial \theta}\right)^x d\theta\,\vec{e}_x + \left(\frac{\partial \vec{e}_\theta}{\partial \theta}\right)^y d\theta\,\vec{e}_y.$$

Performing the differentiations,

$$d\vec{e}_r = 0\,dr + \left(-\sin\theta\,\vec{e}_x + \cos\theta\,\vec{e}_y\right) d\theta \tag{6.15}$$

$$d\vec{e}_\theta = \left(-\sin\theta\,\vec{e}_x + \cos\theta\,\vec{e}_y\right) dr - r\left(\cos\theta\,\vec{e}_x + \sin\theta\,\vec{e}_y\right) d\theta. \tag{6.16}$$

In order to compare with Eqs. (6.11) and (6.12) we must decompose the vectors $d\vec{e}_r$ and $d\vec{e}_\theta$ in the polar coordinate system. In Eq. (6.15) and in the first term of Eq. (6.16) we recognize by means of Eq. (6.14), that $-\sin\theta\,\vec{e}_x + \cos\theta\,\vec{e}_y = (1/r)\,\vec{e}_\theta$, and in the second term of Eq. (6.16) that $\cos\theta\,\vec{e}_x + \sin\theta\,\vec{e}_y = \vec{e}_r$. We see from this

$$d\vec{e}_r = (1/r)\,d\theta\,\vec{e}_\theta,$$

$$d\vec{e}_\theta = (1/r)\,dr\,\vec{e}_\theta + (-r)\,d\theta\,\vec{e}_r. \tag{6.17}$$

Comparing with Eqs. (6.11) and (6.12) we find the following non-vanishing Christoffel symbols

$$\Gamma^\theta{}_{r\theta} = \frac{1}{r}, \quad \Gamma^\theta{}_{\theta r} = \frac{1}{r}, \quad \text{and} \quad \Gamma^r{}_{\theta\theta} = -r. \tag{6.18}$$

This is in accordance with the geometrical calculations above.

The generalization of these results to the simplest 3-dimensional curvilinear coordinate system, called cylindrical coordinates, is rather trivial. This coordinate system will therefore only be briefly mentioned, and we will then go on to calculate the Christoffel symbols in spherical coordinates, which is of great importance in several applications of the general theory of relativity.

The extension of plane polar coordinates to cylindrical coordinates is made by introducing a third coordinate z with straight coordinate axis normal to the plane of the polar coordinates. The basis vector field $\vec{e}_z$ is a constant vector field. All the vectors $\vec{e}_z$ of this field are unit vectors and have the same direction. This is illustrated in Fig. 6.7.

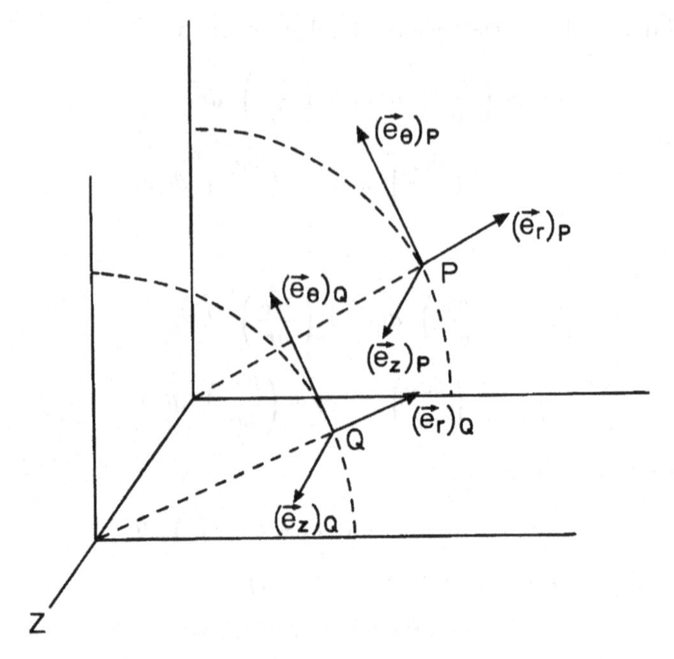

Fig. 6.7 The basis vectors of cylindrical coordinates

None of the basis vectors are changed by a displacement in the z direction. The changes of the basis vectors $\vec{e}_r$ and $\vec{e}_\theta$ have no components along $\vec{e}_z$. And the vector $\vec{e}_z$ does not change by any displacement in the (r, θ) plane. These properties of the basis vectors imply that all the Christoffel symbols with at least one z index are zero. Thus the extension of the plane polar coordinate system to a three-dimensional cylindrical coordinate system introduces no new Christoffel symbols.

6.3 Spherical coordinates

Figure 6.8 invites the reader to imagine a sphere. What is supposed to be seen is marked with full lines. Lines in the interior of the sphere are dotted except some angles. At an arbitrary point P we have drawn three basis vectors; $\vec{e}_r$ along the radial direction from the centre of the sphere and outwards, $\vec{e}_\varphi$ is pointing in a direction corresponding to the variation of longitude on the surface of the Earth, and $\vec{e}_\theta$ in a direction corresponding to the variation of latitude. The hatched two small right-angled triangles converging on the centre of the sphere have sides marked x, y, z, and r and angles φ and θ. Note that the spherical coordinates used in mathematical physics are different from those used on a globus. On the Earth the poles are at 90 degree latitude. The coordinate θ, however, is so that the poles are at $\theta = 0$, and the Equator at $\theta = 90$ degrees.

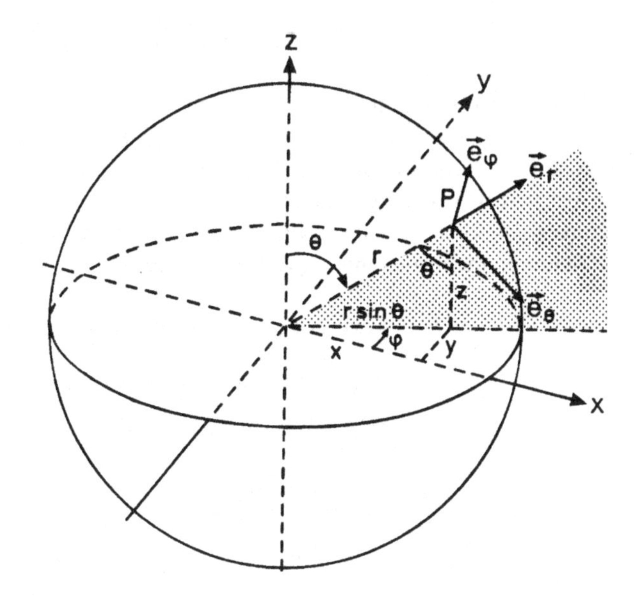

Fig. 6.8 Spherical coordinates

Using Eq. (4.1a) for sinus and cosinus, and inspecting the mentioned triangles we see that

$$\cos\varphi = \frac{x}{r\ \sin\theta}, \quad \sin\varphi = \frac{y}{r\ \sin\theta}, \quad \text{and} \quad \cos\theta = \frac{z}{r}.$$

Thus, the coordinate transformation between the spherial coordinates r, θ, and φ and the Cartesian coordinates x, y, and z is

$$\begin{aligned} & x = r\ \sin\theta\ \cos\varphi, \quad y = r\ \sin\theta\ \sin\varphi, \quad \text{and} \\ & z = r\ \cos\theta. \end{aligned} \tag{6.19}$$

We now use the transformation formula (5.37) and find that the basis vectors of the spherical coordinate system are given by

$$\vec{e}_r = \frac{\partial x}{\partial r}\vec{e}_x + \frac{\partial y}{\partial r}\vec{e}_y + \frac{\partial z}{\partial r}\vec{e}_z \tag{6.20a}$$

$$\vec{e}_\theta = \frac{\partial x}{\partial \theta}\vec{e}_x + \frac{\partial y}{\partial \theta}\vec{e}_y + \frac{\partial z}{\partial \theta}\vec{e}_z \tag{6.20b}$$

$$\vec{e}_\varphi = \frac{\partial x}{\partial \varphi}\vec{e}_x + \frac{\partial y}{\partial \varphi}\vec{e}_y + \frac{\partial z}{\partial \varphi}\vec{e}_z \tag{6.20c}$$

In short, using the Einstein summation convention

$$\vec{e}_\mu = \frac{\partial x^m}{\partial x^\mu}\vec{e}_m \quad \text{where} \quad \begin{cases} x^m \in \{x, y, z\}, \\ x^\mu \in \{r, \theta, \varphi\}. \end{cases} \tag{6.21}$$

From Eqs. (6.20) it is clear that the spherical coordinate system is more complex than the Cartesian, the plane polar, and the cylindrical systems. However, systems with spherical symmetry are the most important ones for us, since we seem to live on a spherical Earth in an isotropic universe. So it is worth the effort to work out the Christoffel symbols in this coordinate system.

We start by calculating the basis vectors $\vec{e}_r$, $\vec{e}_\theta$, and $\vec{e}_\varphi$ from Eq. (6.20), using the expressions for x, y, and z in Eq. (6.19). Performing the partial differentiations and substituting the results into Eqs. (6.20) we get

$$\vec{e}_r = \sin\theta\,\cos\varphi\,\vec{e}_x + \sin\theta\,\sin\varphi\,\vec{e}_y + \cos\theta\vec{e}_z \tag{6.22a}$$

$$\vec{e}_\theta = r\,\left(\cos\theta\,\cos\varphi\,\vec{e}_x + \cos\theta\,\sin\varphi\,\vec{e}_y - \sin\theta\vec{e}_z\right) \tag{6.22b}$$

$$\vec{e}_\varphi = r\,\left(-\sin\theta\,\sin\varphi\,\vec{e}_x + \sin\theta\,\cos\varphi\,\vec{e}_y\right) \tag{6.22c}$$

Differentiating once more, using the product rule (2.24), leads to

$$\begin{aligned} d\vec{e}_r = &\left(\cos\theta\,\cos\varphi\,\vec{e}_x + \cos\theta\,\sin\varphi\,\vec{e}_y - \sin\theta\,\vec{e}_z\right) d\theta \\ &+ \left(-\sin\theta\,\sin\varphi\,\vec{e}_x + \sin\theta\,\cos\varphi\,\vec{e}_y\right) d\varphi \end{aligned} \tag{6.23a}$$

$$\begin{aligned} d\vec{e}_\theta = &\left(\cos\theta\,\cos\varphi\,\vec{e}_x + \cos\theta\,\sin\varphi\,\vec{e}_y - \sin\theta\,\vec{e}_z\right) dr \\ &+ \left(-r\,\sin\theta\,\cos\varphi\,\vec{e}_x - r\,\sin\theta\,\sin\varphi\,\vec{e}_y - r\,\cos\theta\,\vec{e}_z\right) d\theta \\ &+ \left(-r\,\cos\theta\,\sin\varphi\,\vec{e}_x + r\,\cos\theta\,\cos\varphi\vec{e}_y\right)\,d\varphi \end{aligned} \tag{6.23b}$$

$$\begin{aligned} d\vec{e}_\varphi = &\left(-\sin\theta\,\sin\varphi\,\vec{e}_x + \sin\theta\,\cos\phi\,\vec{e}_y\right) dr \\ &+ \left(-r\,\cos\theta\,\sin\varphi\,\vec{e}_x + r\,\cos\theta\,\cos\varphi\,\vec{e}_y\right)\,d\theta \\ &- \left(r\,\sin\theta\,\cos\varphi\,\vec{e}_x + r\,\sin\theta\,\sin\varphi\,\vec{e}_y\right)\,d\varphi \end{aligned} \tag{6.23c}$$

In order to calculate the Christoffel symbols from Eq. (6.13) we must express the right-hand side of Eqs. (6.23) as linear combinations of the vectors $\vec{e}_r$, $\vec{e}_\theta$, and $\vec{e}_\varphi$, not of $\vec{e}_x$, $\vec{e}_y$, and $\vec{e}_z$. Comparing the terms inside the parenthesis in the expression for $d\vec{e}_r$ with the vector $\vec{e}_\theta$ as given in Eq. (6.22), we see that these terms are equal to $(1/r)\,\vec{e}_\theta$. Similarly, the terms in the expression for $d\vec{e}_r$ in front of $d\varphi$ is equal to $(1/r)\,\vec{e}_\varphi$. Thus

$$d\vec{e}_r = (1/r)\,d\theta\,\vec{e}_\theta + (1/r)\,d\varphi\,\vec{e}_\varphi. \tag{6.24}$$

Comparing now the three terms in front of dr, $d\theta$, and $d\varphi$ in the expression for $d\vec{e}_\theta$ with the vectors $\vec{e}_r$, $\vec{e}_\theta$, $\vec{e}_\varphi$ in Eq. (6.22) leads to

$$d\vec{e}_\theta = (1/r)\,dr\vec{e}_\theta + (-r)\,d\theta\,\vec{e}_r + \frac{\cos\theta}{\sin\theta}d\varphi\,\vec{e}_\varphi. \tag{6.25}$$

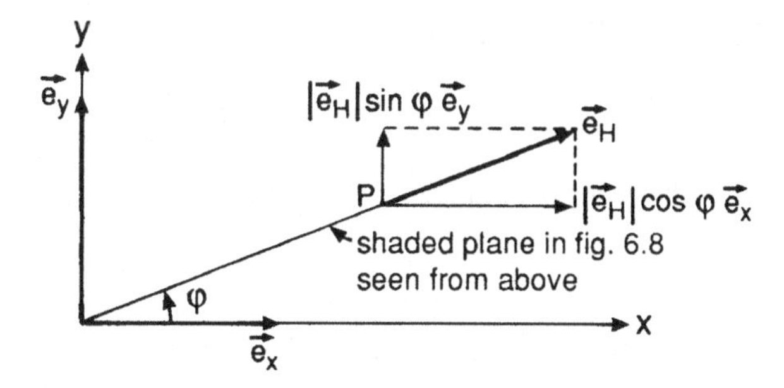

Fig. 6.9 The shaded plane in Fig. 6.8 seen from above

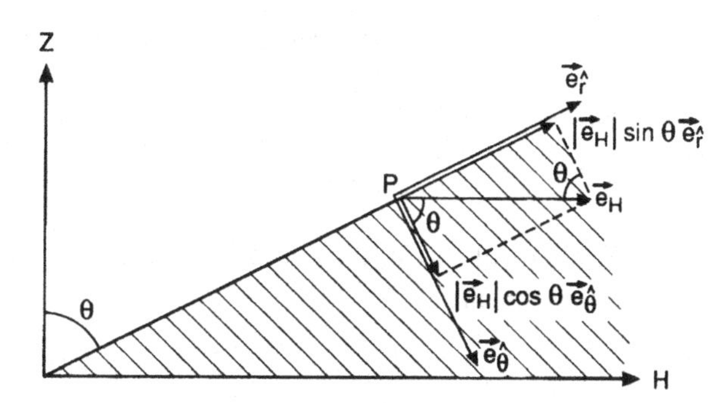

Fig. 6.10 The shaded plane in Fig. 6.8 seen from a horizontal position

This procedure also works for the two first lines in the expression for $d\vec{e}_\varphi$. They are, respectively, $(1/r)\, dr\, \vec{e}_\varphi$ and $\frac{\cos\theta}{\sin\theta}\, d\theta \vec{e}_\varphi$. But the term $-r\ \sin\theta\ (\cos\varphi\, \vec{e}_x + \sin\varphi\, \vec{e}_y)$ in front of $d\varphi$ in the expression for $d\vec{e}_\varphi$ is not simply proportional to one of the vectors $\vec{e}_r$, $\vec{e}_\theta$, or $\vec{e}_\varphi$.

We shall now find an expression for $\cos\varphi\, \vec{e}_x + \sin\varphi\, \vec{e}_y$ in terms of $\vec{e}_r$ and $\vec{e}_\theta$, by simple geometrical reasoning. Consider the shaded plane in Fig. 6.8. Imagine that you look down on this plane from a position far up at the z axis. Then the plane will look like a line to you, as shown in Fig. 6.9.

Let $\vec{e}_H$ be a horizontal unit vector in the shaded plane. From Fig. 6.9 and the usual formula for vector addition follows

$$\vec{e}_H = |\vec{e}_H| \cos\varphi\, \vec{e}_x + |\vec{e}_H| \sin\varphi\, \vec{e}_y = \cos\varphi\, \vec{e}_x + \sin\varphi\, \vec{e}_y .$$

Imagine now that you are positioned in front of the shaded plane and are looking horizontally towards the plane. In Fig. 6.10 we have drawn the plane as seen from this position together with appropriate unit vectors $\vec{e}_{\hat{r}}$, $\vec{e}_{\hat{\theta}}$ and the horizontal vector $\vec{e}_H$.

From Fig. 6.10 is seen that

$$\vec{e}_H = |\vec{e}_H| \sin\theta\, \vec{e}_{\hat{r}} + |\vec{e}_H| \cos\theta \vec{e}_{\hat{\theta}} = \sin\theta\, \vec{e}_{\hat{r}} + \cos\theta\, \vec{e}_{\hat{\theta}} .$$

Calculating the magnitudes of $\vec{e}_r$ and $\vec{e}_\theta$ from

$$|\vec{e}_r| = \left(\vec{e}_r \cdot \vec{e}_r\right)^{1/2} \quad \text{and} \quad |\vec{e}_\theta| = \left(\vec{e}_\theta \cdot \vec{e}_\theta\right)^{1/2}$$

by means of Eq. (6.22), we find

$$|\vec{e}_r| = 1 \quad \text{and} \quad |\vec{e}_\theta| = r.$$

Thus the corresponding unit vectors are

$$\vec{e}_{\hat{r}} = \vec{e}_r \quad \text{and} \quad \vec{e}_{\hat{\theta}} = (1/r)\,\vec{e}_\theta.$$

Inserting this in the expression for $\vec{e}_H$ we have

$$\vec{e}_H = \sin\theta\,\vec{e}_r + (1/r)\cos\theta\,\vec{e}_\theta.$$

Equalizing our two expressions for $\vec{e}_H$ gives

$$\cos\varphi\,\vec{e}_x + \sin\varphi\,\vec{e}_y = \sin\theta\,\vec{e}_r + (1/r)\,\cos\theta\,\vec{e}_\theta.$$

Inserting this for the expression inside the parenthesis in the last term of $d\vec{e}_\varphi$ in Eq. (6.23), we obtain

$$\begin{aligned} d\vec{e}_\varphi = {} & (1/r)\,dr\,\vec{e}_\varphi + \frac{\cos\theta}{\sin\theta}\,d\theta\,\vec{e}_\varphi \\ & + \left(-r\,\sin^2\theta\right)\,d\varphi\,\vec{e}_r + (-\sin\theta\,\cos\theta)\,d\varphi\,\vec{e}_\theta. \end{aligned} \tag{6.26}$$

According to Eq. (6.13) the expressions for $d\vec{e}_r$, $d\vec{e}_\theta$, and $d\vec{e}_\varphi$ are (including only non-vanishing terms)

$$d\vec{e}_r = \Gamma^\theta{}_{r\theta}\,d\theta\,\vec{e}_\theta + \Gamma^\varphi{}_{r\varphi}\,d\varphi\,\vec{e}_\varphi \tag{6.27a}$$

$$d\vec{e}_\theta = \Gamma^\theta{}_{\theta r}\,dr\,\vec{e}_\theta + \Gamma^r{}_{\theta\theta}\,d\theta\,\vec{e}_r + \Gamma^\varphi{}_{\theta\varphi}\,d\varphi\,\vec{e}_\varphi \tag{6.27b}$$

$$\begin{aligned} d\vec{e}_\varphi = {} & \Gamma^\varphi{}_{\varphi r}\,dr\,\vec{e}_\varphi + \Gamma^\varphi{}_{\varphi\theta}\,d\theta\,\vec{e}_\varphi + \Gamma^r{}_{\varphi\varphi}\,d\varphi\,\vec{e}_r \\ & + \Gamma^\theta{}_{\varphi\varphi}\,d\varphi \end{aligned} \tag{6.27c}$$

Comparing, term by term, with the expressions in Eqs. (6.24), (6.25), and (6.26) we find the following non-vanishing Christoffel symbols

$$\begin{aligned} &\Gamma^\theta{}_{r\theta} = \frac{1}{r}, \qquad \Gamma^\varphi{}_{r\varphi} = \frac{1}{r}, \qquad \Gamma^\theta{}_{\theta r} = \frac{1}{r}, \\ &\Gamma^r{}_{\theta\theta} = -r, \qquad \Gamma^\varphi{}_{\theta\varphi} = \frac{\cos\theta}{\sin\theta}, \qquad \Gamma^\varphi{}_{\varphi r} = \frac{1}{r}, \\ &\Gamma^\varphi{}_{\varphi\theta} = \frac{\cos\theta}{\sin\theta}, \qquad \Gamma^r{}_{\varphi\varphi} = -r\,\sin^2\theta, \\ &\Gamma^\theta{}_{\varphi\varphi} = -\sin\theta\,\cos\theta. \end{aligned} \tag{6.28}$$

Of the 27 Christoffel symbols with different indices that exist in 3-dimensional coordinate systems, these nine are the only ones that are different from zero in the spherical coordinate system.

6.4 Symmetry of the Christoffel symbols

The total differential of a basis vector field is given in terms of the partial derivatives, in the same way as in Eq. (2.61) for a scalar field,

$$d\vec{e}_\mu = \frac{\partial \vec{e}_\mu}{\partial x^\alpha} dx^\alpha .$$

Comparing with Eq. (6.13) we get

$$\frac{\partial \vec{e}_\mu}{\partial x^\alpha} = \vec{e}_\nu \, \Gamma^\nu{}_{\mu\alpha} . \tag{6.29}$$

We now decompose the basis vectors in a local Cartesian coordinate system. Then the basis vectors are given by Eq. (6.21). Differentiating this equation gives

$$\frac{\partial \vec{e}_\mu}{\partial x^\alpha} = \frac{\partial^2 x^m}{\partial x^\alpha \partial x^\mu} \vec{e}_m . \tag{6.30}$$

From Eqs. (6.29) and (6.30) follow

$$\vec{e}_\nu \Gamma^\nu{}_{\mu\alpha} = \frac{\partial^2 x^m}{\partial x^\alpha \partial x^\mu} \vec{e}_m .$$

Since [see Eq. (2.66)]

$$\frac{\partial^2 x^m}{\partial^\mu \partial x^\alpha} = \frac{\partial^2 x^m}{\partial x^\alpha \partial x^\mu},$$

it follows that

$$\Gamma^\nu{}_{\mu\alpha} = \Gamma^\nu{}_{\alpha\mu} . \tag{6.31}$$

The Christoffel symbols are symmetric in their subscripts.

Chapter 7
Covariant differentiation

In this chapter a new sort of differentiation, called *covariant differentiation*, will be introduced. The new concept will prove to be of fundamental importance, making it possible to formulate *coordinate invariant* mathematical expressions for the laws of nature.

Why coordinate invariance? Because the laws of nature operated before humans constructed coordinate systems! We have to use *language* expressing the laws, and the formulation of the laws should not depend upon the choice of coordinate systems. As long as we distinguish between (a) the motion of a body that is *left to itself* subject to no forces (a body showing inertial motion), and (b) the motion of a body under the *influence* of gravity, we are slaves of coordinate systems. Why? Because (a) is said to be rectilinear and uniform, whereas (b) is said to be curvilinear and nonuniform. And that is said to be so on false grounds: By means of suitably choosing curvilinear and nonuniformly moving (i.e. accelerated) systems of reference and coordinate systems, *any* motion can be described as rectilinear and uniform *or* curvilinear and nonuniform. The wildest movements can be described as locally straight. Einstein says that these different descriptions are equally valid, In short: give up the distinction between (a) and (b) as absolute, use it as relative, or better, relational. If it suits you, you may announce that the Sun travels around the Earth. Unfortunately the transformation mathematics is not easy, but in 1912 Einstein learned it from his friend Marcel Grossmann, and it formed an indispensable tool which he eventually used in a masterly way when he formulated the general theory of relativity.

7.1 Variation of vector components

Suppose we measure the wind at different points along our path in the landscape. The strength and direction of the wind varies from point to point. Let us describe the velocity of the wind by a vector field $\vec{A}(\lambda)$, where λ is an invariant, i.e. coordinate

Ø. Grøn and A. Næss, *Einstein's Theory: A Rigorous Introduction for the Mathematically Untrained*, DOI 10.1007/978-1-4614-0706-5_7,

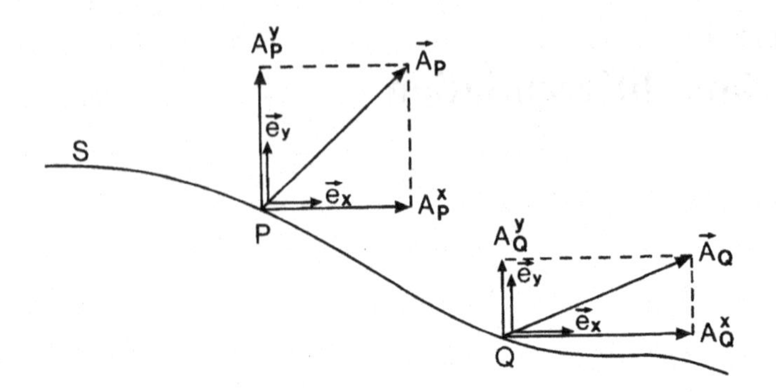

Fig. 7.1 The vector field $\vec{A}$ and its decomposition at two points P and Q on a curve S

independent, parameter along our path, which we treat as a curve. The parameter λ could for example be the path length along the curve, or the time measured with a clock we carry with us, i.e. our proper time. Mathematically the curve is given in parametric form (see Sect. 3.1) by specifying the coordinates as functions of the parameter, $x^\alpha = x^\alpha(\lambda)$.

Imagine that each point of the path is equipped with a wind measuring apparatus, so that the strength and direction of the wind may be measured at every point. Then we go along the path S from a point P to a neighbouring point Q, with parameters λ_P and λ_Q. As decomposed in a Cartesian coordinate system we have

$$\vec{A}_P = A_P{}^x \vec{e}_x + A_P{}^y \vec{e}_y \quad \text{and} \quad \vec{A}_Q = A_Q{}^x \vec{e}_x + A_Q{}^y \vec{e}_y.$$

The vectors and their components are illustrated in Fig. 7.1.

The Cartesian basis vectors are without subscripts P and Q, since they are constant. The change of the vector field from P to Q is denoted by $(\Delta\vec{A})_{PQ}$, and is given by

$$(\Delta\vec{A})_{PQ} = \vec{A}_Q - \vec{A}_P.$$

In the case that Q is indefinitely close to P the change of the vector field is found by differentiation with respect to the curve parameter

$$\frac{d\vec{A}}{d\lambda} = \lim_{\Delta\lambda\to 0}\frac{\Delta\vec{A}}{\Delta\lambda} = \frac{d\left(A^x\vec{e}_x\right)}{d\lambda} + \frac{d\left(A^y\vec{e}_y\right)}{d\lambda}.$$

The quantity $d\vec{A}/d\lambda$ is the *covariant directional derivative* of $\vec{A}$ along the curve. Using the formula for differentiation of a product, $(uv)' = u'v + uv'$, we get

$$\frac{d\vec{A}}{d\lambda} = \frac{dA^x}{d\lambda}\vec{e}_x + A^x\frac{d\vec{e}_x}{d\lambda} + \frac{dA^y}{d\lambda}\vec{e}_y + A^y\frac{d\vec{e}_y}{d\lambda}. \tag{7.1}$$

However, since the Cartesian basis vectors are constant, this equation is reduced to

$$\frac{d\vec{A}}{d\lambda} = \frac{dA^x}{d\lambda}\vec{e}_x + \frac{dA^y}{d\lambda}\vec{e}_y. \tag{7.2}$$

This may be written in condensed form by means of the Einstein summation convention,

$$\frac{d\vec{A}}{d\lambda} = \frac{dA^m}{d\lambda}\vec{e}_m, \quad m \in \{x_1, x_2\} \quad \text{with} \quad \begin{cases} x_1 = x, \\ x_2 = y. \end{cases} \tag{7.3}$$

Usually the vector components are given as functions of the coordinates in a region. And along the curve where the directional derivative of the vector field is to be calculated, the coordinates are given as functions of the parameter λ. The derivative $dA^m/d\lambda$ is then calculated by using the chain rule of differentiation. First we calculate the partial derivative of A^m with respect to a coordinate, then we multiply by the derivative of the coordinate with respect to λ. This is done for all the coordinates. And finally the products are added. In the present case (compare with Eq. (2.61) with $f = A^m$),

$$\frac{dA^m}{d\lambda} = \frac{\partial A^m}{\partial x}\frac{dx}{d\lambda} + \frac{\partial A^m}{\partial y}\frac{dy}{d\lambda}.$$

Using Einstein's summation convention this is written

$$\frac{dA^m}{d\lambda} = \frac{\partial A^m}{\partial x^n}\frac{dx^n}{d\lambda}. \tag{7.4}$$

Equation (7.3) then takes the form

$$\frac{d\vec{A}}{d\lambda} = \frac{\partial A^m}{\partial x^n}\frac{dx^n}{d\lambda}\vec{e}_m. \tag{7.5}$$

We have now come to a salient point. Equation (7.5) is not valid for vectors decomposed in curved coordinate systems. In that case the step from Eq. (7.1) to (7.2) is not valid. This is due to the fact that in general the basis vectors are not constant vectors, but vary from point to point. Taking account of the variability of the basis vectors, and using the formula (2.22) for differentiation of the product between a scalar function and a vector function (in this case a basis vector field), we obtain

$$\frac{d\vec{A}}{d\lambda} = \underbrace{\frac{d(A^\mu \vec{e}_\mu)}{d\lambda}}_{\text{total change}} = \underbrace{\frac{dA^\mu}{d\lambda}\vec{e}_\mu}_{\text{change of components}} + \underbrace{A^\mu \frac{d\vec{e}_\mu}{d\lambda}}_{\text{change of basis vectors}}. \tag{7.6}$$

This equation gives the covariant directional derivative of a vector along a curve, as decomposed in an arbitrary coordinate system. Writing the equation as

$$\frac{dA^\mu}{d\lambda}\vec{e}_\mu = \frac{d\vec{A}}{d\lambda} - A^\mu \frac{d\vec{e}_\mu}{d\lambda},$$

we see that in general the change of the components of a vector field with position has two independent contributions; one from the variation of the vector field itself, and one from the change of basis vector field in an arbitrary coordinate system.

How then do we calculate $d\vec{e}_\mu/d\lambda$? In Eq. (6.13) Elwin Christoffel comes to our rescue. From this equation follows that the derivative of the basis vectors with respect to the curve parameter λ is

$$\frac{d\vec{e}_\mu}{d\lambda} = \Gamma^\nu{}_{\mu\alpha}\frac{dx^\alpha}{d\lambda}\vec{e}_\nu. \tag{7.7}$$

We are now able to answer those generalists who thought, while reading the last chapter: this is for specialists, not for us. Our answer is that if, in the first place, we take the drastic step of introducing curvilinear coordinates, then we have at least to be able to understand the changes of vectors and their components as referred to such coordinate systems. And the description of all such changes necessarily makes use of the Christoffel symbols.

7.2 The covariant derivative

Equation (7.6) offers us an expression of the derivative of a vector along a curve. We shall now find a more powerful expression, which is essential in the mathematical development of the theory of relativity. The mathematics may look difficult in the sections that follow, and one may wonder if it is necessary to go through all of this. Yes, we must, if we are to say honestly that we have gone through all of the mathematics needed to arrive at Einstein's field equations.

Replacing A^m by A^μ and n by ν in Eq. (7.4), we obtain

$$\frac{dA^\mu}{d\lambda} = \frac{\partial A^\mu}{\partial x^\nu}\frac{dx^\nu}{d\lambda}.$$

Equation (3.16), which defines the components of a tangent vector $\vec{u}$ may be written

$$u^\nu = \frac{dx^\nu}{d\lambda}. \tag{7.8}$$

Therefore

$$\frac{dA^\mu}{d\lambda} = \frac{\partial A^\mu}{\partial x^\nu}u^\nu = u^\nu\frac{\partial A^\mu}{\partial x^\nu}. \tag{7.9}$$

Hence, the directional derivative of a scalar function f along a vector $\vec{u} = u^\nu \vec{e}_\nu$ can be expressed by the components of the vector and the partial derivatives as follows

$$\frac{df}{d\lambda} = u^\nu \frac{\partial f}{\partial x^\nu}.$$

Einstein introduced a convenient shorthand for the partial derivative of a function f, for instance the function A^μ. The first part of the right-hand side of Eq. (7.9) he liked to write using a comma

$$\frac{\partial A^\mu}{\partial x^\nu} \equiv A^\mu{}_{,\nu}. \tag{7.10}$$

Consequently Eq. (7.9) can be written

$$\frac{dA^\mu}{d\lambda} = A^\mu{}_{,\nu}\, u^\nu. \tag{7.11}$$

From Eqs. (7.7) and (7.8) we obtain

$$\frac{d\vec{e}_\mu}{d\lambda} = \Gamma^\nu{}_{\mu\alpha}\, u^\alpha\, \vec{e}_\nu. \tag{7.12}$$

Substituting Eqs. (7.11) and (7.12) into (7.6) we obtain

$$\frac{d\vec{A}}{d\lambda} = A^\mu{}_{,\nu}\, u^\nu\, \vec{e}_\mu + A^\mu\, \Gamma^\nu{}_{\mu\alpha}\, u^\alpha\, \vec{e}_\nu. \tag{7.13}$$

Our manipulations have disturbed the distribution of summation indices. It is convenient to stick to our old indices, so that the common factors u and $\vec{e}$ have equal indices in both terms. We therefore let $\mu \to \alpha$, $\nu \to \mu$, and $\alpha \to \nu$ in the last term. This has the desired effect: We can use a parenthesis which units the two terms into one,

$$\begin{aligned} \frac{d\vec{A}}{d\lambda} &= A^\mu{}_{,\nu}\, u^\nu\, \vec{e}_\mu + A^\alpha\, \Gamma^\mu{}_{\alpha\nu}\, u^\nu\, \vec{e}_\mu \\ &= (A^\mu{}_{,\nu} + A^\alpha\, \Gamma^\mu{}_{\alpha\nu})\, u^\nu\, \vec{e}_\mu. \end{aligned} \tag{7.14}$$

Einstein has a useful shorthand notation of the whole of what is inside the parenthesis:

$$A^\mu{}_{;\nu} \equiv A^\mu{}_{,\nu} + A^\alpha\, \Gamma^\mu{}_{\alpha\nu}. \tag{7.15}$$

We end up with a new, very potent formula for the derivative of a vector along a curve, as expressed through its components in an arbitrary coordinate system,

$$\frac{d\vec{A}}{d\lambda} = A^\mu{}_{;\nu}\, u^\nu\, \vec{e}_\mu. \tag{7.16}$$

As mentioned above $d\vec{A}/d\lambda$ is the covariant directional derivative of $\vec{A}$ along the curve. The quantity $A^{\mu}{}_{;\nu}$ in Eq. (7.16) is called the *covariant derivative* of the vector component A^{μ}, and $A^{\mu}{}_{;\nu}\, u^{\nu}$ is the covariant directional derivative of the vector component A^{μ} in the direction of the tangent vector $\vec{u} = u^{\nu}\vec{e}_{\nu}$ of the curve.

We started Ch. 6 with a small narrative: the choreographer who despairs entering a room where everything is bulging. The dance he observes is not like the dance he has ever choreographed. The dance itself corresponds to the vector $\vec{A}$, and what the choreographer observes when he enters the bulging room, is a *component* of the dance (the vector) as referred to the strange room. The bulging of the room corresponds to the variability of the basis vectors. Having received help from Christoffel for some hours, the choreographer is able to choreograph the dance in a room-independent way—"replacing commas by semicolons", and thereby obtaining a covariant description of the movements of the dancers, i.e. component expressions valid in every coordinate system. He can now anticipate how the dance will appear in an arbitrarily bulging room.

The force of a mathematical formalism depends very much upon an economical notation. Equation (7.16) looks simple just because of Einstein's clever notational inventions. If written out it is rather complicated. Even in the simplest, two-dimensional case the expansion of the equation, as written in the first line of Eq. (7.14), contains twelve terms

$$
\begin{aligned}
\frac{d\vec{A}}{d\lambda} = \; & \frac{\partial A^1}{\partial x^1}\frac{dx^1}{d\lambda}\vec{e}_1 + A^1\,\Gamma^1{}_{11}\frac{dx^1}{d\lambda}\vec{e}_1 \\
& + A^2\,\Gamma^1{}_{21}\frac{dx^1}{d\lambda}\vec{e}_1 + \frac{\partial A^1}{\partial x^2}\frac{dx^2}{d\lambda}\vec{e}_1 \\
& + A^1\,\Gamma^1{}_{12}\frac{dx^2}{d\lambda}\vec{e}_1 + A^2\,\Gamma^1{}_{22}\frac{dx^2}{d\lambda}\vec{e}_1 \\
& + \frac{\partial A^2}{\partial x^1}\frac{dx^1}{d\lambda}\vec{e}_2 + A^1\,\Gamma^2{}_{11}\frac{dx^1}{d\lambda}\vec{e}_2 \\
& + A^2\,\Gamma^2{}_{21}\frac{dx^1}{d\lambda}\vec{e}_2 + \frac{\partial A^2}{\partial x^2}\frac{dx^2}{d\lambda}\vec{e}_2 \\
& + A^1\,\Gamma^2{}_{12}\frac{dx^2}{d\lambda}\vec{e}_2 + A^2\,\Gamma^2{}_{22}\frac{dx^2}{d\lambda}\vec{e}_2.
\end{aligned}
\tag{7.17}
$$

In three-dimensional space there would be 36 terms, and in four-dimensional spacetime 80 terms, and in n dimensions $n^2(n+1)$ terms. In comparison the general form in Eq. (7.16) for the derivative of a vector is economical and elegant—or, we are tempted to say: beautiful.

The geometrical interpretation of Eq. (7.17) may be stated as follows. The expression $A^{\mu}{}_{;\nu}\, u^{\nu}$ is the μ component of the derivative of the vector $\vec{A}$, giving the μ component of the change of the vector field $\vec{A}$ by an infinitesimal displacement in the direction of the curve. On the other hand, $A^{\mu}{}_{,\nu}u^{\nu}$ describes the change of A^{μ} by such

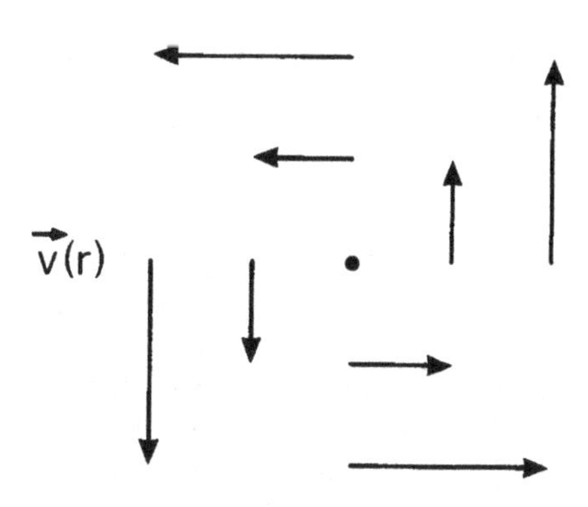

Fig. 7.2 A vector field with vanishing radial component

a displacement. Hence, while $A^{\mu}{}_{,\nu}$ is a derivative acting on the vector component A^{μ}, this is not so in the case of $A^{\mu}{}_{;\nu}$. In this respect the usual expression: $A^{\mu}{}_{;\nu}$ is the covariant derivative of the vector-component A^{μ}, is somewhat misleading. The covariant derivative essentially represents a differentiation of the whole vector $\vec{A}$, not only of its components. In fact $A^{\mu}{}_{;\nu}$ may be different from zero even if the component A^{μ} vanishes at all positions. This conceptually not trivial point is illustrated by Example 7.1 below.

Example 7.1. Consider a vector field $\vec{A}$, in which the vectors are directed along concentric circles (see Fig. 7.2), and have a magnitude equal to the distance from the centre of the circles, i.e. in polar coordinates

$$\vec{A} = r\,\vec{e}_{\theta}. \tag{7.18}$$

None of the arrows have a radial component. Therefore $A^r = 0$ everywhere, so that $A^r{}_{,\theta} = 0$, but from Eq. (7.15) we get in the present case

$$A^r{}_{;\theta} = A^{\theta}\,\Gamma^r{}_{\theta\theta}.$$

Inserting $\Gamma^r{}_{\theta\theta} = -r$ from Eq. (6.18) and $A^{\theta} = r$ from Eq. (7.18), we get

$$A^r{}_{;\theta} = -r\,A^{\theta} = -r^2.$$

Thus the covariant derivative of A^r with respect to θ is different from zero in spite of the fact that A^r is zero everywhere. This means that there is a change of the vector field itself (not of the component A^r) under a dispacement in the $\vec{e}_{\theta}$-direction.

Example 7.2. In this example we shall show how the component expression of the acceleration of a particle in a curved coordinate system, say in plane polar coordinates, is contained in the expression for the covariant derivative. We shall consider the ordinary Newtonian case and use Newtonian time t as parameter along the path of the particle. Differentiation with respect to time will be denoted by an overdot.

In plane polar coordinates the component expression of the velocity is

$$\vec{v} = \frac{dr}{dt}\,\vec{e}_r + \frac{d\theta}{dt}\,\vec{e}_\theta = \dot{r}\,\vec{e}_r + \dot{\theta}\,\vec{e}_\theta.$$

The acceleration is

$$\vec{a} = \frac{d\vec{v}}{dt}. \tag{7.19}$$

Equation (7.13) shall now be applied to calculate the expression for the acceleration in plane polar coordinates. Inserting the left-hand of Eq. (7.11) in the first term of Eq. (7.13), and using Latin indices since only spatial components are involved, this equation takes the form

$$\frac{d\vec{A}}{d\lambda} = \frac{dA^k}{d\lambda}\,\vec{e}_k + \Gamma^k{}_{ij}\,A^i u^j\,\vec{e}_k. \tag{7.20}$$

Replacing the parameter λ by the time t, and the vectors $\vec{A}$ and $\vec{u}$ by the velocity $\vec{v}$, we get from Eqs. (7.19) and (7.20),

$$\vec{a} = \left(\dot{v}^k + \Gamma^k{}_{ij} v^i v^j\right)\vec{e}_k. \tag{7.21}$$

There are only three non-vanishing Christoffel symbols, given in Eq. (6.18), in a plane polar coordinate system. Hence, in the present case Eq. (7.21) reduces to

$$\vec{a} = \left(\ddot{r} + \Gamma^r{}_{\theta\theta}\dot{\theta}^2\right)\vec{e}_r + \left(\ddot{\theta} + 2\Gamma^\theta{}_{r\theta}\dot{r}\,\dot{\theta}\right)\vec{e}_\theta,$$

where we have used the symmetry of the Christoffel symbols. Inserting the values of the Christoffel symbols from Eq. (6.18), we get

$$\vec{a} = \left(\ddot{r} - r\,\dot{\theta}^2\right)\vec{e}_r + \left(\ddot{\theta} + \frac{2}{r}\dot{r}\dot{\theta}\right)\vec{e}_\theta.$$

The unit vectors correspnding to the basis vectors $\vec{e}_r$ and $\vec{e}_\theta$ are $\vec{e}_{\hat{r}} = \vec{e}_r$ and $\vec{e}_{\hat{\theta}} = (1/r)\,\vec{e}_\theta$. They represent an orthonormal basis field in the plane polar coordinate system. The "physical components" of the acceleration vector appear when the acceleration is decomposed in the orthonormal basis field,

$$\vec{a} = \left(\ddot{r} - r\,\dot{\theta}^2\right)\vec{e}_{\hat{r}} + \left(r\,\ddot{\theta} + 2\dot{r}\,\dot{\theta}\right)\vec{e}_{\hat{\theta}}. \tag{7.22}$$

In most books on mechanics this expression for the acceleration in plane polar coordinates, is deduced by geometrical reasoning specially adapted to the particular problem. As shown in Example 7.2, it is neatly contained in the covariant formalism, which may be applied to all sorts of coordinate systems.

7.3 Transformation of covariant derivatives

In the old days, when space was believed to be flat, one could always introduce Cartesian coordinates, which meant that a formulation of the laws of physics valid in arbitrary coordinates could be dispensed with. When Einstein understood that gravitation should be described in terms of curved spacetime, he also saw that he had to formulate the laws of nature in a form valid in arbitrary curved coordinate systems. He knew that vectors do not change under a coordinate transformation, only their components. Maybe a description in terms of vectors might do?

In such a formulation of the theory the calculations are performed in terms of the components of the vectors. Since the laws, for example Newton's 2nd law, are expressed by differential equations, one needs to differentiate the vector components. If the derivative of a vector component is not itself a vector component, we are in trouble with our vector formulation of the theory.

Vector components have one index. However, the partial derivative of a vector component, say $A^{\mu}{}_{,\nu}$, has two indices. Hence it is not a vector component. So, a vector formulation of the theory does not fulfil Einstein's requirements. Einstein needed a theory of tensors of rank higher than one (vectors are tensors of rank one).

The components of tensors of rank 2 have 2 indices, just like the partial derivatives of vector components. Maybe the partial derivatives of the components of a vector make up a tensor of rank 2? Let us investigate this. Imagine that we have two coordinate systems $\{x^{\mu}\}$ and $\{x^{\mu'}\}$, where the marked coordinates are given as functions of the unmarked ones, and vice versa, in a coordinate transformation $x^{\mu'} = x^{\mu'}(x^{\mu})$ and inversely $x^{\mu} = x^{\mu}(x^{\mu'})$. These expressions can be differentiated, and $\partial x^{\mu}/\partial x^{\mu'}$ and $\partial x^{\mu'}/\partial x^{\mu}$ are the elements of the transformation matrix and its inverse. From the chain rule for differentiation follows that partial derivatives transform as follows

$$\frac{\partial}{\partial x^{\nu'}} = \frac{\partial x^{\nu}}{\partial x^{\nu'}} \frac{\partial}{\partial x^{\nu}}.$$

Using this together with Eq. (5.10) for transforming vector components, we get

$$\begin{aligned} A^{\mu'}{}_{,\nu'} &= \frac{\partial A^{\mu'}}{\partial x^{\nu'}} = \frac{\partial x^{\nu}}{\partial x^{\nu'}} \frac{\partial A^{\mu'}}{\partial x^{\nu}} \\ &= \frac{\partial x^{\nu}}{\partial x^{\nu'}} \frac{\partial}{\partial x^{\nu}} \left(\frac{\partial x^{\mu'}}{\partial x^{\mu}} A^{\mu} \right). \end{aligned} \tag{7.23}$$

The product rule (2.24) gives

$$\frac{\partial}{\partial x^{\nu}} \left(\frac{\partial x^{\mu'}}{\partial x^{\mu}} A^{\mu} \right) = \frac{\partial x^{\mu'}}{\partial x^{\mu}} \frac{\partial A^{\mu}}{\partial x^{\nu}} + \frac{\partial^2 x^{\mu'}}{\partial x^{\nu} \partial x^{\mu}} A^{\mu}.$$

Inserting this into Eq. (7.23) we arrive at

$$A^{\mu'}{}_{,\nu'} = \frac{\partial x^{\nu}}{\partial x^{\nu'}} \frac{\partial x^{\mu'}}{\partial x^{\mu}} A^{\mu}{}_{,\nu} + \frac{\partial x^{\nu}}{\partial x^{\nu'}} \frac{\partial^2 x^{\mu'}}{\partial x^{\nu}\, \partial x^{\mu}} A^{\mu}. \tag{7.24}$$

This shows that the partial derivatives of vector components do not transform as tensor components, but in a much more complicated way due to the presence of the last term. In other words, they do not form a tensor. Even if all $A^{\mu}{}_{,\nu}$ are zero, the $A^{\mu'}{}_{,\nu'}$ will in general not be zero because of the second term.

In order to be able to formulate the laws of nature in a coordinate independent way, we need a new sort of differentiation, such that the derivative of a vector component is the component of a tensor of rank 2. The most obvious candidate is the covariant derivative.

The transformation formula for the covariant derivative of a vector component follows most easily from Eq. (7.16). Using the transformation formulae (5.10) and (5.37) for vector components and basis vectors, respectively, we get

$$\frac{d\vec{A}}{d\lambda} = A^{\mu'}{}_{;\nu'}\, u^{\nu'}\, \vec{e}_{\mu'} = A^{\mu}{}_{;\nu}\, u^{\nu}\, \vec{e}_{\mu} = A^{\mu}{}_{;\nu} \frac{\partial x^{\nu}}{\partial x^{\nu'}}\, u^{\nu'} \frac{\partial x^{\mu'}}{\partial x^{\mu}}\, \vec{e}_{\mu'}.$$

Since the succession of the factors does not matter when we calculate by means of Einstein's summation convention, we obtain

$$A^{\mu'}{}_{;\nu'} = \frac{\partial x^{\mu'}}{\partial x^{\mu}} \frac{\partial x^{\nu}}{\partial x^{\nu'}} A^{\mu}{}_{;\nu}.$$

This is just the transformation formula for the mixed components of a tensor of rank 2. Hence, the covariant derivative is the answer to our search for a suitable derivative in a tensor formulation of a physical theory.

7.4 Covariant differentiation of covariant tensor components

Equation (7.15) gives the covariant derivative of the *contravariant* vector components. We shall also need to know the formula for the covariant derivative of the *covariant* components of a vector. In order to deduce this formula it is useful to start with the the simplest possible quantity: a scalar. Such a quantity has no direction. If one differentiates a scalar quantity, for instance a temperature field, the variation of the basis vectors of the coordinate system does not matter. Therefore the covariant derivative of a scalar function is defined as the ordinary partial derivative. For the temperature field $T(x^1, x^2, x^3)$ the covariant derivative of T with respect to x^1, is

$$T_{;x^1} = T_{,x^1} = \frac{\partial T}{\partial x^1}.$$

The dot product of a vector by itself is the square of the magnitude of the vector. This is a scalar function. According to Eq. (5.78) this function is given by

$$\vec{A} \cdot \vec{A} = A_\mu A^\mu.$$

Since the rule (2.24) for differentiating a product is valid both for partial and covariant derivatives, we get

$$\left(A_\mu A^\mu\right)_{;\nu} = A_{\mu;\nu}\, A^\mu + A_\mu\, A^\mu{}_{;\nu}$$

and

$$\left(A_\mu A^\mu\right)_{,\nu} = A_{\mu,\nu}\, A^\mu + A_\mu\, A^\mu{}_{,\nu}.$$

Using that the covariant derivative of a scalar function, like $A_\mu A^\mu$, is equal to the partial derivative, we obtain

$$A_{\mu;\nu}\, A^\mu + A_\mu\, A^\mu{}_{;\nu} = A_{\mu,\nu}\, A^\mu + A_\mu\, A^\mu{}_{,\nu}.$$

Substituting for $A^\mu{}_{;\nu}$ from Eq. (7.14) we get

$$A_{\mu;\nu}\, A^\mu + A_\mu\, \left(A^\mu{}_{,\nu} + A^\alpha \Gamma^\mu{}_{\alpha\nu}\right) = A_{\mu,\nu}\, A^\mu + A_\mu\, A^\mu{}_{,\nu}.$$

Subtracting $A_\mu A^\mu{}_{,\nu} + A_\mu\, A^\alpha\, \Gamma^\mu{}_{\alpha\nu}$ on each side leads to

$$A_{\mu;\nu}\, A^\mu = A_{\mu,\nu}\, A^\mu - A_\mu A^\alpha \Gamma^\mu{}_{\alpha\nu}.$$

Finally, exchanging the names of the summation indices μ and α in the last term gives

$$A_{\mu;\nu}\, A^\mu = A_{\mu\,,\,\nu}\, A^\mu - A_\alpha\, A^\mu\, \Gamma^\alpha{}_{\mu\nu} = \left(A_{\mu\,,\,\nu} - A_\alpha\, \Gamma^\alpha{}_{\mu\nu}\right) A^\mu.$$

In order that this shall be valid for arbitrary vectors $\vec{A} = A^\mu\, \vec{e}_\mu$, the factor in front of A^μ must be equal for every A^μ. Therefore

$$A_{\mu\,;\,\nu} = A_{\mu\,,\,\nu} - A_\alpha\, \Gamma^\alpha{}_{\mu\nu}. \tag{7.25}$$

This is the equation for the *covariant derivative of covariant vector components.*

In the theory of relativity the properties of matter are represented by a so-called energy-momentum tensor of rank two. Energy and momentum conservation is described by putting the covariant divergence of this tensor equal to zero. Therefore we shall need to be able to calculate the covariant derivatives of the components of a tensor of rank two. They are given by the following formulae for the contravariant, mixed and covariant components, respectively:

$$T^{\mu\nu}{}_{;\,\beta} = T^{\mu\nu}{}_{,\,\beta} + T^{\alpha\nu}\, \Gamma^\mu{}_{\alpha\beta} + T^{\mu\alpha}\, \Gamma^\nu{}_{\alpha\beta}, \tag{7.26a}$$

$$T^\mu{}_{\nu\,;\,\beta} = T^\mu{}_{\nu\,,\,\beta} + T^\alpha{}_\nu\, \Gamma^\mu{}_{\alpha\beta} - T^\mu{}_\alpha\, \Gamma^\alpha{}_{\nu\beta}, \tag{7.26b}$$

$$T_{\mu\nu\,;\,\beta} = T_{\mu\nu\,,\,\beta} - T_{\alpha\nu}\, \Gamma^\alpha{}_{\mu\beta} - T_{\mu\alpha}\, \Gamma^\alpha{}_{\nu\beta}. \tag{7.26c}$$

7.5 Christoffel symbols expressed by the metric tensor

In chapter 6 we calculated the Christoffel symbols from the changes of the coordinate basis vectors with position. In chapter 5 the metric tensor was introduced, and it was mentioned that it is of fundamental importance in the theory of relativity, and contains the information needed to calculate the curvature of spacetime. In this section we shall go one step further towards the calculation of curvature from the components of the metric tensor, by deducing how the Christoffel symbols can be calculated from the metric. Note, however, that there are non vanishing Christoffel symbols even in flat spacetime as described in terms of curved coordinate systems. The Christoffel symbols characterize the geometrical properties of the coordinate system, not of spacetime itself. It will be shown in chapter 9 how the curvature of spacetime can be calculated from the Christoffel symbols and their derivatives.

Consider the unit tensor of rank 2, whose mixed components are equal to the Kronecker symbols, defined in Eq. (5.15). Let us calculate the covariant derivative of the components. Using Eq. (7.26b) we get

$$\begin{aligned}\delta^{\mu}{}_{\nu\,;\,\beta} &= \delta^{\mu}{}_{\nu\,,\,\beta} + \delta^{\alpha}{}_{\nu}\,\Gamma^{\mu}{}_{\alpha\beta} - \delta^{\mu}{}_{\alpha}\,\Gamma^{\alpha}{}_{\nu\beta} \\ &= \delta^{\mu}{}_{\nu\,,\,\beta} + \Gamma^{\mu}{}_{\nu\beta} - \Gamma^{\mu}{}_{\nu\beta} = \delta^{\mu}{}_{\nu\,,\,\beta} = 0,\end{aligned}$$

where the last equality follows since the partial derivatives of the numbers 1 and 0 vanish. Thus the unit tensor of rank 2 is a constant tensor.

As we saw in Eq. (5.79) the mixed components of the metric tensor are equal to the Kronecker symbols. Hence the metric tensor may be written (see Sect. 5.5)

$$g = \delta^{\mu}{}_{\nu}\,\vec{e}_{\mu} \otimes \vec{e}^{\nu},$$

i.e. it is just the unit tensor of rank 2. This is a constant tensor, meaning that the derivative of the tensor along an arbitrary curve with paramete λ vanishes,

$$\frac{dg}{d\lambda} = 0. \tag{7.27}$$

The covariant derivative of the tensor components of a tensor of arbitrary rank are defined in the same way as Eq. (7.16) for vector components. The covariant derivatives of the mixed and covariant components of the metric tensor, for example, are given by

$$\begin{aligned}\frac{dg}{d\lambda} &= \delta^{\mu}{}_{\nu;\beta}\,u^{\beta}\,\vec{e}_{\mu} \otimes \vec{e}^{\nu} = g_{\mu\nu;\beta}u^{\beta}\,\vec{e}^{\mu} \otimes \vec{e}^{\nu}, \\ u^{\beta} &= \frac{dx^{\beta}}{d\lambda}.\end{aligned} \tag{7.28}$$

From Eqs. (7.27) and (7.28) follows

$$g_{\mu\nu\,;\,\beta} = 0.$$

Using Eq. (7.26c) this equation takes the form

$$g_{\mu\nu\,;\,\beta} = g_{\mu\nu\,,\,\beta} - g_{\alpha\nu}\,\Gamma^{\alpha}{}_{\mu\beta} - g_{\mu\alpha}\,\Gamma^{\alpha}{}_{\nu\beta},$$

which gives

$$g_{\mu\nu\,,\,\beta} = g_{\alpha\nu}\,\Gamma^{\alpha}{}_{\mu\beta} + g_{\mu\alpha}\,\Gamma^{\alpha}{}_{\nu\beta}. \tag{7.29a}$$

Relabelling we have

$$g_{\mu\beta\,,\,\nu} = g_{\alpha\beta}\,\Gamma^{\alpha}{}_{\mu\nu} + g_{\mu\alpha}\,\Gamma^{\alpha}{}_{\beta\nu} \tag{7.29b}$$

and

$$g_{\nu\beta\,,\,\mu} = g_{\alpha\beta}\,\Gamma^{\alpha}{}_{\nu\mu} + g_{\nu\alpha}\,\Gamma^{\alpha}{}_{\beta\mu}. \tag{7.29c}$$

Taking Eq. (7.29a) + Eq. (7.29b) – Eq. (7.29c) we get

$$\begin{aligned} g_{\mu\nu,\beta} + g_{\mu\beta,\mu} - g_{\nu\beta,\mu} = {} & g_{\alpha\nu}\Gamma^{\alpha}{}_{\mu\beta} + g_{\mu\alpha}\Gamma^{\alpha}{}_{\nu\beta} + g_{\alpha\beta}\Gamma^{\alpha}{}_{\mu\nu} \\ & + g_{\mu\alpha}\Gamma^{\alpha}{}_{\beta\nu} - g_{\alpha\beta}\Gamma^{\alpha}{}_{\nu\mu} - g_{\nu\alpha}\Gamma^{\alpha}{}_{\beta\mu}. \end{aligned}$$

Due to the symmetry of the Christoffel symbols and the metric tensor, the first and the last terms at the right-hand side cancel, the third and the fifth terms cancel, and the second and fourth terms are equal. Exchanging the left and right hand sides we thus get

$$2g_{\mu\alpha}\,\Gamma^{\alpha}{}_{\nu\beta} = g_{\mu\nu\,,\,\beta} + g_{\mu\beta\,,\,\nu} - g_{\nu\beta\,,\,\mu}.$$

Multiplying by $g^{\tau\mu}$, as defined in Eq. (5.70), dividing by 2, and using the equation

$$g^{\tau\mu}\,g_{\mu\alpha}\,\Gamma^{\alpha}{}_{\nu\beta} = \delta^{\tau}{}_{\alpha}\,\Gamma^{\alpha}{}_{\nu\beta} = \Gamma^{\tau}{}_{\nu\beta},$$

provides us with

$$\Gamma^{\tau}{}_{\nu\beta} = \frac{1}{2}\,g^{\tau\mu}\,\left(g_{\mu\nu\,,\,\beta} + g_{\mu\beta\,,\,\nu} - g_{\nu\beta\,,\,\mu}\right), \tag{7.30}$$

which is the desired expression. It is the expression you will find most often for calculating the Christoffel symbols, if you go to a library and look rapidly through books on general relativity.

There are in fact two kinds of Christoffel symbols. The ones given by Eq. (7.30) are called Christoffel symbols of the second kind. The Christoffel symbols of the first kind, $\Gamma_{\alpha\mu\nu}$, are defined by

$$\Gamma_{\alpha\mu\nu} \equiv g_{\alpha\tau}\Gamma^{\tau}{}_{\mu\nu}.$$

Inserting the expression (7.30) and using Eq. (5.74) we get

$$\begin{aligned}\Gamma_{\alpha\mu\nu} &= \frac{1}{2} g_{\alpha\tau} g^{\tau\beta} \left(g_{\beta\mu,\nu} + g_{\beta\nu,\mu} - g_{\mu\nu,\beta}\right) \\ &= \frac{1}{2} \delta_{\alpha}{}^{\beta} \left(g_{\beta\mu,\nu} + g_{\beta\nu,\mu} - g_{\mu\nu,\beta}\right).\end{aligned}$$

In the summation over β only terms with $\beta = \alpha$ are different from zero, since $\delta_{\alpha}{}^{\beta} = 0$ for $\beta \neq \alpha$. Thus we get

$$\Gamma_{\alpha\mu\nu} = \frac{1}{2} \left(g_{\alpha\mu,\nu} + g_{\alpha\nu,\mu} - g_{\mu\nu,\alpha}\right). \tag{7.31}$$

These are the Christoffel symbols of the first kind. They will be used in Sect. 11.2.

Chapter 8
Geodesics

'Geodesy' comes from Greek $\gamma\eta$, Earth, and $\delta\alpha i\omega$, divide, i.e. 'Earth division'. 'Geodesic' will be used in a rather special geometric sense in the following, but it will be related to the old problem of measuring the shortest path on the curved surface of the Earth. From the Euclidean geometry of a plane surface, we know that the shortest path between two points is the straigth line between the points on the surface. However, on the spherical surface of the Earth you cannot find straight paths, so the shortest path between two points is the *straightest possible* path on the surface between the points. Such paths are called *geodesic curves*.

In this chapter we shall give a generally valid and precise mathematical definition of the concept 'geodesic curve'. Such curves can be defined in two conceptually different, but mathematically equivalent ways; either as the shortest (or the longest—in spacetime) curve between two points, or as the straightest possible curve. We choose to define a geodesic curve as the straigthest possible curve.

In order to approach a mathematical precision of the concept 'straightest possible' we note that the tangent vectors of a straight line on a plane, have the same direction. This means that the tangent vector field of a straight line consists of vectors that are connected by parallel transport. In flat space, or on a plane, this has an intuitively obvious meaning: if you move a tangent vector along the curve, without changing the direction of the vector, it will arrive at a new place on the curve and cover (coincide with) a tangent vector at this place.

8.1 Generalizing 'flat space concepts' to 'curved and flat space concepts'

As a prerequisite for a precise definition of 'straightest possible' we shall in the next section define the concept 'parallel transport' of a vector in a covariant way. The definition will be generally valid in curved as well as flat spaces. In this section we shall introduce a powerful and simple method for making such general

Ø. Grøn and A. Næss, *Einstein's Theory: A Rigorous Introduction for the Mathematically Untrained*, DOI 10.1007/978-1-4614-0706-5_8,

definitions. The way we proceed here, is typical for the way that one can generalize a mathematical concept, with a known equation valid in Cartesian coordinates in flat space, to a more general concept, useful also in curved spaces: We start with the known equation valid in flat space. This is generally not a tensor equation. Then we construct a *tensor equation* reducing to our first equation in a Cartesian coordinate system.

If we are to accept the resulting equation as a proper generalization of the original equation, we must know that there is only one tensor equation which reduces to the original one in Cartesian coordinates. We need the following theorem: *If a tensor vanishes in one particular coordinate system, then it vanishes in every coordinate system.*

This is expressed mathematically by the transformation law Eq. (5.13) for the vector components, $u^{\mu'} = (\partial x^{\mu'}/\partial x^{\mu})\, u^{\mu}$. If every u^{μ} vanish, then every $u^{\mu'}$ vanish also.

That this theorem is valid for the special case that the tensor is a vector, is intuitively clear. We are used to think of a vector as a coordinate-independent arrow with fixed length and direction. If a vector vanishes in one particular coordinate system, then there is no arrow, and a transformation to a new coordinate system does not create a vector. For a tensor of arbitrary rank the intuitive basis is no longer there. But the component concept is essentially the same as for vectors. If a tensor vanishes in *one* coordinate system, then all the components of the tensor in this coordinate system vanish, and the transformation laws for tensor components, Eq. (5.80), then secures that *all* the components of the tensor vanish in an arbitrary coordinate system. This proves our theorem: *tensors of arbitrary rank are coordinate independent quantities, just like vectors.*

This means that there is only *one* tensor generalization of a non-tensorial equation valid in a Cartesian coordinate system.

8.2 Parallel transport: unexpected difficulties

Imagine that we pick a piece of paper formed like an arrow, and put it so that it covers the vector $\vec{A}$ on Fig. 8.1. We will let the arrow slide along the boundary of the triangle in such a way that its direction is not changed. If we succeed in parallel

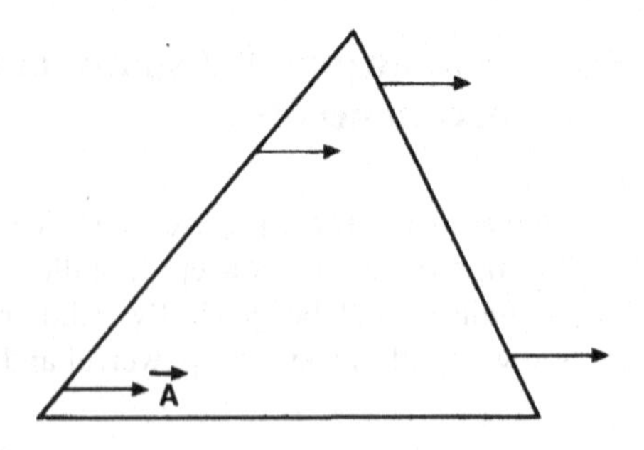

Fig. 8.1 Parallel transport on a flat surface

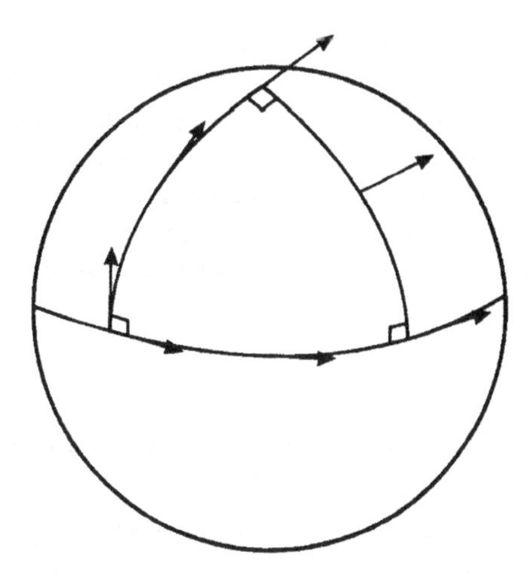

Fig. 8.2 Parallel transport on a sphere

transporting the arrow around the triangle, it will return to the original position with just the same orientation as it had when it left this position.

Let us now imagine that we are two-dimensional creatures ('flatlanders') on a great spherical surface. Unavoidably it sounds like a paradox to ask for this, since it seems to require that we may perceive ourselves simultaneously as two-dimensional creatures on a surface, and as three-dimensional creatures looking at ourselves on that surface. Imagine, furthermore, that the spherical surface is covered with a coordinate system like that on a globe, such that we may define latitude and longitude on the sphere. Assume that we are somewhere on the Equator, with a straight arrow. Letting the arrow point along the Equator we then proceed to walk eastwards to a certain longitude. Then we turn northwards all the way to the North Pole. After a long rest we turn southwards along a latitude further west, so that we reach Equator at just the point we started from. All the time we have kept the direction of the arrow fixed. But, as suggested in Fig. 8.2, the result of this transport is not what we expect from our parallel transport of an arrow on a plane.

We see, in Fig. 8.2, that when the arrow arrives at the point of departure, after the round trip, it is pointing in a direction which deviates from its direction at the start of the trip.

It may be objected that the drawing is misleading. The drawing of a curved surface is possible only if we imagine that the surface is embedded in a flat (Euclidean) 3-dimensional space. One might suspect that our conclusion that the direction of an arrow is changed during a round trip on a spherical surface is, somehow, an illusory result of our reasoning from a three-dimensional point of view. That the result is real enough, however, also for two-dimensional creatures on the surface, will be proved mathematically in the next chapter.

Before proceeding to a mathematical definition of parallel transport, we owe it to the reader to comment a little on the above thought experiment. How can we keep the direction of the arrow constant? On the Earth there is a magnetic field determining the direction of a compass needle. This leads outside the core of the theory of relativity, which concerns space, time and *gravitation*, not electromagnetism. What we therefore need, is a compass of inertia. It was noted by the great mathematician Kurt Gödel that such a compass exists, and it is remarkably simple. A pendulum is a 'compass of inertia'. Its swinging plane remains unchanged relative to the direction of the stars. The swinging plane of the pendulum determines a fixed direction of reference.

8.3 Definition of parallel transport

We shall first describe parallel transport of vectors in a Cartesian coordinate system in flat space. Let a curve pass through the space. The curve is described mathematically by means of a parameter (curve-coordinate) λ. Assume that there is a vector field $\vec{A}$ in the space. If the vectors of the field are connected by parallel transport, they all have the same direction. We shall also demand that vectors connected by parallel transport have the same length. Said in an intuitive way: If two vectors in flat space are connected by parallel transport, they are identical except for their position. In other words: If the vectors of a vector field are connected by parallel transport, then the vector field is constant. And the derivative of a constant vector field vanishes. Thus

$$\frac{d\vec{A}}{d\lambda} = 0. \tag{8.1}$$

The left-hand side is the directional derivative of the vector field along the curve. When we give a parametric description of the curve we specify the coordinates as functions of the curve parameter λ (see chapter 3). So we may use the chain rule for differentiation, Eq. (2.31), and write (using Latin indices for components in a Cartesian coordinate system)

$$\frac{d\vec{A}}{d\lambda} = \frac{\partial A^m}{\partial x^i}\frac{dx^i}{d\lambda} = A^m{}_{,i}\, u^i .$$

In the last step we have introduced Einstein's comma notation (7.10) for partial derivatives, and the Eq. (3.16) for the components of tangent vectors. Equation (8.1) can now be written

$$A^m{}_{,i}\, u^i = 0. \tag{8.2}$$

This is the equation for parallel-transport of vectors, a s expressed in a Cartesian coordinate system.

As we know from the transformation equation (7.24) for partial derivatives of vector components, Eq. (8.2) is not a tensor equation. It is not valid, as anequation for parallel transport, in arbitrary coordinate systems. But from the results of chapter 7 we know how to generalize this equation to one that is generally valid. We just have to replace ordinary partial derivatives by covariant derivatives. Due to the ingenious notation introduced by Einstein, this can be performed just by replacing the comma in Eq. (8.2) by a semicolon. In this way we obtain the covariant equation for parallel transport of vectors,

$$A^{\mu}{}_{;\nu}\, u^{\nu} = 0. \tag{8.3}$$

Substituting the expression (7.15) for the covariant derivative, the equation takes the form

$$A^{\mu}{}_{,\nu}\, u^{\nu} + A^{\alpha}\, \Gamma^{\mu}{}_{\alpha\nu}\, u^{\nu} = 0.$$

Using Eq. (7.11) and placing A^{α} between $\Gamma^{\mu}{}_{\alpha\nu}$ and u^{ν}, we get

$$\frac{dA^{\mu}}{d\lambda} + \Gamma^{\mu}{}_{\alpha\nu}\, A^{\alpha}\, u^{\nu} = 0. \tag{8.4}$$

In this general form the equation for parallel transport is suitable for two most important applications. In the next section we shall see that the equation for geodesic curves is closely related to it, and in the next chapter the equation will be taken as a point of departure for the definition of the Riemann curvature tensor.

8.4 The general geodesic equation

The definition of a 'straightest possible curve' may be stated as follows: A curve is said to be *straightest possible* if and only if the tangent vectors of the curve are connected by parallel transport. Such curves shall be called *geodesic curves.*

According to this definition the equation of a geodesic curve is obtained from Eq. (8.3) just by replacing the arbitrary vector $\vec{A}$ by the tangent vector $\vec{u}$ of the curve. This gives *the equation of a geodesic curve on covariant form*

$$u^{\mu}{}_{;\nu}\, u^{\nu} = 0, \tag{8.5}$$

which may be stated with words as follows: the covariant directional derivative of the tangent vector field of a geodesic curve, in the direction of the curve, is equal to zero.

Using Eq. (8.4) this equation can be written as

$$\frac{du^{\mu}}{d\lambda} + \Gamma^{\mu}{}_{\alpha\nu}\, u^{\alpha}\, u^{\nu} = 0 \tag{8.6}$$

or

$$\frac{d^2x^\mu}{d\lambda^2} + \Gamma^\mu{}_{\alpha\nu}\frac{dx^\alpha}{d\lambda}\frac{dx^\nu}{d\lambda} = 0. \tag{8.7}$$

An equation expressed in terms of tensors, (and not their components), is said to have an *invariant form*. The invariant form of the geodesic equation is

$$\frac{d\vec{u}}{d\lambda} = 0. \tag{8.8}$$

Note that each of the component Eqs. (8.5)–(8.7), represents a set of n equations in an n-dimensional space; one for each value of the index μ. This is understood to be the case also for Eq. (8.8), although this fact is not visible in this equation. A vector equation in four-dimensional spacetime represents four component equations.

We shall give two illustrations; geodesics on an Euclidean plane, and geodesics on a spherical surface.

Example 8.1. Let x and y be Cartesian coordinates on the plane. Then all the Christoffel symbols are equal to zero, and Eq. (8.6) is reduced to

$$\frac{du^x}{d\lambda} = 0 \quad \text{and} \quad \frac{du^y}{d\lambda} = 0.$$

These equations are integrated in just the same way as the corresponding equations of motion of a free particle in section 3.7. Since the derivative of a constant is zero, we get

$$u^x = k_1 \quad \text{and} \quad u^y = k_2,$$

where k_1 and k_2 are constants. Hence

$$\frac{dx}{d\lambda} = k_1 \quad \text{and} \quad \frac{dy}{d\lambda} = k_2.$$

Integrating once more, using Eq. (3.33) with the substitutions $x \to \lambda$, $p = 1$, and $C = x_0$ in the first equation and $C = y_0$ in the second, we find

$$x = x_0 + k_1\lambda \quad \text{and} \quad y = y_0 + k_2\lambda.$$

This is the parametric equation of a straight line (see Eq. (3.15) with $t = \lambda$, $v^x = k_1$, and $v^y = k_2$).

Example 8.2. Next we shall answer the question: what kind of curves on a spherical surface are geodesic? We consider a surface with radius R, and introduce spherical angular coordinates $x^1 = \theta$ and $x^2 = \phi$ on the surface. In this case Eq. (8.7) represents the following two equations (note that the symmetry of the Christoffel symbols has been used)

$$\frac{d^2\theta}{d\lambda^2} + \Gamma^{\theta}{}_{\theta\theta}\left(\frac{d\theta}{d\lambda}\right)^2 + 2\,\Gamma^{\theta}{}_{\theta\phi}\frac{d\theta}{d\lambda}\frac{d\phi}{d\lambda} + \Gamma^{\theta}{}_{\phi\phi}\left(\frac{d\phi}{d\lambda}\right)^2 = 0 \tag{8.9}$$

$$\frac{d^2\phi}{d\lambda^2} + \Gamma^{\phi}{}_{\theta\theta}\left(\frac{d\theta}{d\lambda}\right)^2 + 2\,\Gamma^{\phi}{}_{\theta\phi}\frac{d\theta}{d\lambda}\frac{d\phi}{d\lambda} + \Gamma^{\phi}{}_{\phi\phi}\left(\frac{d\phi}{d\lambda}\right)^2 = 0. \tag{8.10}$$

To some readers the transition from Eq. (8.7) to Eqs. (8.9) and (8.10) may for a moment seem slightly magical. But note that α and ν in the Christoffel symbol of Eq. (8.7) are dummy indexes. We have to expand and summarize with the substitutions $\theta\theta$, $\theta\phi$, and $\phi\phi$ for $\alpha\nu$. With $x^{\mu} = \theta$ we get Eq. (8.9), including four Christoffel symbols, but the symmetry $\theta\phi = \phi\theta$ reduces this to 3. That is the reason for the factor 2 in the third term of each equation. We are sorry to admit that the magic vanishes. Note, also, that in the present case the invariant parameter λ is the arclength along the geodesic curve.

From Eq. (6.28) we find that the only non-vanishing of these Christoffel symbols are

$$\Gamma^{\phi}{}_{\theta\phi} = \frac{\cos\theta}{\sin\theta} \quad \text{and} \quad \Gamma^{\theta}{}_{\phi\phi} = -\sin\theta\,\cos\theta.$$

Consequently Eqs. (8.9) and (8.10) reduce to

$$\frac{d^2\theta}{d\lambda^2} - \sin\theta\,\cos\theta\left(\frac{d\phi}{d\lambda}\right)^2 = 0 \tag{8.11}$$

and

$$\frac{d^2\phi}{d\lambda^2} + 2\frac{\cos\theta}{\sin\theta}\frac{d\theta}{d\lambda}\frac{d\phi}{d\lambda} = 0. \tag{8.12}$$

In order to find the curves represented by these equations, we note that a geodesic curve must be the intersection between a plane and the spherical surface, since it is the straightest possible curve. Obviously such a curve must have circular shape. Since there is no preferred direction on the surface we may choose our coordinate system such that the intersecting circle has a constant value of the angle θ, i.e. so that it is parallel to the 'equatorial circle' of the coordinate system. Thus $d^2\theta/d\lambda^2 = 0$. Then the second term of Eq. (8.11) must be equal to zero, in order that this equation shall be fulfilled. Since $d\phi/d\lambda \neq 0$, this demands $\sin\theta\,\cos\theta$ to equal zero, which is possible only if the angle $\theta = \pi/2$ (in the interval from $\theta = 0$ (the North Pole) to $\theta = \pi/2$ (the Equator)). Hence the geodesic curve is a great circle along the Equator. It remains to show that this solution also satisfies Eq. (8.12). With $\theta = \pi/2$ the second term in Eq. (8.12) vanishes and the equation is reduced to $d^2\phi/d\lambda^2 = 0$, giving $d\phi/d\lambda = K$, where K is a constant. Measuring the angle ϕ in radians, we get $K = 1/R$, and $\phi = \lambda/R$, which is the longitude measured along the equatorial circle.

We have now found that the equatorial *great circle* is the only curve with $\theta =$ constant that solves the geodesic equations. Since the orientation of our coordinate system is arbitrary, we conclude that on a spherical surface the geodesic curves are great circles.

The rather obvious result of Example 8.2 may lead the reader to think that after all we have shot only a crow with our cannon. However, the main point we have illustrated with this example is that *equations (8.5)–(8.8) describe geodesic curves, not only on flat space relative to arbitrary coordinate systems, but in curved space as well.*

The equation will be of vital importance in the theory of relativity. According to this theory spacetime is curved, and free particles move along geodesic curves in spacetime (see Ch. 12).

8.5 Local Cartesian and geodesic coordinate systems

A useful result is the following. At an arbitrary point P in curved spacetime one can construct a coordinate system in a small region around P such that the metric has the Minkowski form given by Eq. (5.121) at P, and the Christoffel symbols vanish at P.

The Minkowski form of the metric is obtained simply by choosing a coordinate system with orthogonal unit basis vectors at P. Coordinates with vanishing Christoffel symbols at P may be constructed as follows. Consider a geodesic curve $\vec{r}(\lambda)$ through P. Let $\vec{u}_P = (u^\mu)_P(\vec{e}_\mu)_P$ be the unit tangent vector of the curve at P, as decomposed in an arbitrary coordinate system. The parameter λ is the path length, or distance, along the curve with $\lambda = 0$ at P. We now introduce a special coordinate system $\{x^\mu\}$ such that at a point Q at a distance λ from P has coordinates

$$x^\mu = (u^\mu)_P\lambda. \tag{8.13}$$

In the new coordinate system, all geodesics through P have equations of the form (8.13). Differentiating twice with respect to λ, and remembering that $(u^\mu)_P$ are the components of a fixed vector, leads to

$$\left(\frac{d^2x^\mu}{d\lambda^2}\right)_P = 0. \tag{8.14}$$

Comparing with the general geodesic equation (8.7), we see that the Christoffel symbols vanish at P.

We have thus succeeded in constructing a coordinate system around P with Minkowski metric and vanishing Christoffel symbols at P. In the following this type of coordinates shall simply be called *local Cartesian coordinates*. From chapter 1 we know that the second derivative of a function describing a curve tells how fast its direction changes, which means that it describes its curvature. Equation (8.14) tells that, as referred to a local Cartesian coordinate system, a geodesic curve in an arbitrary space is straight. In this sense a geodesic curve is the 'straightest possible' curve. From our knowledge of curves and distances on flat surfaces it is tempting to conclude that the geodesic curve is also the shortest path between the given points P_1 and P_2. However, when it comes to geodesic curves in spacetime, you shall be presented with a great surprise. The geodesic curves of spacetime have the *longest* path lengths between two events (see Sect. 5.14).

Chapter 9
Curvature

9.1 The curvature of plane curves

'Spacetime is curved'. It is, of course, not easy to understand adequately what is meant by that sentence as it occurs in the general theory of relativity. There are two principal axes of 'precisation', one leads into pure mathematics, the other into physics and cosmology. We shall start with the mathematical.[1]

One of the simplest properties of a space is its dimension, which was defined in Sect. 5.1. If you wonder how you can find out the number of dimensions of a space, you may just count how many positions in different 'orthogonal' directions you need in order to specify a point in the space. The number you arrive at is the dimension of the space. For example, if you want to make an appointment, you have to specify three positions, the latitude, the longitude and the height in order to fix a point in space, and one number in order to fix the point of time. Thus spacetime is four-dimensional. If you want to specify a point on a curve one position suffices. Thus a curve may be considered as a one-dimensional space.

The simplest case of a one-dimensional space is a straight line. It has no curvature. A straight line may be called an Euclidean one-dimensional space. The next simplest case is a curve in a plane, i.e. a plane curve. Let us consider curves in the xy-plane, and assume that the curves are graphs of differentiable functions, $y = y(x)$. In Sect. 2.3 it was mentioned that there is a connection between the second derivative of a function and the curvature of its graph. We shall now find that connection.

A definition of the curvature of a plane curve is needed. Qualitatively the curvature is an expression of how fast the slope of a curve changes as we move

[1]A formulation U is more precise than a formulation T if and only if the set of different interpretations of U is a genuine subset of the interpretations of T. In other words, if U is more precise than T, there is a higher definiteness of meaning and less ambiguity in U than in T. The more precise formulations, $U_1, U_2, \ldots$, are called precisations.

Ø. Grøn and A. Næss, *Einstein's Theory: A Rigorous Introduction for the Mathematically Untrained*, DOI 10.1007/978-1-4614-0706-5_9,

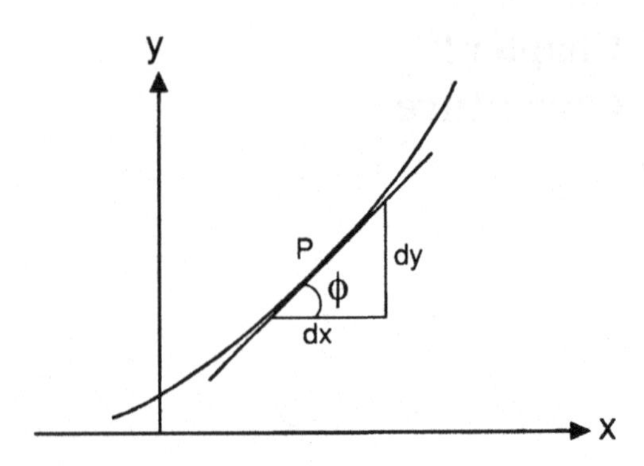

Fig. 9.1 A curve and its tangent

along the curve. In Fig. 9.1 we have drawn a part of a curve in a region around a point P on the curve. Let ϕ be the angle that the tangent to the curve at P makes with the x-axis.

From Eq. (4.3) and the figure is seen that

$$\tan\phi = \frac{dy}{dx}.$$

According to the definition (2.4) of the derivative it then follows that

$$y' = \tan\phi. \tag{9.1}$$

Let ds be a displacement along the curve corresponding to a displacement dx along the x-axis and a displacement dy along the y-axis. The curvature K of a plane curve is defined as the change of the slope angle ϕ per unit displacement along the curve

$$K \equiv \frac{d\phi}{ds}. \tag{9.2}$$

The relationship between the displacements ds, dx, and dy is given by the Pythagorean theorem.

$$(ds)^2 = (dx)^2 + (dy)^2.$$

Putting $(dx)^2$ outside a parenthesis we get

$$(ds)^2 = \left[1 + \frac{(dy)^2}{(dx)^2}\right](dx)^2.$$

Using a rule from the calculus of fractions, $a^2/b^2 = (a/b)^2$, we obtain

$$(ds)^2 = \left[1 + \left(\frac{dy}{dx}\right)^2\right](dx)^2 = \left(1 + y'^2\right)(dx)^2.$$

Taking the positive square root on both sides of the equality sign we have

$$ds = \sqrt{1 + y'^2}\, dx.$$

Inserting this in the definition (9.2) leads to

$$K = \frac{1}{\sqrt{1 + y'^2}} \frac{d\phi}{dx} = \frac{\phi'}{\sqrt{1 + y'^2}}. \tag{9.3}$$

Here ϕ' may be expressed in terms of y'' by differentiating Eq. (9.1). Using Eq. (4.27) and the chain rule, Eq. (2.31), since ϕ is a function of x , we find

$$y'' = (\tan \phi)' = \left(1 + \tan^2 \phi\right) \phi'.$$

Substituting for $\tan \phi$ from Eq. (9.1) leads to

$$y'' = \left(1 + y'^2\right) \phi'.$$

Dividing by $1 + y'^2$ on both sides gives

$$\phi' = \frac{y''}{1 + y'^2}.$$

Inserting this into Eq. (9.3) and using that $\sqrt{1 + y'^2}\,(1 + y'^2) = (1 + y'^2)^{3/2}$, we finally arrive at the expression for the curvature of a plane curve

$$K = \frac{y''}{\left(1 + y'^2\right)^{3/2}}. \tag{9.4}$$

The second derivative of y is not quite the same as the curvature of the graph $y = y(x)$. While y'' is the rate of change of slope (i.e. tangens to the slope angle) per unit distance in the x-direction, the curvature K is the rate of change of slope angle per unit distance along the curve.

Let us again consider the parabola $y = x^2$ of Fig. 2.4. This function has $y' = 2x$ and $y'' = 2$. Thus, its curvature is $K = 2/(1 + 4x^2)^{3/2}$. This expression shows that the curvature of the parabola is largest, equal to 2, at $x = 0$, and decreases towards zero for large values of x. Far away from the y-axis the parabola approaches a straight line. Looking at Fig. 2.4 you see that this is in accordance with the shape of the curve. Note also that if we had tried to define the curvature as the rate of change of slope per unit distance along the x-axis, i.e. as y'', we would have found a constant curvature, equal to 2 for the parabola. A constant curvature corresponds to a circular arc, which clearly does not correspond to the shape of the parabola.

9.2 The curvature of surfaces

Let us inspect a simple surface; the *surface* of a cylinder. Suppose the cylinder you look at is of paper and that you roll it out like you would do with a rolled-up newspaper fetched by your dog. If there were drawings of triangles and other figures on the cylinder they would all retain their shapes and lengths. There would be no trace of distortions or deformations. The geometry of a cylindrical surface is the same as that of a flat paper. The right triangle drawn on a paper with lengths 3, 4, and 5 cm, retain these quantities when we roll the paper into a cylinder. The theorem of Pythagoras holds good also on the curved cylindrical surface. The sum of the angles remains 180 degrees. Two-dimensional creatures on a cylinder develop the plane Euclidean geometry. They have no concept of a cylinder as we see it. But of course a sufficiently long *straight* journey on the surface may lead back to where they started. This is an astonishing fact for the cylindrians, showing that the topology of their world is different from that of an Euclidean plane.

As three-dimensional creatures we may curl the paper cylinder more and more. The diameter of the cylinder gets smaller and smaller, the paper curves 'more and more' per cm. The 5 cm straight line of our drawn triangle is eventually changed into a spiral. But the cylindrians, as *mathematicians*, may not notice anything whatsoever. The geometrical measurements of the cylindrians show no difference. Technically we say that the *intrinsic curvature* of the cylindrical surface is zero. Embedded in a three-dimensional space, it has curvature which is greater or smaller. The curvature of a two-dimensional surface as measured when embedded in a three-dimensional space is called the *extrinsic curvature.*

In general relativity we are concerned with curved four-dimensional spacetime, without reference to any higher-dimensional flat space in which spacetime could be imagined to be embedded. Thus by 'curvature of spacetime' is always meant 'intrinsic curvature'. We might speak of the extrinsic curvature of spacetime, embedded in for example a ten-dimensional Euclidean space, but there is no need for such an exploit.

We shall deduce an expression showing how the intrinsic curvature of a two-dimensional surface can be found by measurements on the surface itself. The circumference and radius of a circle can be measured by means of a ruler. The quotient between these quantities is equal to 2π on a plane. We shall find how the intrinsic curvature of a surface is given by the deviation of this quotient from 2π.

Consider a small part of any surface, for instance a part of the surface of a dome which surrounds a point P. By making the neighborhood around P small enough, the form of the considered mathematical surface will depart as little as one pleases from a spherical form. This is illustrated in Fig. 9.2.

From this figure and the definition of an angle as measured in radians (see Sect. 4.1) we see that

$$\alpha = \frac{r}{R}.$$

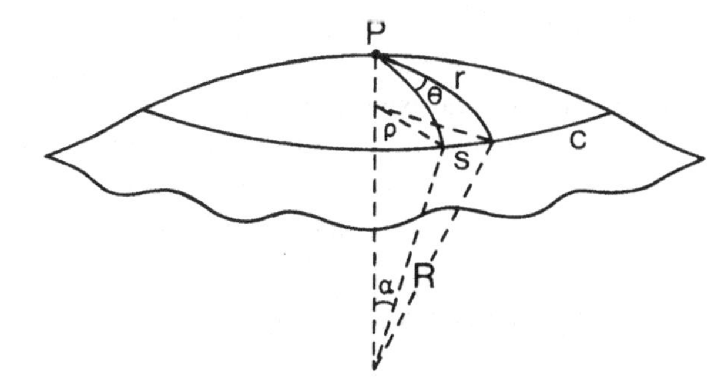

Fig. 9.2 Surface of a dome

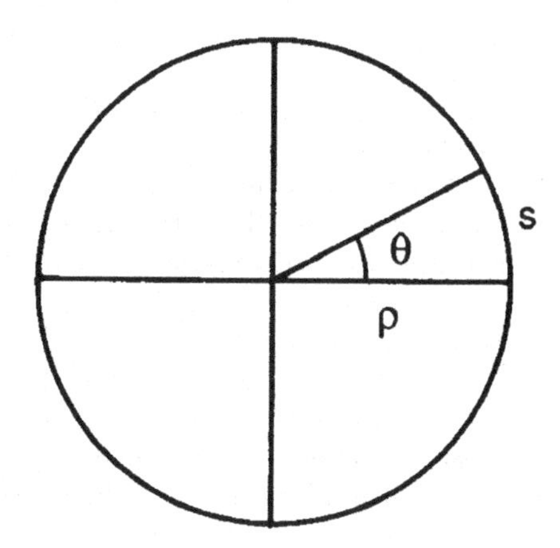

Fig. 9.3 The dome seen from above

The flatter the dome, the larger is R, and the smaller is the angle α. Furthermore

$$\sin\alpha = \frac{\rho}{R}.$$

Thus

$$\rho = R\,\sin\alpha = R\,\sin(r/R).$$

Also, as shown more clearly in Fig. 9.3

$$\theta = \frac{s}{\rho},$$

where s is the distance along the circle C extended by the angle θ. We get

$$s = \theta\rho = \theta\,R\,\sin(r/R). \tag{9.5}$$

This expression shall be used to find the curvature at a point P of the surface depicted on Fig. 9.2. In general the curvature of a surface is a function of the position on the surface. The curvature at P is a local quantity, generally different from the curvature at other points on the surface. Therefore, we shall need to develop an expression for s, valid for very small values of r. This is performed by means of a series expansion of the expression (9.5).

The first two terms of the MacLaurin series for $\sin x$ is (see Sect. 4.1.1)

$$\sin x = x - \frac{1}{6}x^3 + \cdots . \tag{9.6}$$

Here $x = r/R$, and we get from Eq. (9.5)

$$s = \theta\, R \left(\frac{r}{R} - \frac{1}{6}\frac{r^3}{R^3} + \cdots \right) = \theta \left(r - \frac{1}{6}\frac{1}{R^2} r^3 + \cdots \right). \tag{9.7}$$

We have obtained our desired expression for s. This will be used to find the curvature of our surface, Fig. 9.2, at P.

Here R is the radius of the sphere that the surface around P is part of. The *extrinsic curvature*—the curvature measured from the outside of the surface—is called 'K_E', and is defined as

$$K_E \equiv \frac{1}{R}.$$

It is called extrinsic because it refers to the radius of curvature R of the two-dimensional surface as embedded in the three-dimensional space external to the surface. Equation (9.7) may now be written as

$$s = \theta \left(r - \frac{1}{6} {K_E}^2 r^3 + \cdots \right).$$

The length ℓ of the circle C around P with radius r on the surface, is

$$\ell = 2\pi \left(r - \frac{1}{6} {K_E}^2 r^3 + \cdots \right), \tag{9.8}$$

since the angle θ around the circumference is 2π. Dividing each side of Eq. (9.8) by 2π we get

$$\frac{l}{2\pi} \approx r - \frac{1}{6} {K_E}^2 r^3.$$

Thus

$$\frac{1}{6} {K_E}^2 r^3 \approx r - \frac{\ell}{2\pi} = \frac{2\pi r - \ell}{2\pi}.$$

Multiplying by 6 and dividing by r^3 we get

$${K_E}^2 \approx \frac{3}{\pi} \frac{2\pi r - \ell}{r^3}.$$

Taking the limit $r \to 0$ we get

$$K_E{}^2 = \frac{3}{\pi} \lim_{r \to 0} \frac{2\pi r - \ell}{r^3}. \tag{9.9}$$

We have now succeeded in expressing the extrinsic curvature of the considered surface in terms of quantities that can be measured by a two-dimensional creature on the surface.

The *intrinsic curvature* K_I of a surface at a point P is given by the expression at the right-hand side of Eq. (9.9). Thus

$$K_I = \frac{3}{\pi} \lim_{r \to 0} \frac{2\pi r - \ell}{r^3}. \tag{9.10}$$

The intrinsic curvature K_I is called the *Gaussian curvature*. From Eqs. (9.9) and (9.10) it follows that for the considered surface $K_I = K_E{}^2$. This extremely simple relation is not easily seen intuitively, and is *not* valid in general. It is due to the special property of the considered surface; that this surface was assumed to be indistinguishable from a spherical cap at a sufficiently small region around a point P. This is not a general property of surfaces. The *cylindrical surface* we mentioned above, for example, is locally similar to a plane. It has $K_E \neq 0$ and $K_I = 0$.

In general the extrinsic curvature of a surface may be characterized as follows. At every point on the surface we consider the straightest possible curves with different directions. We then find the curvature of these curves. They curve in the direction normal to the surface. Generally the curvature varies with the direction of the curve (except in the isotropic case that we considered above). The maximal and minimal curvatures are called the 'principal curvatures' of the surface, and are denoted by κ_1 and κ_2. The *extrinsic curvature* is defined by

$$K_E \equiv \frac{1}{2} (\kappa_1 + \kappa_2). \tag{9.11}$$

The *intrinsic* (Gaussian) curvature is given by

$$K_I \equiv \kappa_1 \kappa_2. \tag{9.12}$$

For a spherical surface $\kappa_1 = \kappa_2$, which gives $K_I = K_E{}^2 = 1/R^2$. In the case of a cylindrical surface $\kappa_1 \neq 0$ and $\kappa_2 = 0$, which gives $K_E = \kappa_1/2 \neq 0$ and $K_I = 0$.

If $K_I \neq 0$, K_I may be either positive or negative. The case we have considered in Fig. 9.2 has $\ell < 2\pi r$, giving $K_I > 0$.

The geometry of such surfaces is called *parabolic*. In the case that $l > 2\pi r$, the spherical surface is replaced by a surface similar to a saddle. In this case $K_I < 0$. The geometry of such surfaces is called *hyperbolic*.

In elementary school the famous π is a symbol for the ratio between the circumference and diameter of a circle, and the value of π is given as $3.14159\cdots$. True or false? On a flat surface: true. But what about circles drawn on curved surfaces? Think of a circle on a spherical surface. Let the centre of the circle be at the North Pole. From the point of view of a Euclidean three-dimensional space in which the spherical surface is embedded, the radius of the circle is a curved path

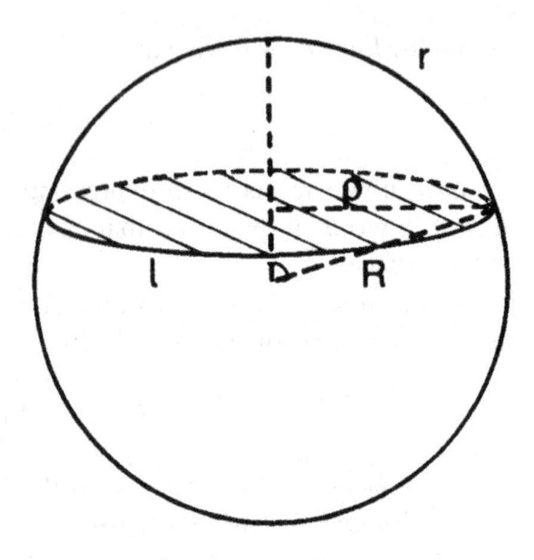

Fig. 9.4 A circle on a spherical surface

on the surface, from the North Pole to the circle. Assume that the radius of the circle is made larger, so that the circle approaches the Equator, and even passes the Equator. Then the circumference of the circle *decreases* while the radius increases. Obviously, the ratio of the length of the circumference and the length of the radius of the circle is not constant when the radius of the circle increases. It is due to the curvature of the surface.

What we have seen is that if one maintains the school-definition of π, then the magnitude of π will depend upon the diameter of the circle that it refers to. This may be maintained, but not the implicit assumption that the circle must be drawn on a flat surface. But both mathematical and physical flatness is an extremely special condition.

Defining the concept of π as quotient between circumference and diameter, its value may vary from zero to infinity, dependent upon the curvature of a surface! We shall in what follows mercifully retain the value of π you always find, by modifying the school definition of π, since it is admittedly not very practical that π might sometimes be equal to 2.15 and sometimes for example equal to 8.50. In order not to violate the near-sacredness of $3.14159\cdots$ we need only to add four words to the definition of π: "on a flat surface". This more precise definition has a consequence with respect to the number of radians we get for the angle around a circle (see Sect. 4.1.1). In order that a radian shall be a geometry-independent measure of an angle, we shall demand that the number of radians around a circle is the same whether it refers to circles on flat or curved surfaces. Then the definition of a radian must also refer to a flat surface. We shall illustrate all this by example 9.1.

Example 9.1. Consider a circle on a spherical surface, as shown in Fig. 9.4.

Here ℓ is the circumference of a circle with radius r on the spherical surface. Let us for a moment retain the 'school-definition' of π mentioned above and apply it to curved space

$$\varpi \equiv \frac{\ell}{2r}. \tag{9.13}$$

The symbol ω also denotes the Greek letter π, but we reserve the π symbol for the flat space value

$$\pi = 3.14159\cdots.$$

The radius of the circle on the shaded plane in Fig. 9.4 is denoted by ρ. Since the geometry is Euclidean on this plane,

$$\ell = 2\pi\rho. \tag{9.14}$$

Inserting Eq. (9.14) into Eq. (9.13) gives

$$\varpi = \frac{\rho}{r\pi}.$$

Now we want to express ϖ as a function of α, the angle at the centre of a globe (see Fig. 9.4). It is seen that

$$\rho = R \sin\alpha$$

so that

$$\varpi = \frac{R\sin\alpha}{r\pi}. \tag{9.15}$$

Since α is measured in radians

$$\alpha = \frac{r}{R}. \tag{9.16}$$

Inserting Eq. (9.16) into Eq. (9.15) gives

$$\varpi(\alpha) = \frac{\sin\alpha}{\alpha}\pi.$$

Let us consider some particular values, remembering that we refer to measurements performed at the North Pole.

For a small circle around the North Pole $\alpha \approx 0$ and the measured value of ϖ according to the 'school-definition', is

$$\lim_{\alpha\to 0}\varpi(\alpha) = \pi \lim_{\alpha\to 0}\frac{\sin\alpha}{\alpha} = \pi \lim_{\alpha\to o}\frac{\alpha}{\alpha} = \pi,$$

where we have used the fact that in the limit $\alpha \to 0$ it is sufficient to retain the first term in the MacLaurin series, Eq. (9.6), for $\sin\alpha$.

For the equatorial circle the angle α is defined with reference to our Euclidean paperplane. Thus the equator corresponds to the angle $\alpha = \pi/2$, giving

$$\varpi(\pi/2) = \frac{\pi}{\pi/2} = \frac{2\pi}{\pi} = 2.$$

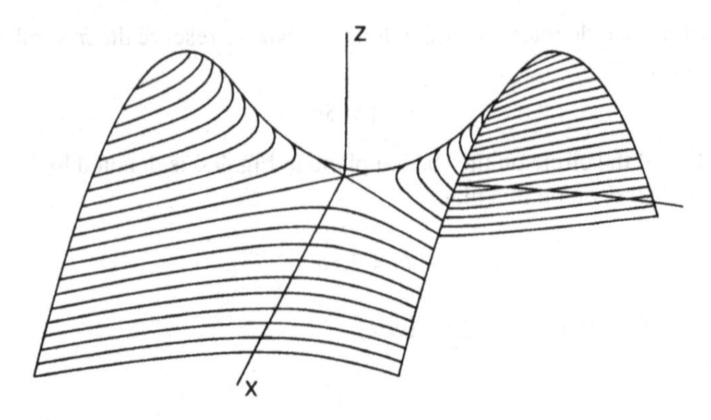

Fig. 9.5 A surface with negative curvature

If our observer at the North Pole measures π using a circle with centre at the North Pole, and with the largest as possible radius, r (see Fig. 9.4), namely a small circle at the South Pole of the surface, he gets

$$\varpi(\pi) = \frac{0}{\pi}\pi = 0.$$

Generally when the curvature of a surface is positive, $\varpi < \pi$.

The simplest surface with negative curvature is a saddle shaped surface, as shown in Fig. 9.5.

A circle drawn on such a surface is shaped like a wave going up and down. The length of such a curve with a given radius is clearly longer than the length of a plane circle with the same radius. This means that the value of $\varpi \equiv \ell/(2r)$ is greater than π on a surface with negative curvature.

When we come to the geometry of three-dimensional spaces—or even four-dimensional spacetimes—the possibility of obtaining a visual image of the curvature of the space has been lost. We must resort to less *intuitive* ways of describing the geometry of such spaces. However, it was discovered by 'the prince of mathematics', Carl Friedrich Gauss—in what he called 'theorema egregium' ('the extraordinary theorem'), that an inhabitant of, say, a three-dimensional space, may perform measurements, *within the three-dimensional space*, which reveals to him the curvature of the space he lives in. This may be achieved by performing analogous measurements to the one we treated in the last paragraph, where the curvature of a surface was found by measuring how the quotient between the circumference of a circle and its diameter deviates from the 'Euclidean' value 3.14.... In the following paragraphs we shall consider one such 'intrinsic' way of measuring the curvature in a space with an arbitrary number of dimensions, namely by measuring the change suffered by a vector when it is parallel transported around a closed path in a curved space.

9.3 Curl

The main topic of this chapter is to introduce a definition of the Riemann curvature tensor. This will be done by calculating the change of a vector by parallel transport around a small closed curve in a curved space. The idea of using such a procedure comes from the considerations in Sect. 8.2, where we found that a vector parallel transported around a triangle on the Earth suffers a change, while on a plane there is no such change.

Before we can proceed with the calculation, we need to make a few definitions. Consider a vector field $\vec{B}$, and a closed curve T. Let $d\vec{r} = dx^\nu \vec{e}_\nu$ be an infinitesimal displacement vector along the curve. The *circulation* C of $\vec{B}$ around the curve is defined as

$$C \equiv \oint \vec{B} \cdot d\vec{r}, \tag{9.17}$$

where the symbol $\oint$ means the integral around the curve. The scalar product $\vec{B} \cdot d\vec{r}$ is the magnitude of the vector $\vec{B}$ times the projection along the vector of an infinitesimal displacement $d\vec{r}$ along the curve. We can say the circulation of a vector field represents the 'flow' of the vector field around a closed curve. If, for example the curve is a stream line in a fluid, and $\vec{B} = \rho\vec{v}$, where ρ is the density of the fluid and $\vec{v}$ its velocity field, the circulation represents the rate of flow of fluid mass around the curve.

We shall now define the *curl* of a vector field. Let $\vec{n}$ be the unit vector normal to a small surface with area ΔS enclosed by the curve T. The component of the curl of a vector $\vec{B}$ in the $\vec{n}$ direction is defined by

$$\text{curl}\,\vec{B} \equiv \vec{n} \lim_{\Delta S \to 0} \frac{1}{\Delta S} \oint \vec{B} \cdot d\vec{r}. \tag{9.18}$$

The curl of a vector field $\vec{B}$ is the circulation density, i.e. the circulation of $\vec{B}$ around small curve T per unit area of the surface enclosed by T. If $\vec{B}$ is the velocity field, $\vec{B} = \vec{v}$, the curl is a local measure of the rate of rotation of a fluid. A fluid with curl free velocity field is said to be irrotational.

Writing $\lim_{\Delta S \to 0} \Delta S = dS$, and multiplying each side of Eq. (9.18) by dS, we have

$$\oint \vec{B} \cdot d\vec{r} = (\text{curl}\,\vec{B})_n \, dS,$$

where $(\text{curl}\,\vec{B})_n$ is the component of curl $\vec{B}$ normal to the surface dS. This equation says that the circulation of a vector field $\vec{B}$ around an indefinitely small curve T is equal to the curl of the vector times the area of the surface enclosed by T. Introducing coordinates $\{x^\alpha, x^\beta\}$, the component of $curl\,\vec{B}$ normal to the surface

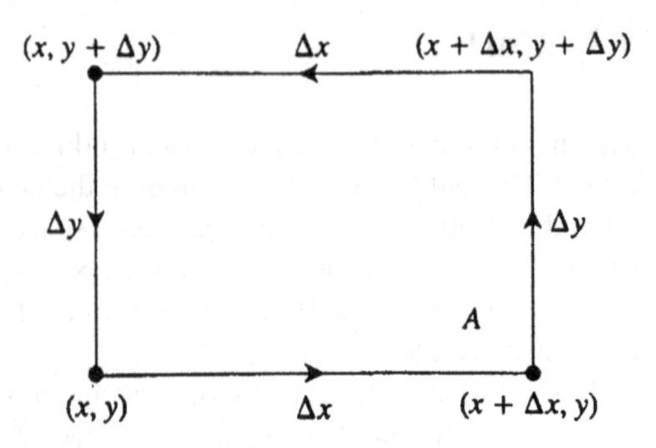

Fig. 9.6 Infinitesimal curve

are written $(\text{curl}\,\vec{B})_{\alpha\beta}$ and the area of the surface $dS^{\alpha\beta}$. Then the circulation of $\vec{B}$ around T can be expressed in the following way

$$\oint \vec{B} \cdot d\vec{r} = (\text{curl}\,\vec{B})_{\alpha\beta}\, dS^{\alpha\beta}. \tag{9.19}$$

We shall now calculate the curl of a vector field $\vec{B}$ in the most simple case of a vector field in a plane covered with a Cartesian coordinate system, i.e. an (x, y) plane. In the following it will be advantageous to express a vector by its covariant components (subscripts), and not by its contravariant components (superscripts). The covariant components of the vector field are $B_x(x, y)$ and $B_y(x, y)$.

Consider a small curve T, enclosing a rectangular surface with area $\Delta S^{xy} = \Delta x\, \Delta y$, as shown in fig. 9.6.

The counterclockwise circulation of $\vec{B}$ around the curve T is the sum of the flow rates along the sides, i.e. of the component of $\vec{B}$ times the length of the side. The flow rates are:

$$\begin{aligned} F_{\text{bottom}} &= B_x(x, y)\, \Delta x, \\ F_{\text{right}} &= B_y(x + \Delta x, y)\, \Delta y, \\ F_{\text{top}} &= -B_x(x, y + \Delta y)\, \Delta x, \\ F_{\text{left}} &= -B_y(x, y)\Delta y. \end{aligned}$$

We add opposite pairs to get, for the top and bottom edges:

$$-\left[B_x(x, y + \Delta y) - B_x(x, y)\right] \Delta x = -\frac{\Delta B_x}{\Delta y}\, \Delta y\, \Delta x$$

and for the right and left sides

$$\left[B_y(x + \Delta x, y) - B_y(x, y)\right] \Delta y = \frac{\Delta B_y}{\Delta x}\, \Delta x\, \Delta y.$$

Adding these expressions gives the circulation around the rectangular curve T

$$C = \left(\frac{\Delta B_y}{\Delta x} - \frac{\Delta B_x}{\Delta y} \right) \Delta x \, \Delta y.$$

Taking the limit of an indefinitely small rectangle, we get from the definition (9.18) of the curl

$$\text{curl}\, \vec{B} = \left(\frac{\partial B_y}{\partial x} - \frac{\partial B_x}{\partial y} \right) \vec{e}_z.$$

Using Einstein's comma notation for partial derivatives, and denoting the z component of the *curl* by $(\text{curl}\, \vec{B})_{xy}$, we have

$$(\text{curl}\, \vec{B})_{xy} = B_{y,x} - B_{x,y}.$$

This expression is valid in a Cartesian coordinate system only. The covariant generalization is obtained by replacing the ordinary partial derivatives by covariant derivatives,

$$(\text{curl}\, \vec{B})_{\alpha\beta} = B_{\beta;\alpha} - B_{\alpha;\beta}. \tag{9.20}$$

However, due to the symmetry (6.31) of the Christoffel symbols one can, in fact, replace the covariant derivatives by ordinary partial derivatives in arbitrary coordinate systems. This can be seen as follows. Inserting the expression (7.25) for the covariant derivative of covariant vector components into Eq. (9.20) we get

$$\begin{aligned}(\text{curl}\, \vec{B})_{\alpha\beta} &= B_{\beta,\alpha} - B_\tau \Gamma^\tau{}_{\beta\alpha} - \left(B_{\alpha,\beta} - B_\tau \Gamma^\tau{}_{\alpha\beta} \right) \\ &= B_{\beta,\alpha} - B_{\alpha,\beta} - B_\tau \left(\Gamma^\tau{}_{\beta\alpha} - \Gamma^\tau{}_{\alpha\beta} \right).\end{aligned}$$

Since the last term vanishes the expression

$$(\text{curl}\, \vec{B})_{\alpha\beta} = B_{\beta,\alpha} - B_{\alpha,\beta}. \tag{9.21}$$

is valid in an arbitrary coordinate system.

9.4 The Riemann curvature tensor

We made appeal to our intuition in chapter eight, to see what would happen to a vector if it was *parallel transported* around a closed curve on a spherical surface, such as, for example, on the surface of a globe. We reached the thought provoking result that when we transport the vector firstly along the equator, then along a certain longitude to the North Pole, and then along another longitude back to the equator

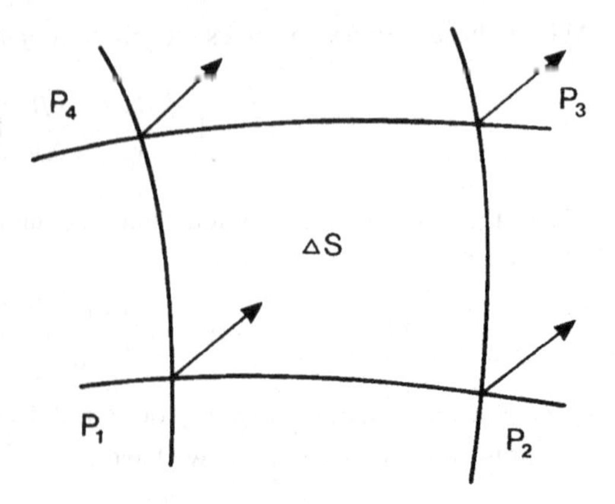

Fig. 9.7 An area on a plane surface

so that it reaches the point of departure, then the vector will not point along the same direction as when it started! If a similar trip is performed on a plane surface, or even on the surface of a cylinder, no change of direction will take place. The change seems to depend in some way upon the curvature of the spherical surface, as distinct from a plane or cylindrical surface. In this connection it may be noted that a cylindrical surface may be 'rolled out' on a plane surface.

In the following we shall search for a consistent geometrical interpretation of the change of direction of a vector due to parallel transport around a closed curve. This will turn out to be rather involved since we are seeking a mathematical expression valid for curved surfaces generally, that is, for an overwhelmingly rich variety of surfaces. You are invited to take part in what might be called 'index gymnastics'. The equations in this section are relatively short and simple, but looking closely you will notice a bewildering change of indices. There is reason to suspect that Einstein, who felt he was a pure physicist, and never a mathematician, disliked this sort of gymnastics. But he saw no way to avoid it in his search for a relativistic theory of gravitation. So he not only learned it and mastered it. He also contributed to this part of the mathematics by inventing useful notation—as for example the Einstein summation convention.

Let us start by inspection of a small area ΔS on a *plane* surface. This area is enclosed by a pair of coordinate curves of an arbitrary curvilinear coordinate system (see Fig. 9.7). Parallel transport on a *plane* surface does not change a vector. If we denote the change of a vector $\vec{A}$ due to parallel transport around a closed curve by $\Delta\vec{A}$, then, in the case of a plane surface, we may safely write

$$\Delta\vec{A} = 0.$$

In this chapter we intend to deduce an expression for $\Delta\vec{A}$ in the case of parallel transport around an *infinitely* small, closed curve on a *curved* surface of any kind. We shall then use the notation $d\vec{A}$ instead of $\Delta\vec{A}$.

In our deduction of the general expression for $d\vec{A}$ we will need to know how the components of $\vec{A}$ change during parallel transport. As made clear in Sect. 8.2, if the vectors of a vector field $\vec{A}$ are connected by parallel transport along a curve, then the vector field is constant along the curve. This is expressed by the vector equation (see Eq. (8.1))

$$d\vec{A} = 0,$$

which holds in curved space as well as in flat space.

There is something strange here. Parallel transporting a vector around a triangle on a spherical surface we found a change $\Delta\vec{A} \neq 0$. However, by adding vanishing changes $d\vec{A}$ around a curve one cannot get a non-vanishing change. The solution of this apparent paradox is that the change of a vector $d\vec{A}$, when the vector is parallel transported an infinitesimal distance dx^μ, vanishes only to first order in dx^μ. Another way of expressing this is to say that the change of a vector field by parallel transport is not visible when we are concerned only with first derivatives of the vector field. But if we calculate to second order in the differentials there *is* such a change.

If the vector changes by parallel transport around a curve, then the components of the vector will change by such a transport, too, since the vector is transported back to the same point it started from, with the same basis vectors. We shall, as announced, calculate the change of the vector components when the vector is parallel transported around an indefinitely small curve. It turns out to be most convenient to calculate the changes dA_μ of the covariant vector components. Parallel transporting a vector $\vec{A}$ a coordinate distance dx^ν we have

$$A_{\mu;\nu} dx^\nu = 0.$$

Inserting the expression (7.25) for the covariant derivatives of the covariant vector components, we get

$$A_{\mu,\nu} dx^\nu = A_\tau \Gamma^\tau{}_{\mu\nu} dx^\nu.$$

Thus

$$A_{\mu,\nu} = A_\tau \Gamma^\tau{}_{\mu\nu}. \tag{9.22}$$

Furthermore, the change of A_μ by parallel transport along dx^ν is

$$dA_\mu = A_\tau \Gamma^\tau{}_{\mu\nu} dx^\nu. \tag{9.23}$$

According to what was said above one might be surprised that we need not include a second derivative term in the expression for dA_μ. However, according to Eq. (9.21), the sum of the first derivatives (the Christoffel symbols) around a curve may be expressed by the second derivatives (the derivatives of the Christoffel symbols) on the surface enclosed by the curve. Therefore the expression (9.23) is sufficiently accurate.

We shall now calculate the total change, ΔA_μ, of A_μ by parallel transport around a closed curve. In other words, we shall integrate dA_μ, as given by Eq. (9.23), around the curve,

$$\Delta A_\mu = \oint dA_\mu .$$

The right-hand side of Eq. (9.23) can be written

$$dA_\mu = B_{\mu\nu}\, dx^\nu , \tag{9.24}$$

where the quantities $B_{\mu\nu}$ are given by

$$B_{\mu\nu} = A_\tau \Gamma^\tau{}_{\mu\nu} . \tag{9.25}$$

We now introduce vectors $\vec{B}_\mu$ with covariant components $B_{\mu\nu}$ in order to write the right-hand side of Eq. (9.24) as a scalar product. Usually vector components have only one index. Here we need two, one index μ to specify which vector we consider, and one index ν to specify the component of the vector (the x component or the y component and so forth). According to Eq. (5.119) the scalar product of two vectors, $\vec{N}$ and $d\vec{r}$, can be expressed in terms of their components in an arbitrary coordinate system as follows

$$\vec{N} \cdot d\vec{r} = N_\nu\, dx^\nu$$

where $d\vec{r} = dx^\mu \vec{e}_\mu$ have been used. In the present case the components N_ν are replaced by the components $B_{\mu\nu}$ of the vectors $\vec{B}_\mu$, and we have

$$dA_\mu = B_{\mu\nu}\, dx^\nu = \vec{B}_\mu \cdot d\vec{r} .$$

Thus the total change of the vector components A_μ around an indefinitely small curve T of arbitrary shape, is equal to

$$\Delta A_\mu = \oint dA_\mu = \oint \vec{B}_\mu \cdot d\vec{r} . \tag{9.26}$$

Comparing with Eq. (9.17) we see that Eq. (9.26) is just the circulation of $\vec{B}_\mu$ around the curve. According to Eq. (9.19) this is equal to a sum of the components of the curl of $\vec{B}_\mu$ times the components of the area $dS^{\alpha\beta}$ of the surface enclosed by the curve. The components of the curl are given by Eq. (9.21). Thus

$$\Delta A_\mu = (\text{curl}\, \vec{B}_\mu)_{\alpha\beta}\, \Delta S^{\alpha\beta} = (B_{\mu\beta,\alpha} - B_{\mu\alpha,\beta})\, \Delta S^{\alpha\beta} .$$

Inserting the expression (9.25) for the components of $\vec{B}_\mu$ we get

$$\Delta A_\mu = \left[\left(A_\tau \Gamma^\tau{}_{\mu\beta}\right)_{,\alpha} - \left(A_\tau \Gamma^\tau{}_{\mu\alpha}\right)_{,\beta} \right] \Delta S^{\alpha\beta} .$$

Differentiating the products $A_\tau \Gamma^\tau{}_{\mu\alpha}$ and $A_\tau \Gamma^\tau{}_{\mu\beta}$ and renaming the summation indices so that τ is replaced by ν in the second and fourth terms, gives

$$\Delta A_\mu = \left(A_{\tau,\alpha}\Gamma^\tau{}_{\mu\beta} + A_\nu \Gamma^\nu{}_{\mu\beta,\alpha} - A_{\tau,\beta}\Gamma^\tau{}_{\mu\alpha} - A_\nu \Gamma^\nu{}_{\mu\alpha,\beta}\right) \Delta S^{\alpha\beta}. \tag{9.27}$$

According to Eq. (9.22) (replacing μ by τ, ν by β and τ by ν) $A_{\tau,\alpha} = A_\nu \Gamma^\nu{}_{\tau\alpha}$ and (replacing μ by τ, ν by α and τ by ν) $A_{\tau,\beta} = A_\nu \Gamma^\nu{}_{\tau\beta}$. Inserting these expressions into Eq. (9.27), we get

$$\Delta A_\mu = \left(A_\nu \Gamma^\nu{}_{\tau\alpha}\, \Gamma^\tau{}_{\mu\beta} - A_\nu \Gamma^\nu{}_{\tau\beta}\, \Gamma^\tau{}_{\mu\alpha} + A_\nu \Gamma^\nu{}_{\mu\beta,\alpha} - A_\nu \Gamma^\nu{}_{\mu\alpha,\beta}\right) \Delta S^{\alpha\beta}.$$

Putting the common factor A_ν outside the parenthesis, and exchanging the factors in the products of the Christoffel symbols (only for aestetic reasons, to get the same succession of the indices α and β in the two first terms and the two last terms), we get

$$\Delta A_\mu = \left(\Gamma^\tau{}_{\mu\beta}\, \Gamma^\nu{}_{\tau\alpha} - \Gamma^\tau{}_{\mu\alpha}\, \Gamma^\nu{}_{\tau\beta} + \Gamma^\nu{}_{\mu\beta,\alpha} - \Gamma^\nu{}_{\mu\alpha,\beta}\right) A_\nu\, \Delta S^{\alpha\beta}. \tag{9.28}$$

The *Riemann curvature tensor* is the fourth rank tensor with components $R^\nu{}_{\mu\alpha\beta}$ defined by

$$R^\nu{}_{\mu\alpha\beta} \equiv \Gamma^\tau{}_{\mu\beta}\, \Gamma^\nu{}_{\tau\alpha} - \Gamma^\tau{}_{\mu\alpha}\, \Gamma^\nu{}_{\tau\beta} + \Gamma^\nu{}_{\mu\beta,\alpha} - \Gamma^\nu{}_{\mu\alpha,\beta}. \tag{9.29}$$

The change of the covariant components a vector $\vec{A}$ by parallel transport around an indefinitely small closed curve enclosing a surface with area $dS^{\alpha\beta}$ may now be written

$$\Delta A_\mu = \frac{1}{2} R^\nu{}_{\mu\alpha\beta}\, A_\nu\, \Delta S^{\alpha\beta}. \tag{9.30}$$

The change of the vector itself is given by the change of its components, (raising the index μ at both sides of the last equality sign)

$$\Delta\vec{A} = \Delta A^\mu \vec{e}_\mu = \frac{1}{2} R^{\nu\mu}{}_{\alpha\beta}\, A_\nu\, \Delta S^{\alpha\beta}\, \vec{e}_\mu, \tag{9.31}$$

since the vector has been parallel transported around a closed curve and thus has come back to the point where its transport started, with the same basis vectors.

Equation (9.31) shows that the change of a vector by parallel transport around a closed path in a curved space, is proportional to the product of the curvature of the space and the area of the surface enclosed by the path.

The result expressed in Eq. (9.31) is purely mathematical. It concerns the curvature of curved surfaces and curved spaces. The 'theorema egregium' of Gauss is contained in Eq. (9.28), since the values of the Christoffel symbols and their derivatives are defined 'intrinsically', i.e. without reference to any higher-dimensional space which the one with curvature $R^{\mu}{}_{\nu\alpha\beta}$ could be embedded in. Due to its mathematical character this theorem needs no empirical support, it tells nothing about the physical world. The geometrical theory can be developed abstractly as a non-geometrical, logical system. The terms 'vector', 'point', 'coordinate', and 'path' are then introduced without any reference to a physical space that we might be said to live in. Drawings on a paper are of no significance for the deductions. Drawings are only of heuristic value.

However, some equations of the abstract system can be made to *correspond* to certain empirically studied physical relationships. If we construct a theory which, for example, through certain equations, successfully predicts how light moves in flat spacetime, and also in curved spacetime, we would talk of a *physical* spacetime, whose geometrical properties we then can investigate empirically. The general theory of relativity is just such a theory. This makes it meaningful to talk about the geometry of *physical* spacetime.

We shall further on be talking about a four-dimensional model of the universe where we are born and presumably are going to die. But the level of abstraction will be astonishing. Abhorrent and frightening to some, awesome to others. The so-called Einstein equations, superceeding Newton's, require skyhigh levels of abstraction.

The question is unavoidable: will humanity never get back to a fairly easily understandable, but grand theory of the universe of the Newtonian kind? For those who hope to see a trend in that direction, the development of this century has been discouraging.

Chapter 10
Conservation laws of classical mechanics

10.1 Introduction

In order to be able to understand Einstein's field equations we should first consider some important concepts of Newtonian physics.

Three fundamental principles of Newtonian physics are:

1. conservation of momentum (mass times velocity)
2. conservation of mass
3. conservation of energy

The general theory of relativity is a theory which is conceptually very different from Newtonian mechanics and gravitational theory. It is often said that Einstein's theory generalizes Newton's theory. This must, however, be understood in a restricted sense; the general theory of relativity contains the predictions of Newton's theory of gravitation, as a limiting case with small velocities (compared to the velocity of light) and weak gravitational fields. Physicists are confident that the general theory of relativity can describe, with great accuracy, also phenomena involving strong gravitational fields, such as very compact stars, and systems with great velocities, like the expanding universe. But to do this a new conceptual framework had to be invented.

Of particular significance is the concept 'spacetime' which was introduced into physics by Hermann Minkowski in a famous speech on September 21, 1908, starting by the words:

> The views of space and time which I wish to lay before you have sprung from the soil of experimental physics, and therein lies their strength. They are radical. Henceforth space by itself, and time by itself, are doomed to fade away into mere shadows, and only a kind of union of the two will preserve an independent reality.

If one wants to apply a theory of gravity to the interior of a star, or to the universe as a whole, one must be able to combine a law of gravitation with a hydrodynamical

Ø. Grøn and A. Næss, *Einstein's Theory: A Rigorous Introduction for the Mathematically Untrained*, DOI 10.1007/978-1-4614-0706-5_10,

formulation of the basic principles stated above. This is because both the interior of stars and the contents of the universe as a whole can today most simply be *modelled* as a kind of fluid or a gas.

For example, the simplest model of the cosmic mass is a fluid or gas without viscosity, called 'perfect fluid'. It has by definition only three properties: mass (energy), pressure (or tension) and motion. These properties are represented by the mass density (mass per unit volume), ρ, the pressure, p, and a velocity field, $\vec{v}$.

In this chapter we shall study slowly moving systems in regions free of gravitational fields. Such systems can be adequately investigated by means of conventional Newtonian dynamics, as formulated in the nineteenth century. However, in order to prepare for the covariant formulation of the fundamental laws of nature in the way they appear in the general theory of relativity, we shall here describe fluids moving in the *four-dimensional* flat spacetime of special relativity, using a tensor formalism that may also be applied in curved spacetime. So, even if the numerical results obtained when the equations of this chapter are applied to hydrodynamical problems, for example to the task of predicting tomorrow's weather, are equal to those of Newton's theory—the spirit of our treatment is due to Einstein and Minkowski.

10.2 Divergence

In order to be able to give a condensed, general mathematical formulation of the fundamental conservation laws, we shall need to become familiar with what is called the divergence. Even if this concept is of a geometrical nature, it will be useful to introduce it by means of a physical example.

Imagine that something is flowing out from a region and something flows into it. In some regions, called 'sources', more is flowing out of it than into it. For 'sinks' the situation is the opposite one.

In the story which unfortunately has contributed in the forming of a negative picture of wolves, namely Little Red Riding Hood, more is moving into the cavern than out of it. The cavern is a sink. On the other hand, the sun or a birth clinic, is a source. More heat or people are flowing out of it than into it.

The *flux* of a quantity through a surface is defined as the product of that quantity and the area of the surface. We define the *divergence* of a quantity at a point P as the net flux per unit volume of that quantity moving out or in from an indefinitely small region surrounding P.

Already at this point one may guess that a conservation law can be expressed mathematically in terms of a vanishing divergence. Our task now is to express the concept 'divergence' mathematically.

Let an indefinitely small region around a point P be equipped with a Cartesian coordinate system, as shown in Fig. 10.1. The point P is located in a vector field

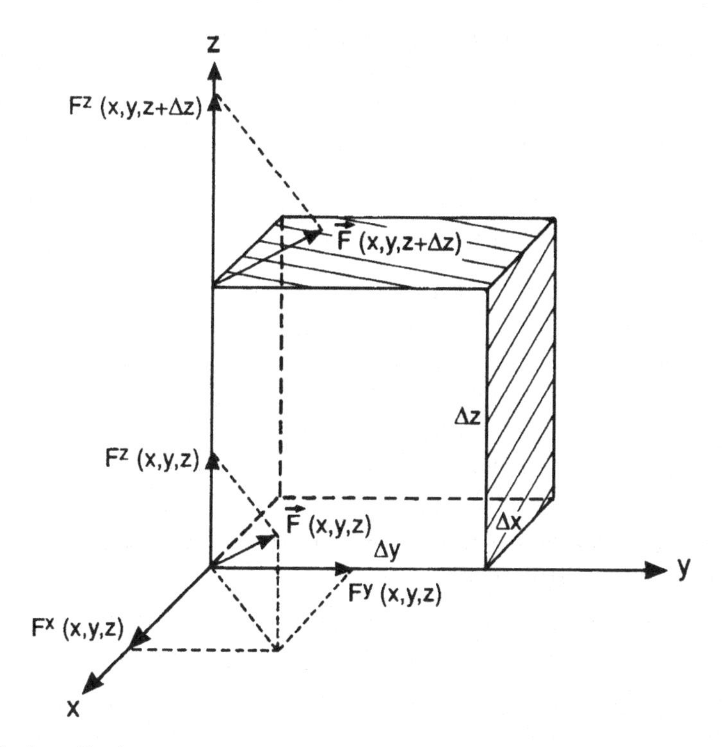

Fig. 10.1 A small volume

$\vec{F}(x, y, z)$. The physical meaning of $\vec{F}$ is of no significance. But if one wants to think of something which is streaming, then one might let $\vec{F}$ represent, for example, the velocity field of a fluid. Our region is now chosen as a parallel-epiped (generalization of the cube, with different lengths of the sides) with side-lengths Δx, Δy, Δz along the coordinate axes.

Consider the z direction. At the point P the z component of the vector field is $F^z(x, y, z)$. A small distance Δz from P along the z axis the z component has changed to $F^z(x, y, z + \Delta z)$. In Fig. 10.1 this component is represented by a longer arrow than the one at P. From our definition then follows that the region is a source of $\vec{F}$ as far as the z direction concerns.

The z component of the net flux of $\vec{F}$ out of the parallel-epiped is

$$F^z(x, y, z + \Delta z)\Delta x \, \Delta y - F^z(x, y, z)\Delta x \, \Delta y.$$

The volume of the parallel-epiped is

$$\Delta V = \Delta x \, \Delta y \, \Delta z.$$

The z component of the net flux per unit volume is

$$\frac{[F^z(x, y, z + \Delta z) - F^z(x, y, z)]\,\Delta x\,\Delta y}{\Delta x\,\Delta y\,\Delta z} = \frac{F^z(x, y, z + \Delta z) - F^z(x, y, z)}{\Delta z}.$$

Adding the x, y, and z components and taking the limit of an indefinitely small region, we find the divergence of $\vec{F}(x, y, z)$ at P

$$\begin{aligned}\operatorname{div} \vec{F} = {} & \lim_{\Delta x \to 0} \frac{F^z(x + \Delta x, y, z) - F^z(x, y, z)}{\Delta x} \\ & + \lim_{\Delta y \to 0} \frac{F^y(x, y + \Delta y, z) - F^y(x, y, z)}{\Delta y} \\ & + \lim_{\Delta z \to 0} \frac{F^z(x, y, z + \Delta z) - F^z(x, y, z)}{\Delta z}.\end{aligned}$$

Comparing this with the definition of partial derivatives of functions of several variables (see Sect. 2.8), we find

$$\operatorname{div} \vec{F} = \frac{\partial F^x}{\partial x} + \frac{\partial F^y}{\partial y} + \frac{\partial F^z}{\partial z}. \tag{10.1}$$

If we use Einstein's summation convention, and the comma notation for partial derivatives, this can be written as

$$\operatorname{div} \vec{F} = \frac{\partial F^i}{\partial x^i} = F^i{}_{,i}. \tag{10.2}$$

The definition of divergence, as formulated above, does not refer to any particular coordinate system. The particular expressions (10.1) and (10.2), however, are only valid in locally Cartesian coordinate systems. We sorely need a coordinate independent generalization of these expressions.

The mathematical apparatus that we have developed makes it nearly miraculously simple to find this generalization. It is simply the tensor-expression which is reduced to Eq. (10.1) in a locally Cartesian coordinate system. The tensor-generalization of partial derivatives of tensor components is found just by replacing the partial derivatives with covariant derivatives. Thus, the tensor expression replacing Eq. (10.2) is found by replacing the comma by a semicolon

$$\operatorname{div} \vec{F} = F^\mu{}_{;\mu}. \tag{10.3}$$

We have also replaced Latin indices referring to Cartesian coordinates in three-dimensional space by Greek indices referring to arbitrary coordinates in spacetime. Note that in spacetime the summation extends over the time coordinate and the three space coordinates.

The generalization of Eq. (10.3) to the divergence of a tensor-field $T^{\mu\nu}$ of rank 2 is a vector with components

$$(\text{div}\, T)^{\mu} = T^{\mu\nu}{}_{;\nu}. \tag{10.4}$$

In general, taking the divergence of a tensor-field reduces its rank with one. For example the divergence of a tensor of rank 2 is a vector, and the divergence of a vector-field is a scalar (i.e. the field of a tensor of rank 0).

If the divergence of a vector field vanishes, one gets a single (scalar) equation

$$F^{\mu}{}_{;\mu} = 0.$$

We started this section with an application of mathematics to physics, in the sense of an illustration. We end this section with a purely mathematical concept. However, the next section starts and ends within physics.

10.3 The equation of continuity

In sections 10.3–10.5 we shall derive the two basic equations governing the motion of fluids moving slowly compared to the velocity of light. This is needed to formulate a relativistic hydrodynamics, which will be used when we are going to apply the general theory of relativity to the task of modelling the large scale structure of our universe.

In Newtonian physics we postulate that mass is conserved. Consider a fluid in a region with volume V enclosed by a surface with area S. Conservation of mass, meaning that no mass is created or destroyed, can then be expressed by the equation

$$\begin{bmatrix}\text{increase of mass within}\\ \text{a volume } V \text{ per unit time}\end{bmatrix} = \begin{bmatrix}\text{net flux of mass across}\\ \text{surface } S \text{ per unit time}\end{bmatrix}.$$

We shall now express this equation in terms of our mathematical language.

Mass density, ρ, is defined as mass per unit volume,

$$\rho = \frac{m}{V},$$

where m is the mass inside a volume V. Hence

$$m = \rho V. \tag{10.5}$$

We have here assumed that V is sufficiently small that the density is constant inside the volume.

The *current density* of a fluid is defined as the mass density of the fluid times its velocity.

Consider the surface of a cube, at rest in a reference frame with a Cartesian coordinate system. The fluid is here flowing in the negative y direction of coordinate

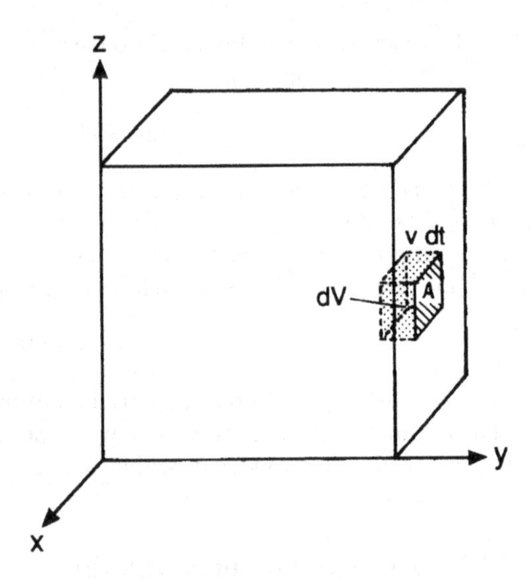

Fig. 10.2 The infinitesimal volume dV

system. The cube is oriented as in Fig. 10.1. During a time interval dt a fluid-element moves a distance $dy = v\,dt$ normal to the surface facing the positive y axis. The fluid having passed this surface during a time interval dt fills a volume $dV = A\,dx = A\,v\,dt$ inside the cube (see Fig. 10.2). The mass entering the cube through this surface during the time dt, is

$$\begin{aligned} d_y\,m &= -\rho\,dV = -\rho\,A\,dy \\ &= -\rho\,A\,v\,dt = -\rho\,v\,A\,dt. \end{aligned} \tag{10.6}$$

The minus sign has been included since a positive velocity v is directed out of the sweep-net, causing a *decrease* of mass inside it. A negative velocity causes an increase of mass inside the cube.

Equation (10.6) shows that the mass which has passed through the surface element during the time dt, is minus the flux of the current density, $\rho\,v$ times dt. The amount of mass, dm, which has passed through all the walls of the cube during a time interval dt is therefore minus the net flux of the fluid's current density through all the walls of the cube times dt.

The *divergence* of a quantity is defined as the net flux of the quantity per unit volume. Thus, the net increase of fluid-mass in a time interval dt, inside the cube with volume V, is

$$dm = -\operatorname{div}(\rho\vec{v})\,V\,dt. \tag{10.7}$$

On the other hand, if the cube has constant volume, the change of mass inside it results in a change of density,

$$d_t\,\rho = \frac{\partial\rho}{\partial t}\,dt.$$

In general the density is a function both of time and position. In the present case we consider the change of density at a fixed place in the fluid. It is the *local* change of density when it varies with the time. This has been marked by a subscript t on the differential, and has made it necessary to use partial derivatives. From Eq. (10.5) follows that the corresponding change of mass inside the cube may be expressed by

$$dm = V\, d_t \rho = \frac{\partial \rho}{\partial t} V\, dt. \tag{10.8}$$

Setting the two expressions (10.7) and (10.8) for dm equal to each other, leads to

$$\frac{\partial \rho}{\partial t} V\, dt = -\operatorname{div}(\rho \vec{v})\, V\, dt$$

or

$$\frac{\partial \rho}{\partial t} = -\operatorname{div}(\rho \vec{v}). \tag{10.9}$$

This is called the *equation of continuity* of classical hydrodynamics, which might better be called 'the equation of mass conservation'. It is the mathematical expression of the assumption that mass is conserved, i.e. that mass is neither created nor destroyed.

Note that mass is not always conserved. For example in nuclear reactions there may be a loss of mass which is transformed into energy. Relativistically one would say that mass-energy is conserved also in such a situation, but in Newton's theory mass conservation and energy conservation are two different assumptions.

10.4 The stress tensor

A so-called 'perfect fluid' is defined as a fluid which (according to Newton's theory) has only four 'dynamical' properties: mass, energy, stresses, and motion. We shall at first discuss how the stresses are represented mathematically in classical fluid dynamics.

A small cube inside a fluid is drawn in Fig. 10.3. A vector pointing orthogonally to a surface is called a *normal vector* of the surface. We now introduce a Cartesian coordinate system so that the sides of the cube are lying along the coordinate-axes. Then the unit normal vectors of the cube's surfaces are $\vec{e}_x$, $\vec{e}_y$, and $\vec{e}_z$. A *stress*-component—force per unit area—acting on a surface with normal vector $\vec{e}_i$ and pointing in the direction of $\vec{e}_j$, is termed t^{ij}. The first superscript indicates the normal vector of the surface on which the stress acts, and the second superscript shows the direction of the force-component. Thus, t^{xx}, t^{xy}, and t^{xz} are, respectively, the x, y, and z components of the stress acting on the surface with

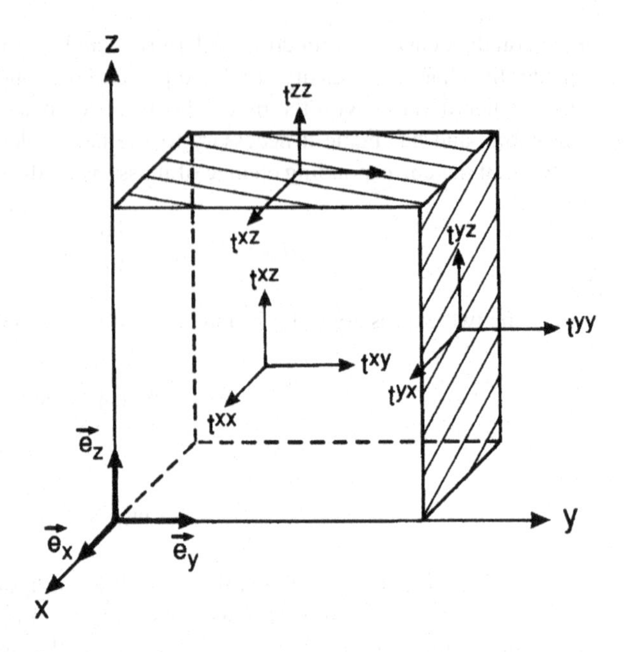

Fig. 10.3 A cube inside a fluid

normal vector $\vec{e}_x$. The stress-component t^{xx} points in the direction orthogonal to the yz plane (having normal-vector $\vec{e}_x$) on which it acts. It is therefore called a *normal* stress. The stress-components t^{xy} and t^{xz} point along this plane. They are called *shear* stresses. The existence of shear stresses depends upon some sort of friction in the fluid. A perfect fluid is friction-free. So, there are no shear stresses in a perfect fluid. A normal-stress pointing towards the surface it acts upon, is called a *pressure*, while a normal-stress pointing away from the surface is called a *tension*.

The nine stress-components acting in an arbitrary fluid make up a *stress tensor* of rank 2. The components are often written in matrix form as follows.

$$t^{ij} = \begin{bmatrix} t^{xx} & t^{xy} & t^{xz} \\ t^{yx} & t^{yy} & t^{yz} \\ t^{zx} & t^{zy} & t^{zz} \end{bmatrix}.$$

The shear stress t^{xy} is the force per unit area acting upon the front side of the cube (see Fig. 10.3), and pointing in the y direction. It tends to give the cube a rotating motion in the counter-clockwise direction about the z axis. In order that this rotational motion shall not increase beyond any limit, it must be counteracted by a force, t^{yx} per unit area, on the right-hand side of the cube, acting in the x direction. If these stress-components shall keep the cube in equilibrium, the condition $t^{xy} = t^{yx}$ must be fulfilled. If $t^{xy} \neq t^{yx}$ these stress components would cause the fluid elements to rotate faster and faster, which is not observed. In a similar way, considering rotations about the x and y axes, leads to the conclusion that *the stress tensor must be symmetric*.

The pressures are usually denoted by p^x, p^y,and p^z and correspond to the three first diagonal components of the stress tensor $t^{xx} = p^x$, $t^{yy} = p^y$, and $t^{zz} = p^z$. In a perfect fluid the pressure is isotropic, $p^x = p^y = p^z = p$. The stress tensor of a perfect fluid with isotropic (same in all directions) pressure is reduced to

$$t^{ij} = p\,\delta^{ij}, \tag{10.10}$$

where the Kronecker symbol may here be thought of as the metric tensor of flat three-space in a Cartesian coordinate system.

10.5 The net surface force acting on a fluid element

In order to find an expression for the resultant surface force acting on our fluid element, we start by considering the force-components in the x-direction. We denote the stress acting on the surface of the cube in Fig. 10.3 with normal vector $\vec{e}_x$ by $\vec{t}^x$. Stress is force per unit area. In other words, force is stress times area. Since the area of the surface is $\Delta x\,\Delta y$, the force acting on this surface is

$$\vec{F}^x = \vec{t}^x\,\Delta y\,\Delta z.$$

This force can be expressed in component form

$$\begin{aligned}\vec{F}^x(x+\Delta x, y, z) = \big[\,t^{xx}(x+\Delta x, y, z)\,\vec{e}_x &+ t^{xy}(x+\Delta x, y, z)\,\vec{e}_y\\ &+ t^{xz}(x+\Delta x, y, z)\,\vec{e}_z\,\big]\,\Delta y\,\Delta z.\end{aligned} \tag{10.11a}$$

The first term represents the force in the x direction acting on the right-hand surface normal to the x axis, with corresponding interpretations of the other terms.

The reason for the arguments $x + \Delta x$ inside the parentheses is that the surface with normal vector $\vec{e}_x$ is located at the position $x + \Delta x$. The surface with normal vector $-\vec{e}_x$, on the other hand, is at the position x (we shall keep an arbitrary x in our expressions, even if $x = 0$ in the drawing), and since this force points in the negative x direction, it is given by

$$\vec{F}^x(x, y, z) = \big[-t^{xx}(x, y, z)\,\vec{e}_x - t^{xy}(x, y, z)\,\vec{e}_y - t^{xz}(x, y, z)\,\vec{e}_z\,\big]\,\Delta y\,\Delta z, \tag{10.11b}$$

In the same way, we find the forces on the remaining four surfaces of our cube

$$\vec{F}^y(x, y+\Delta y, z) = \big[\, t^{yx}(x, y+\Delta y, z)\,\vec{e}_x + t^{yy}(x, y+\Delta y, z)\,\vec{e}_y + t^{yz}(x, y+\Delta y, z)\,\vec{e}_z \big]\, \Delta x\, \Delta z, \tag{10.11c}$$

$$\vec{F}^y(x, y, z) = \big[-t^{yx}(x, y, z)\,\vec{e}_x - t^{yy}(x, y, z)\,\vec{e}_y - t^{yz}(x, y, z)\,\vec{e}_z \big]\, \Delta y\, \Delta z, \tag{10.11d}$$

$$\vec{F}^z(x, y, z+\Delta z) = \big[\, t^{zx}(x, y, z+\Delta z)\,\vec{e}_x + t^{zy}(x, y, z+\Delta z)\,\vec{e}_y + t^{zz}(x, y, z+\Delta z)\,\vec{e}_z \big]\, \Delta x\, \Delta y, \tag{10.11e}$$

$$\vec{F}^z(x, y, z) = \big[-t^{zx}(x, y, z)\,\vec{e}_x - t^{zy}(x, y, z)\,\vec{e}_y - t^{zz}(x, y, z)\,\vec{e}_z \big]\, \Delta x\, \Delta y. \tag{10.11f}$$

We now compute the x component (the six terms of Eqs. (10.11a)–(10.11f) in front of the basis vector $\vec{e}_x$) of the resultant surface force $\vec{S}$ acting on all the surfaces of the cube. It is due to the normal stresses acting on the $\pm\vec{e}_x$ surfaces and the shear stresses acting on the $\pm\vec{e}_y$ and $\pm\vec{e}_z$ surfaces,

$$\begin{aligned} S^x = &\,[t^{xx}(x+\Delta x, y, z) - t^{xx}(x, y, z)]\, \Delta y\, \Delta z \\ &+ [t^{yx}(x, y+\Delta y, z) - t^{yx}(x, y, z)]\, \Delta x\, \Delta z \\ &+ [t^{zx}(x, y, z+\Delta z) - t^{zx}(x, y, z)]\, \Delta x\, \Delta y. \end{aligned}$$

Putting Δx outside the first bracket, Δy outside the second, and Δz outside the third, this equation may be rewritten as

$$\begin{aligned} S^x = \Bigg[&\frac{t^{xx}(x+\Delta x, y, z) - t^{xx}(x, y, z)}{\Delta x} \\ &+ \frac{t^{yx}(x, y+\Delta y, z) - t^{yx}(x, y, z)}{\Delta y} \\ &+ \frac{t^{zx}(x, y, z+\Delta z) - t^{zx}(x, y, z)}{\Delta z} \Bigg] \Delta x\, \Delta y\, \Delta z. \end{aligned}$$

Considering the limit of an indefinitely small cube, and using the definition of partial derivatives (see Sect. 2.8), this equation takes the form

$$S^x = \left(\frac{\partial t^{xx}}{\partial x} + \frac{\partial t^{yx}}{\partial y} + \frac{\partial t^{zx}}{\partial z} \right) V, \tag{10.12}$$

where $V = \Delta x\, \Delta y\, \Delta z$ is the volume of the cube. Using Einstein's summation convention this may be written

$$S^x = \frac{\partial t^{jx}}{\partial x^j}\, V.$$

Similar equations are valid for the y and z components of the force $\vec{S}$. Denoting an arbitrary force component by S^i, we obtain

$$S^i = \frac{\partial t^{ji}}{\partial x^j}\, V.$$

Using the symmetry of the stress tensor, $t^{ji} = t^{ij}$, we get

$$S^i = \frac{\partial t^{ij}}{\partial x^j}\, V.$$

Comparing with the expression (10.4) for the divergence of a tensor of rank 2, we find that

$$\vec{S} = V \operatorname{div} T,$$

where T is the stress tensor. This equation shows that the net surface force per unit volume on a fluid element is equal to the divergence of the stress tensor.

In the case of a perfect fluid the only stress is the pressure, and the stress tensor reduces to the form in Eq. (10.10). Inserting this in Eq. (10.12) leads to

$$\begin{aligned} S^x &= -\left[\frac{\partial(p\,\delta^{xx})}{\partial x} + \frac{\partial(p\,\delta^{yx})}{\partial y} + \frac{\partial(p\,\delta^{zx})}{\partial z}\right] V \\ &= -\left[\frac{\partial(p \times 1)}{\partial x} + \frac{\partial(p \times 0)}{\partial y} + \frac{\partial(p \times 0)}{\partial z}\right] V = -\frac{\partial p}{\partial x}\, V. \end{aligned}$$

Thus, the x component of the net pressure force per unit volume on the (infinitely small) fluid element is

$$s^x \equiv \frac{S^x}{V} = -\frac{\partial p}{\partial x}.$$

The y and z components of this force are given by similar expressions, but with partial derivatives with respect to y and z, respectively. The resulting equation for the net pressure force per unit volume on a fluid element is

$$\vec{s} = -\frac{\partial p}{\partial x}\,\vec{e}_x - \frac{\partial p}{\partial y}\,\vec{e}_y - \frac{\partial p}{\partial z}\,\vec{e}_z. \tag{10.13}$$

The *gradient* of a function $f(x, y, z)$ is denoted by $\operatorname{grad} f$ and is defined as a vector which has the following component form in a Cartesian coordinate system,

$$\operatorname{grad} f \equiv \nabla f \frac{\partial f}{\partial x}\,\vec{e}_x + \frac{\partial f}{\partial y}\,\vec{e}_y + \frac{\partial f}{\partial z}\,\vec{e}_z. \tag{10.14}$$

where we have introduced the so-called gradient operator (for a general definition, see appendix A), which in a Cartesian coordinate system takes the form[1]:

$$\nabla = \frac{\partial}{\partial x}\vec{e}_x + \frac{\partial}{\partial y}\vec{e}_y + \frac{\partial}{\partial z}\vec{e}_z. \tag{10.15}$$

The pressure gradient is a vector pointing in the direction in which the pressure increases fastest, and with a magnitude equal to the rate of change of the pressure in this direction. Equation (10.13) says that the net pressure force per unit volume on a fluid element has a magnitude equal to that of the pressure gradient, and points in the opposite direction of the pressure gradient,

$$\vec{s} = -\nabla p.$$

Hence the net pressure force on a fluid element points in the direction of maximally *decreasing* pressure. In index notation the equation for the i component of this force takes the form

$$s^i = -\frac{\partial p}{\partial x^i}. \tag{10.16}$$

This expression will appear below in the equation of motion of a fluid.

10.6 The material derivative

The velocity of a fluid element may change for two different reasons. Think of a fluid element in a river. Firstly, the velocity of the water in the river may increase with time due to for example heavy rain, and, secondly, the fluid element may move into a narrow region of the river with a greater flow velocity.

This is a typical situation for all sorts of fields, not only velocity fields. Imagine you are measuring the temperature of a gas, while you are moving through it with a thermometer. The temperature you measure may change because it is evening and the air cools towards the night, and it may change because you move northwards toward cooler regions. In other words the temperature measured by a moving thermometer may change both because the temperature field changes with time, and because it is inhomogeneous, i.e. the temperature is different at different places.

These changes are represented mathematically by two sorts of derivatives, called the *local derivative* and the *convective derivative*, respectively. Their sum is called the *total* derivative. Their definition will appear in a natural way when we consider the total differential of a field $f(x, y, z, t)$ (scalar field or vector field—that

[1]For a general definition of the gradient operator, see Appendix A

does not matter) which is a function of the time coordinate and the three spatial coordinates (see chapter 2)

$$\begin{aligned} df &= \frac{\partial f}{\partial t}\,dt + \frac{\partial f}{\partial x}\,dx + \frac{\partial f}{\partial y}\,dy + \frac{\partial f}{\partial z}\,dz \\ &= dt\,\frac{\partial f}{\partial t} + dx\,\frac{\partial f}{\partial x} + dy\,\frac{\partial f}{\partial y} + dz\,\frac{\partial f}{\partial z}. \end{aligned}$$

The rate of change of the field is

$$\frac{df}{dt} = \frac{\partial f}{\partial t} + \frac{dx}{dt}\frac{\partial f}{\partial x} + \frac{dy}{dt}\frac{\partial f}{\partial y} + \frac{dz}{dt}\frac{\partial f}{\partial z}. \tag{10.17}$$

Here $\frac{dx}{dt}$, $\frac{dy}{dt}$, and $\frac{dz}{dt}$ are the x, y, and z components of the velocity of the *measuring apparatus* through the fluid. The rate of change of f with time which this apparatus measures, is $\frac{df}{dt}$. This quantity is the *total* derivative of f. The term $\frac{\partial f}{\partial t}$ represents the rate of change of f that would be measured by an apparatus at rest. It is the *local* derivative of f. The three remaining terms in Eq. (10.17) represent the rate of change of f that would have been measured by the moving apparatus due to the position dependence of f in a stationary fluid, i.e. in a fluid where this property does not change with time. It is the *convective* derivative of f.

If the measuring apparatus moves with the fluid, so that

$$\frac{dx}{dt} = v^x, \quad \frac{dy}{dt} = v^y, \quad \text{and} \quad \frac{dz}{dt} = v^z,$$

where v^x, v^y, and v^z are the components of the fluid velocity, then the total derivative is denoted by $\frac{Df}{Dt}$ and is called the *material derivative*. Thus

$$\frac{Df}{Dt} = \frac{\partial f}{\partial t} + v^x\,\frac{\partial f}{\partial x} + v^y\,\frac{\partial f}{\partial y} + v^z\,\frac{\partial f}{\partial z}.$$

Using Einstein's summation convention this may be written

$$\frac{Df}{Dt} = \frac{\partial f}{\partial t} + v^j\,\frac{\partial f}{\partial x^j}.$$

The material derivative of the velocity field of a fluid is obtained by substituting for f the i component of the velocity field, $f = v^i$. This gives

$$\frac{Dv^i}{Dt} = \frac{\partial v^i}{\partial t} + v^j\,\frac{\partial v^i}{\partial x^j}. \tag{10.18}$$

10.7 The equation of motion

Newton's second law says that the vector sum of the forces acting on a fluid element is equal to the mass of the fluid element times its acceleration, $\vec{a}$. The forces acting on an element of a fluid may be separated into two parts in a natural way. On the one hand there are the so-called 'body forces', $\vec{G}$, acting on the whole of the fluid element. A prominent example is the force of gravity or weight of a fluid element. On the other hand there are forces $\vec{S}$ acting upon the surface enclosing the fluid element, for example pressure forces. Newton's second law as applied to a fluid element with mass m thus may be written

$$\vec{G} + \vec{S} = m\,\vec{a}.$$

In hydrodynamics it is most useful to consider the force *per unit volume* on a fluid element. Also one prefers to exchange the two sides of the equation, in order to emphasize that the acceleration of a fluid element is resulting from the forces acting on it. So, we write the equation of motion as

$$\rho\,\frac{D\vec{v}}{Dt} = \rho\,\vec{g} + \vec{s},$$

where $\vec{s}$ is the pressure per unit volume introduced in Eq. (10.13), and $\vec{g}$ is the acceleration of gravity. The i component of this equation is

$$\rho\,\frac{Dv^i}{Dt} = \rho\,g^i + s^i.$$

Substituting for $D\,v^i/D\,t$ from Eq. (10.18) and for s^i from Eq. (10.16) leads to the standard form of the equation of motion for a perfect fluid, called *Euler's equations of motion*,

$$\rho\left(\frac{\partial v^i}{\partial t} + v^j\,\frac{\partial v^i}{\partial x^j}\right) = \rho\,g^i - \frac{\partial p}{\partial x^i}. \tag{10.19}$$

This completes our Newtonian deduction of the hydrodynamical equation of mass conservation (10.9) and the corresponding equation of motion.

10.8 Four-velocity

According to Newton's conceptions a body moves through three-dimensional space. There exists a universal time t, and the velocity of the body is the rate of change of position with time in this space. As decomposed in a Cartesian coordinate system the velocity is

$$\vec{v} = v^x\,\vec{e}_x + v^y\,\vec{e}_y + v^z\,\vec{e}_z = \frac{dx}{dt}\,\vec{e}_x + \frac{dy}{dt}\,\vec{e}_y + \frac{dz}{dt}\,\vec{e}_z$$

or

$$\vec{v} = v^i \, \vec{e}_i = \frac{dx^i}{dt} \, \vec{e}_i \, . \tag{10.20}$$

Since the metric in an ordinary Cartesian coordinate system is (see Eq. (5.15) for the definition of the Kronecker symbol, and Eq. (5.60) for the metric in a two-dimensional Cartesian coordinate system)

$$g_{ij} = \delta_{ij} ,$$

the square of the magnitude of a usual three-velocity is

$$\begin{aligned} |\vec{v}|^2 &= \vec{v} \cdot \vec{v} = v_i v^i = g_{ij} v^i v^j = \delta_{ij} v^i v^j \\ &= (v^x)^2 + (v^y)^2 + (v^z)^2 = \left(\frac{dx}{dt}\right)^2 + \left(\frac{dy}{dt}\right)^2 + \left(\frac{dz}{dt}\right)^2 . \end{aligned}$$

The magnitude $|\vec{v}|$ of the velocity in three-space is called *speed*, and has no direction.

We shall now go on to define the velocity of a body in spacetime. Since spacetime is four-dimensional, vectors in spacetime must have four components. They are therefore called *four-vectors*.

Furthermore time is not universal according to the special theory of relativity. The Lorentz transformation, Eq. (5.98), imply that people moving relative to each other will not agree as to which events are measured as simultaneous. And their clocks do not go equally fast (see Ch. 5).

The velocity of a body in spacetime is called the *four-velocity*. The components of the four-velocity should be given by expressions similar to those of an ordinary velocity in three-space (see Eq. (10.20)); rate of change of coordinates with time. In order that these expressions shall indeed transform as vector-components, the coordinates must be differentiated with respect to an invariant time appropriate for the particular body which is to be studied. In chapter 5 we defined such an invariant time; the *proper time* of the body, $d\tau$. It is the time shown by a clock that moves together with the body. So, we define the four-velocity of a body by

$$\vec{u} \equiv \frac{dx^\mu}{d\tau} \, \vec{e}_\mu . \tag{10.21}$$

Let us consider a particle moving in flat spacetime with Cartesian coordinates (x, y, z, t). An infinitesimal distance vector in spacetime, $d\vec{r} = dx^\mu \vec{e}_\mu$, then has components dx, dy, dz, and dt. The first three components are measured in metres and the last one in seconds. Thus, the components of the distance vector in a space direction and the time direction seem to be measured in different units. This cannot be correct. The components of a vector are just the projections of the vector onto chosen directions. It must be possible to measure all the components of a vector in the same unit. Thus, in connection with four-vectors the time coordinate t is not a suitable coordinate. We must introduce a time coordinate x^4 that is measured in metres! In order to obtain a change from seconds to metres we must multiply by metres per second, i.e.

$$\text{metre} = \frac{\text{metre}}{\text{second}}\,\text{second}.$$

Thus we must multiply the time coordinate by a velocity. In order that the new coordinate x^4 shall represent time, and nothing else, this velocity must be a universal constant. There is only one such velocity, the velocity of light, c. The unique choice of time coordinate in connection with four-vectors is therefore

$$x^4 = ct.$$

The square of the magnitude of a four-velocity is

$$\begin{aligned} |\vec{u}|^2 &= |\vec{u}\cdot\vec{u}| = |u_\mu u^\mu| \\ &= |g_{\mu\nu}u^\mu u^\nu| = \left|g_{\mu\nu}\frac{dx^\mu}{d\tau}\frac{dx^\nu}{d\tau}\right|. \end{aligned} \tag{10.22}$$

Combining this with the general expression for the line element, Eq. (5.120), we get

$$|\vec{u}|^2 = \left|\frac{ds^2}{d\tau^2}\right|.$$

According to the definition (5.115) of the proper time interval, the line element may be written

$$ds^2 = -c^2 d\tau^2.$$

Inserting this in Eq. (10.22) gives

$$|\vec{u}|^2 = |-c^2| = c^2.$$

The magnitude of the four-velocity of an arbitrary body, its 'speed' in spacetime, is therefore, surprisingly

$$|\vec{u}| = c. \tag{10.23}$$

This equation shows that all bodies move with the same constant speed in spacetime. Nothing is at rest. A body at a fixed position in space moves in the time direction, meaning that the body's distance in spacetime from the Big Bang event representing the creation of our universe, increases steadily.

Looking at Eq. (10.21) we find that $u^4 = dx^4/d\tau = c(dt/d\tau)$. Hence, the four-velocity has a time component, which is equal to c times the rate of change of coordinate time with proper time. However, the *direction* of the velocity of a material body in spacetime differs from that of light. Light has a greater velocity component in the spatial direction than any particle.

Consider a particle moving through a coordinate system with an ordinary (three-dimensional) speed v. Then the relation between the coordinate-time and the proper-time of the particle is given in Eq. (5.117). Inserting this into the expression for the spatial components of the four-velocity we obtain

$$u^i = \frac{dx^i}{d\tau} = \frac{1}{\sqrt{1 - v^2/c^2}} \frac{dx^i}{dt} = \frac{v^i}{\sqrt{1 - v^2/c^2}}.$$

These four-velocity components can clearly be greater than c. It seems that bodies can move faster than light, after all. However, these four-velocity components are the rate of change of (spatial) position with *proper time*. And the proper-time measured on a clock moving with the velocity of light does not change at all. So the spatial four-velocity components of light are infinitely great.

The time component of the four-velocity is

$$u^t = c\,\frac{dt}{d\tau} = \frac{c}{\sqrt{1 - v^2/c^2}},$$

where we have used Eq. (5.117) for the relativistic time dilation. Consequently the component-form of the four-velocity is

$$\vec{u} = \gamma\,\left(v^x\,\vec{e}_x + v^y\,\vec{e}_y + v^z\,\vec{e}_z + c\,\vec{e}_t\right)$$

where

$$\gamma = 1/\sqrt{1 - v^2/c^2}.$$

In the case of a slowly moving body, $v^i \ll c$, so $\gamma \to 1$, and the components of the four-velocity is, with good approximation, given by

$$u^\mu = (v^x, v^y, v^z, c). \tag{10.24}$$

10.9 Newtonian energy-momentum tensor of a perfect fluid

In order to obtain a unified representation of *all* the properties of a perfect fluid (see Sect. 10.4), the remaining properties—mass, energy and motion—must also be represented by a symmetric tensor of rank 2. The quantity combining these properties in Newton's theory is simply a scalar function; *the kinetic energy* of a fluid element. The kinetic energy per unit volume is

$$E_{\text{kinetic}} = \frac{1}{2}\rho\,v^2, \tag{10.25}$$

where ρ is the mass density of the fluid, and v is the speed of the fluid element.

Since we are going to describe fluids in four-dimensional spacetime, not in Newton's three-dimensional space, we must form a sort of generalization of

Eq. (10.25) in terms of the *four-velocity* $\vec{u} = u^{\mu}\vec{e}_{\mu}$ of the fluid-element. Because this generalization shall be a tensor of rank 2, it cannot involve just the magnitude of the four-velocity. However, the products $u^{\mu} u^{\nu}$ transform as components of a tensor of rank 2. Since the succession of the vector components does not change the values of the products, this tensor is symmetric.

The density of the fluid is defined as the mass of the fluid per unit volume. As applied to our fluid-element, it is the mass of the fluid-element divided by its volume. In Newton's theory the density of the fluid is invariant against a transformation from one reference frame to another. This is not the case in the special theory of relativity, due to the Lorentz contraction of a fluid element, making its volume less, and the relativistic mass increase, making the mass greater, the faster the fluid element moves through three-space. However, the density of the fluid as measured by an observer following its motion, is just the same as if the fluid were at rest. It is called the *proper density* of the fluid, and is equal to the Newtonian density, ρ. The proper density of the fluid is an invariant quantity.

Using the theory of relativity, Einstein deduced in 1905 a relationship between the energy E of a system and its mass m, namely $E = m\,c^2$. Just as $x^4 = c\,t$ and t is essentially the same quantity, a time coordinate, the equation $E = m\,c^2$ means that according to the theory of relativity energy and mass are essentially the same physical quantity. The only difference is that they are measured in different units. We need not then introduce both mass and energy to represent this quantity in the theory of relativity.

We now introduce a symmetric tensor of rank 2, with components $\theta^{\mu\nu}$, representing the mass/energy and motion of a fluid. These components are defined by

$$\theta^{\mu\nu} \equiv \rho\, u^{\mu} u^{\nu}.$$

This expression represents the components of the so-called kinetic energy-momentum tensor. The four-velocity components of a slowly moving fluid is given in Eq. (10.24). For such a fluid the kinetic energy-momentum tensor takes the form

$$\theta^{\mu\nu} = \begin{bmatrix} \rho v^x v^x & \rho v^x v^y & \rho v^x v^z & \rho c\, v^x \\ \rho v^y v^x & \rho v^y v^y & \rho v^y v^z & \rho c\, v^y \\ \rho v^z v^x & \rho v^z v^y & \rho v^z v^z & \rho c\, v^z \\ \rho\, c\, v^x & \rho\, c\, v^y & \rho\, c\, v^z & \rho\, c^2 \end{bmatrix}. \qquad (10.26)$$

Before we can define a total energy-momentum density tensor, we must generalize the three-space stress tensor t^{ij} to a corresponding spacetime tensor $S^{\mu\nu}$. This tensor is defined by demanding that its only non-vanishing components *in a local rest frame of the fluid* are equal to the components of the 'Newtonian' stress tensor t^{ij}. The components of the tensor $S^{\mu\nu}$ in an arbitrary inertial reference frame are then calculated by Lorentz transforming the tensor components from the rest-frame to the arbitrary frame. However, we shall not need to write down these expressions.

The *total-energy momentum density tensor* (later called 'the energy-momentum tensor') is defined by

$$T^{\mu\nu} \equiv \theta^{\mu\nu} + S^{\mu\nu}.$$

Inserting the components of $S^{\mu\nu}$ from the 'rest frame values' in Eq. (10.10), and the components of $\theta^{\mu\nu}$ from Eq. (10.26), we get the components for the energy-momentum tensor for the case of a perfect fluid moving slowly in a (four-dimensional) Cartesian coordinate system (with $x^4 = c\,t$),

$$T^{ij} = \rho\, v^i\, v^j + p\,\delta^{ij}, \tag{10.27a}$$

$$T^{4i} = \rho\, c\, v^i, \tag{10.27b}$$

and

$$T^{44} = \rho\, c^2. \tag{10.27c}$$

This may be called the Newtonian energy-momentum tensor for a perfect fluid.

10.10 Mathematical formulation of the basic conservation laws

We shall now study the divergence, interpreted physically, of the 'Newtonian' energy-momentum tensor of a perfect fluid given in Eq. (10.27). Then we shall be able to understand the physical meaning of stating that this divergence vanishes.

In the present application we let $T^{\mu\nu}$ be decomposed in a locally Cartesian coordinate system comoving with an inertial reference frame. Then the covariant derivatives are reduced to ordinary partial derivatives, and the time component of ${T^{\mu\nu}}_{;\nu}$ is

$${T^{4\nu}}_{;\nu} = {T^{44}}_{,4} + {T^{4i}}_{,i}.$$

Substituting for ${T^{44}}_{,4}$ and ${T^{4i}}_{,i}$ from Eq. (10.27), while noting that

$$T^{44}_{,4} = \frac{\partial T^{44}}{c\,\partial t},$$

and dividing each term by c, we get

$$T^{4\nu}_{,\nu} = \frac{\partial \rho}{\partial t} + \frac{\partial(\rho\, v^i)}{\partial x^i}.$$

In the Newtonian approximation the equation

$${T^{4\nu}}_{;\nu} = 0,$$

takes the form

$$\frac{\partial \rho}{\partial t} + \frac{\partial (\rho\, v^i)}{\partial x^i} = 0,$$

or

$$\frac{\partial \rho}{\partial t} + \mathrm{div}(\rho\,\vec{v}) = 0. \tag{10.28}$$

This is just the equation of mass conservation in hydrodynamics, Eq. (10.9). In view of the 'equivalence' of the measurable quantities mass and energy, represented by Einstein's famous equation $E = m\,c^2$, this can also be interpreted as an expression of the law of energy-conservation.

The significance of $E = m\,c^2$ as representing a *unifying* concept of fundamental importance in physics should be emphasized. In classical Newtonian mechanics there are four basic conservation laws; conservation of mass, energy, momentum and angular momentum. The mass of a body is the quantity m that appears both in Newton's second law and in Newton's law of gravitation. Energy appears in different forms, for example kinetic energy related to the motion of a body, and potential energy related to the position of a body in a force field. Momentum is mass times velocity, and angular momentum is a rotational analogue of momentum, in which mass far from the axis of rotation contributes mostly. Its conservation is made visible when a figure skater draws her arms towards her body and thereby increases her angular velocity in a pirouette.

In 1918 Emmy Noether proved the remarkable fact that conservation of energy, momentum and angular momentum are consequences of a more basic requirement. The laws of physics should be completely independent of the location of an experiment in space and time, and of the orientation of the experimental arrangement. Expressed more technically: three of the four basic conservation of Newtonian mechanics are consequences of the requirement that the laws of nature should be invariant against translations in space, and in time, and under rotations in space. Such invariance properties are often called *symmetries* in physics. In other words, three of the four basic conservation laws of Newtonian physics follows from symmetries of space and time. Conservation of mass is usually postulated separately (although it can also be considered as implicit in the law of conservation of momentum).

We now proceed to investigate the spatial components of $T^{\mu\nu}{}_{;\nu}$. Let us consider the i component $T^{i\nu}{}_{;\nu}$. It takes the form

$$T^{i\nu}{}_{;\nu} = T^{i4}{}_{,4} + T^{ij}_{,j}\,.$$

Substituting from Eq. (10.27), and noting that

$$\begin{aligned} T^{i4}{}_{,4} &= \frac{\partial T^{i4}}{\partial x^4} = \frac{\partial T^{i4}}{c\,\partial t} = \frac{\partial\,(\rho\,c\,v^i)}{c\,\partial t} \\ &= \frac{c\,\partial\,(\rho\,v^i)}{c\,\partial t} = \frac{\partial\,(\rho\,v^i)}{\partial t} = \frac{\partial}{\partial t}\,(\rho\,v^i), \end{aligned}$$

we get

$$T^{i\nu}{}_{;\nu} = \frac{\partial}{\partial t}\left(\rho v^i\right) + \frac{\partial}{\partial x^j}\left(\rho v^i v^j + p\,\delta^{ij}\right).$$

The equation

$$T^{i\nu}{}_{;\nu} = 0$$

results in

$$\frac{\partial}{\partial t}\left(\rho v^i\right) + \frac{\partial}{\partial x^j}\left(\rho v^i v^j + p\,\delta^{ij}\right) = 0. \tag{10.29}$$

Making use of the properties of the Kronecker symbol, the last term is simplified to

$$\frac{\partial}{\partial x^j}\left(p\,\delta^{ij}\right) = \frac{\partial p}{\partial x^i}. \tag{10.30}$$

Subtracting Eq. (10.30) from Eq. (10.29) leads to

$$\frac{\partial}{\partial t}\left(\rho v^i\right) + \frac{\partial}{\partial x^i}\left(\rho v^i v^j\right) = -\frac{\partial p}{\partial x^i}, \tag{10.31}$$

which expresses the Newtonian law of conservation of momentum. The right hand side is the pressure force per unit volume acting on a fluid element, as given in Eq. (10.16).

We shall now show more explicitly the relation of this equation to Newton's second law, which says that mass times acceleration is equal to the sum of the forces. Thus, we must rewrite Eq. (10.31) so that the left-hand side is replaced by 'mass times acceleration'. In order to do this, we differentiate the second term of Eq. (10.31). Using the rule for differentiating a product we get

$$\frac{\partial}{\partial x^j}\left(\rho v^i v^j\right) = v^i \frac{\partial}{\partial x^j}\left(\rho v^j\right) + \rho v^j \frac{\partial v^i}{\partial x^j}.$$

Applying the equation of mass conservation in the form (10.28) to the first term at the right-hand side we find

$$\frac{\partial}{\partial x^j}\left(\rho v^i v^j\right) = -v^i \frac{\partial \rho}{\partial t} + \rho v^j \frac{\partial v^i}{\partial x^j}. \tag{10.32}$$

Inserting Eq. (10.32) into Eq. (10.31) leads to

$$v^i \frac{\partial \rho}{\partial t} + \rho \frac{\partial v^i}{\partial t} - v^i \frac{\partial \rho}{\partial t} + \rho v^j \frac{\partial v^i}{\partial x^j} = -\frac{\partial p}{\partial x^i}$$

or

$$\rho\left(\frac{\partial v^i}{\partial t} + v^j \frac{\partial v^i}{\partial x^j}\right) = -\frac{\partial p}{\partial x^i}. \tag{10.33}$$

This is Euler's equation of motion of a fluid.

It is not like Eq. (10.19), however. There is no term with the acceleration of gravity here. Fortunately, the reason for this discrepancy is simple to find, and simple to 'repair'. The effect of gravity has not been included in the present section because we decided to simplify the equations by decomposing the tensors and vectors in a Cartesian coordinate system comoving with an inertial reference frame. This made it possible to replace covariant differentiation by ordinary partial differentiation. This was sufficient for the present purposes; namely to formulate the Newtonian laws of hydrodynamics in a way that may be generalized, to include the effects of arbitrarily strong gravitational fields and relativistic velocities. Such generalization may be obtained by the receipt: replace ordinary partial derivatives by covariant derivatives!

Later we shall see that certain Christoffel symbols that automatically appear when we use covariant derivatives represent, in a unified way, inertial forces experienced in an accelerated reference frame, and gravitational forces due to mass concentrations.

We have now seen: In spacetime $T^{\mu\nu}{}_{;\nu} = 0$ represents four equations. The time component is the equation of continuity, representing the law of conservation of mass, or energy. The three spatial components represent the law of conservation of momentum. Together with the equation of mass conservation this leads to the equation of motion of a fluid element. Thus, the basic conservation laws of hydrodynamics, and the equation of motion of fluids, are all contained in the physical interpretation of the covariant equations $T^{\mu\nu}{}_{;\nu} = 0$.

10.11 Relativistic energy-momentum of a perfect fluid

The expression (10.27) for the components of the energy-momentum tensor of a perfect fluid moving slowly in a Cartesian coordinate system shall now be generalized to a relativistically valid expression. We must then replace the components v^i of the ordinary velocity by the components u^μ of the four velocity of the fluid elements. Note also that δ^{ij} are the components of the metric tensor in the Cartesian coordinate system. This must be replaced by the components $g^{\mu\nu}$ of the metric in an arbitrary coordinate system in spacetime. The relativistic generalization of the expression for T^{ij} in Eq. (10.27) therefore seems to be

$$\bar{T}^{\mu\nu} = \rho u^\mu u^\nu + p g^{\mu\nu}. \tag{10.34}$$

In a local inertial rest frame of the fluid at an arbitrary point in spacetime the values of the components of the energy-momentum tensor are found by inserting $v^i = 0$ in Eq. (10.27). Denoting the rest frame values by marked indices, we get

$$T^{i'j'} = p\,\delta^{i'j'}, \quad T^{i'4'} = 0, \quad \text{and} \quad T^{4'4'} = \rho c^2.$$

However, calculating the rest frame values from Eq. (10.34) by inserting $u^{\mu'} = (c, 0, 0, 0)$, $g^{i'j'} = \delta^{i'j'}$, $g^{i'4'} = 0$, and $g^{4'4'} = -1$, we get

$$\bar{T}^{i'j'} = p\,\delta^{i'j'}, \quad \bar{T}^{i'4'} = 0, \quad \text{and} \quad \bar{T}^{4'4'} = \rho c^2 - p.$$

Due to the wrong value of $\bar{T}^{4'4'}$ Eq. (10.34) cannot be the correct expression for the energy-momentum tensor. We must add a term that contributes with p to $T^{4'4'}$, but does not contribute to the components $T^{i'j'}$ and $T^{i'4'}$. The simplest such term is $(p/c^2)\, u^\mu u^\nu$. Adding this term to Eq. (10.34) we obtain an expression which we can use to define the components of the relativistic energy-momentum tensor of a perfect fluid,

$$T^{\mu\nu} \equiv \left(\rho + p/c^2\right) u^\mu u^\nu + p\, g^{\mu\nu}. \tag{10.35}$$

We see from this expression that the energy-momentum tensor is symmetric, $T^{\nu\mu} = T^{\mu\nu}$.

Chapter 11
Einstein's field equations

11.1 A new conception of gravitation

According to Einstein's theory of gravitation there is no gravitational field of forces. The notion of a gravitational field of force is replaced by that of curved spacetime.

Because no force is needed, the theory is often said to be purely geometric. Einstein is said to have 'geometrized' gravitation. This is somewhat misleading because his theory is one of physics making use of empirically vulnerable terms. In geometry, understood as a mathematical, not a physical science, there is no empirical vulnerability. We would like, instead, to say that Einstein's theory of gravitation is purely 'kinematical', i.e. purely movement oriented, emphasizing the absence of any gravitational *force* in this theory. Matter moves according to the curvature of space and space is curved by matter.

The great relativist J. A. Wheeler formulated the aphorism: "matter tells space how to curve, and space tells matter how to move".

The mathematical expressions of this are Einstein's field equations and the geodesic equation. They replace Newton's gravitational law that 'tells' how matter creates a gravitational field of force, and Newton's second law, that 'tells' how a particle responds to this field. The famous slogan of Wheeler is, as most slogans, likely to be misunderstood; the 'telling' demands no force. To certain physical properties in a region there *correspond* certain geometrical properties of spacetime. A logical relation of correspondence!

In order to avoid thinking in terms of force and influence, we may propose the following alternative to Wheeler's slogan: with mass space is curved, and so are the paths of the mass.

It is difficult to avoid the concept of 'force' when we *talk* about gravitational phenomena. For example, people often say that mass *causes* spacetime curvature, and that this curvature *causes* the path of a free particle to deviate from a straight line in ordinary three-dimensional space. According to Newtonian gravitational theory there are causal relationships involving forces. General relativity *is* nevertheless, a causal theory of gravitation and spacetime. But it is free of forces. The causal

Ø. Grøn and A. Næss, *Einstein's Theory: A Rigorous Introduction for the Mathematically Untrained*, DOI 10.1007/978-1-4614-0706-5_11,

character of the theory is related to the mathematical expression of the theory. In the weak field approximation general relativity is similar to Maxwell's electromagnetic theory. If the Sun changes its shape, for example, it emits gravitational waves moving with the velocity of light. Still, as far as gravitation is concerned, Einstein's God does no pushing; everything flows freely. Gravitation is of a kinematical character, it is not an interacting force.

Einstein's field equations relate spacetime curvature to matter and radiation, which is characterized by an energy-momentum tensor. As noted in section 10.11, it is a symmetric tensor of rank 2. The field equations of Einstein say that a certain curvature tensor of spacetime is proportional to the energy-momentum tensor of the matter and energy that is present. Like the energy-momentum tensor this curvature tensor must be a symmetric tensor of rank 2, since two tensors that are proportional to each other must be of the same rank and have the same symmetries.

Since the Riemann curvature tensor is of rank 4, Einstein had to find a different curvature tensor for the left-hand side of his field equations.

One may perhaps wonder: Why not construct an energy-momentum tensor of rank 4? Then the field equations could have the form: 'The Riemann curvature tensor is proportional to the energy-momentum tensor'. The reason that Einstein did not follow this route is that *it would not make physical sense*. A simple illustration may be useful. Temperature is a quantity that may be specified by means of just *one* number (and an expression of a unit, for example 'degree Celsius'). So it does not make much sense to introduce for example two numbers, say a temperature vector with two components. In a similar way there are enough components in a symmetric tensor of rank 2 to specify *all* the kinematical and dynamical properties of matter. Introducing an energy-momentum tensor of rank 4 would only make sense if *all* geometrical properties of a curved space corresponded to the physical properties of matter. So far our observations make us believe that this is not the case.

11.2 The Ricci curvature tensor

There is a way to construct a tensor of rank 2 from one of rank 4, or more generally, a tensor of rank $n-2$ from one of rank n. The procedure is called *contraction*. Consider, as an example a tensor S of rank 2. The contravariant components are $S^{\mu\nu}$. Contraction is defined by the following operations. One index is lowered and then made equal to an upper index. The two equal indices are then regarded as summation indices, by an extension of Einstein's summation convention, and the indicated summation is performed. Therefore, the contraction tensor of the energy-momentum tensor, which is of rank 2 and has components $T^{\mu\nu}$, is the scalar quantity (that is tensor of rank 0)

$$T \equiv T^{\mu}{}_{\mu} = T^{1}{}_{1} + T^{2}{}_{2} + T^{3}{}_{3} + T^{4}{}_{4}. \tag{11.1}$$

Another example, which will be used below, is the contraction of the metric tensor. We start with the components $g^{\mu\nu}$. Lowering the last index (see Eq. (5.74)) gives $g^{\mu\alpha} g_{\nu\alpha} = \delta^{\mu}{}_{\nu}$, where $\delta^{\mu}{}_{\nu}$ is the Kronecker symbol. We put μ equal to ν and thereby reduce information. Then we summarize and get:

$$\delta^{\nu}{}_{\nu} = \delta^{1}{}_{1} + \delta^{2}{}_{2} + \delta^{3}{}_{3} + \delta^{4}{}_{4} = 1 + 1 + 1 + 1 = 4. \tag{11.2}$$

Thus the contraction of the metrical tensor is just the number 4. The only information left is the number of dimensions. If other tensors of rank 2 are contracted, we do not in general obtain merely a definite number, but a scalar function, such as the scalar T in Eq. (11.1).

We are searching for a curvature tensor of rank 2 by contracting our wellknown Riemann curvature tensor of rank 4. We start with the mixed components $R^{\mu}{}_{\nu\alpha\beta}$. This tensor has three lower indices. Does this mean that we can obtain three different contracted curvature tensors, depending upon which lower index we put equal to μ?

On the next two pages we shall show that the Riemann tensor possesses certain symmetries that make the contraction of the Riemann tensor a curvature tensor of rank 2 unique, and reduce the number of independent components of the Riemann tensor in four-dimensional spacetime from 256 to 20. This is of great significance as to the possibility of obtaining a useful geometrical theory of space, time and gravitation. Therefore we find it justifiable to make the reader completely competent to understand the mathematical basis of these symmetries.

We now write the components of the curvature tensor with reference to a local Cartesian coordinate system (Sect. 8.5). In such a coordinate system the Christoffel symbols vanish. Then the terms with products of Christoffel symbols in Eq. (9.29) vanish, and the expression for the components of Riemann's curvature tensor is reduced to (where we have exchanged the indices μ and ν in Eq. (9.29))

$$R^{\mu}{}_{\nu\alpha\beta} = \Gamma^{\mu}{}_{\nu\beta,\alpha} - \Gamma^{\mu}{}_{\nu\alpha,\beta}. \tag{11.3}$$

From this equation we get

$$\begin{aligned} R^{\mu}{}_{\nu\beta\alpha} &= \Gamma^{\mu}{}_{\nu\alpha,\beta} - \Gamma^{\mu}{}_{\nu\beta,\alpha} \\ &= -\left(\Gamma^{\mu}{}_{\nu\beta,\alpha} - \Gamma^{\mu}{}_{\nu\alpha,\beta}\right) = -R^{\mu}{}_{\nu\alpha\beta}. \end{aligned} \tag{11.4}$$

The Riemann curvature tensor is antisymmetric in its last two indices. Since the symmetries of a tensor are independent of the coordinate system, this result is valid in general.

Lowering the index μ in each term of Eq. (11.3) we get

$$R_{\mu\nu\alpha\beta} = \Gamma_{\mu\nu\beta,\alpha} - \Gamma_{\mu\nu\alpha,\beta}. \tag{11.5}$$

From this equation we can deduce a symmetry relation fulfilled by the Riemann tensor, as follows

$$R_{\mu\nu\alpha\beta} + R_{\mu\beta\nu\alpha} + R_{\mu\alpha\beta\nu} = \Gamma_{\mu\nu\beta,\alpha} - \Gamma_{\mu\nu\alpha,\beta} + \Gamma_{\mu\beta\alpha,\nu} - \Gamma_{\mu\beta\nu,\alpha} + \Gamma_{\mu\alpha\nu,\beta} - \Gamma_{\mu\alpha\beta,\nu}.$$

We now apply the symmetry of the Christoffel symbols, $\Gamma_{\mu\alpha\beta}=\Gamma_{\mu\beta\alpha}$, and exchange the succession of the terms

$$R_{\mu\nu\alpha\beta} + R_{\mu\beta\nu\alpha} + R_{\mu\alpha\beta\nu} = \Gamma_{\mu\nu\beta,\alpha} - \Gamma_{\mu\nu\beta,\alpha} + \Gamma_{\mu\beta\alpha,\nu} - \Gamma_{\mu\beta\alpha,\nu} + \Gamma_{\mu\nu\alpha,\beta} - \Gamma_{\mu\nu\alpha,\beta}.$$

All the terms cancel each other. Consequently

$$R_{\mu\nu\alpha\beta} + R_{\mu\beta\nu\alpha} + R_{\mu\alpha\beta\nu} = 0. \tag{11.6}$$

This is the second symmetry relation fulfilled by the Riemann tensor.

According to Eq. (7.31) the Christoffel symbols of the first kind are given by

$$\Gamma_{\mu\nu\alpha} = \frac{1}{2}\left(g_{\mu\nu,\alpha} + g_{\mu\alpha,\nu} - g_{\nu\alpha,\mu}\right). \tag{11.7}$$

We are now going to insert this expression into Eq. (11.5). Since the derivatives of the Christoffel symbols, containing first derivatives of the metric, are present in Eq. (11.5), we shall get second derivatives of the metric in the expression of the curvature. This may not be too surprising, if we recall from section 2.3 that the curvature of a graph of a function, $y = f(x)$, is proportional to the second derivative of the function.

The following notation shall be used for the second derivatives of the tensor components $g_{\mu\nu}$

$$g_{\mu\nu,\beta\alpha} \equiv \left(g_{\mu\nu,\beta}\right)_{,\alpha}.$$

Consider the first term at the right-hand side of Eq. (11.5), $\Gamma_{\mu\nu\beta,\alpha}$. Here we shall differentiate the Christoffel symbol given by Eq. (11.7), but with α replaced by β. Then we get

$$\begin{aligned}\Gamma_{\mu\nu\beta,\alpha} &= \frac{1}{2}\left(g_{\mu\nu,\beta} + g_{\mu\beta,\nu} - g_{\nu\beta,\mu}\right)_{,\alpha} \\ &= \frac{1}{2}\left(g_{\mu\nu,\beta\alpha} + g_{\mu\beta,\nu\alpha} - g_{\nu\beta,\mu\alpha}\right).\end{aligned} \tag{11.8}$$

In the same way, only exchanging α and β,

$$\Gamma_{\mu\nu\alpha,\beta} = \frac{1}{2}\left(g_{\mu\nu,\alpha\beta} + g_{\mu\alpha,\nu\beta} - g_{\nu\alpha,\mu\beta}\right). \tag{11.9}$$

Subtracting Eq. (11.9) from Eq. (11.8) and using Eq. (11.5), we find

$$R_{\mu\nu\alpha\beta} = \frac{1}{2}\left(g_{\mu\nu,\beta\alpha} + g_{\mu\beta,\nu\alpha} - g_{\nu\beta,\mu\alpha} - g_{\mu\nu,\alpha\beta} - g_{\mu\alpha,\nu\beta} + g_{\nu\alpha,\mu\beta}\right). \tag{11.10}$$

Since the succession of differentiation of two consecutive partial differentiations does not influence the result, $g_{\mu\nu,\alpha\beta} = g_{\mu\nu,\beta\alpha}$. Therefore the first and the fourth terms at the right-hand side of Eq. (11.10) cancel each other, and we are left with the expression (valid in a local Cartesian coordinate system)

$$R_{\mu\nu\alpha\beta} = \frac{1}{2}\left(g_{\mu\beta,\nu\alpha} - g_{\nu\beta,\mu\alpha} - g_{\mu\alpha,\nu\beta} + g_{\nu\alpha,\mu\beta}\right). \tag{11.11}$$

Exchanging the indices μ and ν in the expression inside the parenthesis leads to a change of sign, i.e.

$$g_{\mu\beta,\nu\alpha} - g_{\nu\beta,\mu\alpha} = -\left(g_{\nu\beta,\mu\alpha} - g_{\mu\beta,\nu\alpha}\right)$$

and

$$-g_{\mu\alpha,\nu\beta} + g_{\nu\alpha,\mu\beta} = -\left(-g_{\nu\alpha,\mu\beta} + g_{\mu\alpha,\nu\beta}\right).$$

Thus, we get

$$R_{\nu\mu\alpha\beta} = -R_{\mu\nu\alpha\beta} \tag{11.12}$$

which shows that the Riemann curvature tensor is antisymmetric in its first two indices, too.

The fourth and last symmetry of the Riemann curvature tensor is also found from the expression (11.11). We start by writing this expression, exchanging the index pairs $\mu\nu$ and $\alpha\beta$

$$R_{\alpha\beta\mu\nu} = \frac{1}{2}\left(g_{\alpha\nu,\beta\mu} - g_{\beta\nu,\alpha\mu} - g_{\alpha\mu,\beta\nu} + g_{\beta\mu,\alpha\nu}\right).$$

Exchanging the first and the last term in the parenthesis, and using the symmetry of the metric tensor and that the successsion of the partial derivatives doesn't matter, so that, for example, $g_{\beta\mu,\alpha\nu} = g_{\mu\beta,\nu\alpha}$, we get

$$R_{\alpha\beta\mu\nu} = \frac{1}{2}\left(g_{\mu\beta,\nu\alpha} - g_{\nu\beta,\mu\alpha} - g_{\mu\alpha,\nu\beta} + g_{\nu\alpha,\mu\beta}\right)$$

which is just the expression (11.11). This shows that the Riemann curvature tensor has the symmetry

$$R_{\mu\nu\alpha\beta} = R_{\alpha\beta\mu\nu}. \tag{11.13}$$

The four symmetries of the Riemann curvature tensor, Eqs. (11.4), (11.6), (11.12) and (11.13), reduce the number of independent components of the Riemann tensor in a four-dimensional spacetime from $4^4 = 256$ to 20, which is of decisive importance for the construction of a curvature tensor which can be used in Einstein's field equations, because it reduces drastically the number of equations needed.

As noted above we need to find a symmetric curvature tensor of rank 2. In order to construct such a tensor, we calculate the contraction of the Riemann curvature tensor. Since the components of this tensor can be written with one superscript and three subscripts, there exist three possible contractions of the Riemann curvature tensor. Let us consider all three possibilities.

We first put $\nu = \mu$, in $R^{\mu}{}_{\nu\alpha\beta}$ and get

$$\begin{aligned} R^{\mu}{}_{\mu\alpha\beta} &= g^{\mu\nu} R_{\nu\mu\alpha\beta} = -g^{\mu\nu} R_{\mu\nu\alpha\beta} = -g^{\nu\mu} R_{\mu\nu\alpha\beta} \\ &= -g^{\mu\nu} R_{\nu\mu\alpha\beta} = -R^{\mu}{}_{\mu\alpha\beta}. \end{aligned} \tag{11.14}$$

Here we first used the rule (5.77) for raising an index, then we used Eq. (11.13), then the symmetry of the metric tensor, and then the freedom to change the name of summation indices. In the present case we have let $\mu \to \nu$ and $\nu \to \mu$. Equation (11.14) implies that $R^{\mu}{}_{\mu\alpha\beta} = 0$, since if something positive is equal to something negative this 'something' must be zero.

Next we contract α with μ. Using Eq. (11.3) we get

$$R^{\mu}{}_{\nu\beta\mu} = -R^{\mu}{}_{\nu\mu\beta}$$

which shows that the tensor obtained by contraction of β with μ is the negative of the tensor obtained by contracting α with μ.

We had three possibilities. Contraction with the first subscript gave zero, and contraction with the third subscript gave the same quantity, but with opposite sign, as contraction with the second subscript. Therefore we get essentially only one tensor of rank 2 by contraction of the Riemann curvature tensor. The contracted tensor is called the *Ricci curvature tensor*. Its components $R_{\mu\nu}$ are defined by

$$R_{\mu\nu} \equiv R^{\alpha}{}_{\mu\alpha\nu}, \tag{11.15}$$

where we have chosen to rename the indices, with μ and ν as free indices, and α as summation index. From the symmetry (11.13) follows that the Ricci tensor is symmetric,

$$R_{\mu\nu} = R_{\nu\mu}. \tag{11.16}$$

Perhaps the Ricci curvature tensor is the one we ultimately are looking for? Does it make sense to postulate that it is proportional to the energy-momentum tensor $T_{\mu\nu}$? It is natural to try it out since it is a symmetric tensor of rank 2. And Einstein actually proposed it, writing down field equations that seemingly appeared unique and extremely simple, namely

$$R_{\mu\nu} = \kappa T_{\mu\nu},$$

where κ is a constant. But for reasons that will be apparent below, he gave it up and ended with a somewhat different tensor.

11.3 The Bianchi identity, conservation of energy, and Einstein's tensor

According to the fundamental laws of dynamics, thermodynamics and electromagnetism the amount of energy and momentum (mass times velocity) do not change for an isolated system. This is often expressed by saying that energy and momentum are conserved quantities.

Einstein demanded from his general theory of relativity that energy and momentum conservation should follow from the field equations. As we have seen in Sect. 10.10, energy and momentum conservation is expressed by vanishing divergence of the energy-momentum tensor. In order that this shall be a consequence of the field equations the curvature tensor must have vanishing divergence, which should be a purely geometric property of the curvature tensor, not the result of anything physical.

We shall now calculate the divergence of the Ricci curvature tensor, and see if it vanishes. This is most easily done by means of a geometric relation, unknown to the young Einstein, called the Bianchi identity.

The deduction of the Bianchi identity is greatly simplified by use of locally Cartesian coordinates. Then the components of the Riemann tensor are given by Eq. (11.3). By partial differentiation of each term of Eq. (11.3) with respect to a coordinate x^{γ}, we get

$$R^{\mu}{}_{\nu\alpha\beta,\gamma} = \Gamma^{\mu}{}_{\nu\beta,\alpha\gamma} - \Gamma^{\mu}{}_{\nu\alpha,\beta\gamma}. \tag{11.17}$$

Replacing α by γ, β by α, and γ by β leads to

$$R^{\mu}{}_{\nu\gamma\alpha,\beta} = \Gamma^{\mu}{}_{\nu\alpha,\gamma\beta} - \Gamma^{\mu}{}_{\nu\gamma,\alpha\beta}. \tag{11.18}$$

Similarly, replacing γ by β, α by γ, and β by α results in

$$R^{\mu}{}_{\nu\beta\gamma,\alpha} = \Gamma^{\mu}{}_{\nu\gamma,\beta\alpha} - \Gamma^{\mu}{}_{\nu\beta,\gamma\alpha}. \tag{11.19}$$

Let us inspect the sum of Eqs. (11.17)–(11.19). In the first term of Eq. (11.17) and the last term of Eq. (11.19) the succession of differentiation is different, but this does not influence the result. Accordingly, these two terms cancel each other when Eqs. (11.17)–(11.19) are added. The same applies to term number two of Eq. (11.17) and term number one of Eq. (11.18), and to term two of Eq. (11.18) and term one of Eq. (11.19). Hence

$$R^{\mu}{}_{\nu\alpha\beta,\gamma} + R^{\mu}{}_{\nu\gamma\alpha,\beta} + R^{\mu}{}_{\nu\beta\gamma,\alpha} = 0. \tag{11.20}$$

Due to the presence of the partial derivatives this is not a tensor equation. It is valid only in a locally Cartesian coordinate system. However, the corresponding generally covariant tensor equation is obtained simply by replacing the

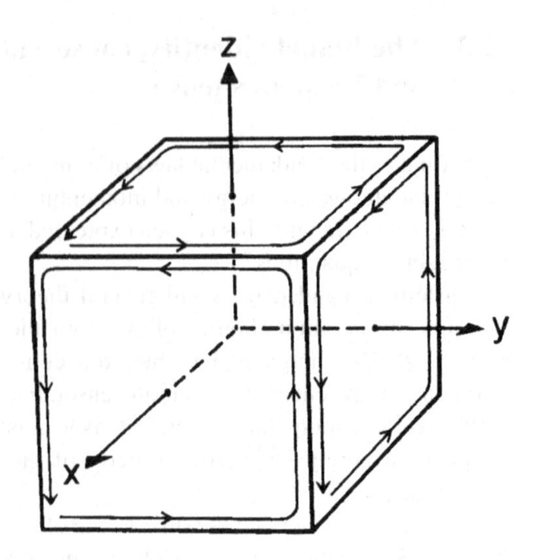

Fig. 11.1 A cube

partial derivatives with covariant derivatives. That this is a valid generalization of Eq. (11.20) follows from the fact that if a tensor expression vanishes in one particular coordinate system, then it vanishes in all other coordinate systems. The final relation is accordingly

$$R^{\mu}{}_{\nu\alpha\beta;\gamma} + R^{\mu}{}_{\nu\gamma\alpha;\beta} + R^{\mu}{}_{\nu\beta\gamma;\alpha} = 0. \tag{11.21}$$

This is the famous *Bianchi identity*.

It is possible to give a geometrical interpretation of the Bianchi identity. In order to arrive at such an interpretation we shall need a two-pages deduction. The geometrical contents of the Bianchi identity will thereby become apparent.

Let us introduce a local Cartesian coordinate system. Consider a coordinate cube as shown in Fig. 11.1.

Imagine that a vector $\vec{A}$ is parallel transported around the front side with normal vector $\vec{e}_x$. This side is at $x + \Delta x$. Since the basis vectors of the locally Cartesian coordinates are constant in the indefinitely small region of the cube (because of vanishing Christoffel symbols, see Sect. 8.5), the change of the vector $\vec{A}$ due to this parallel transport is given by the change of the vector components (see Eq. (9.30)),

$$(\Delta A^{\mu})_{x+\Delta x} = R^{\mu}{}_{\nu yz}(x + \Delta x, y, z)\, A^{\nu}\, \Delta y\, \Delta z.$$

Parallel transport of $\vec{A}$ around the opposite surface gives a similar contribution, except that now the sign is reversed and the evaluation takes place at x rather than at $x + \Delta x$,

$$(\Delta A^{\mu})_{x} = -R^{\mu}{}_{\nu yz}(x, y, z)\, A^{\nu}\, \Delta y\, \Delta x.$$

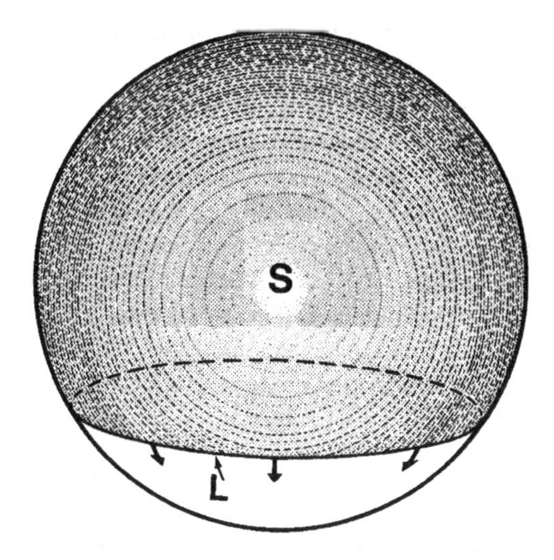

Fig. 11.2 A surface S with boundary L

The sum of the contributions from parallel transport of $\vec{A}$ around these two surfaces gives, in the case of a vanishingly small cube $\Delta x \to 0$,

$$\begin{aligned}
&\lim_{\Delta x \to 0} \left[(\Delta A^{\mu})_{x+\Delta x} + (\Delta A^{\mu})_{x}\right] \\
&= \lim_{\Delta x \to 0} \left[R^{\mu}{}_{\nu yz}(x+\Delta x, y, z) - R^{\mu}{}_{\nu yz}(x, y, z)\right] A^{\nu}\, \Delta y\, \Delta z \\
&= \lim_{\Delta x \to 0} \frac{R^{\mu}{}_{\nu yz}(x+\Delta x, y, z) - R^{\mu}{}_{\nu yz}(x, y, z)}{\Delta x} A^{\nu}\, \Delta x\, \Delta y\, \Delta z \\
&= \frac{\partial R^{\mu}{}_{\nu yz}}{\partial x} A^{\nu}\, \Delta x\, \Delta y\, \Delta z = \frac{\partial R^{\mu}{}_{\nu yz}}{\partial x} A^{\nu}\, \Delta V,
\end{aligned}$$

where $\Delta V \equiv \Delta x\, \Delta y\, \Delta z$ is the volume of the coordinate cube. The change of the vector $\vec{A}$ when it is parallel transported around all six boundary surfaces of the cube is

$$\Delta A^{\mu} = \left(\frac{\partial R^{\mu}{}_{\nu yz}}{\partial x} + \frac{\partial R^{\mu}{}_{\nu zx}}{\partial y} + \frac{\partial R^{\mu}{}_{\nu xy}}{\partial z}\right) A^{\nu}\, \Delta V,$$

or, making use of Einstein's comma notation for partial derivatives,

$$\Delta A^{\mu} = \left(R^{\mu}{}_{\nu yz,x} + R^{\mu}{}_{\nu zx,y} + R^{\mu}{}_{\nu xy,z}\right) A^{\nu}\, \Delta V. \tag{11.22}$$

In order to find the value of this change of the vector $\vec{A}$ we shall use a geometrical argument. Consider the shaded part S of the spherical surface in Fig. 11.2.

The circle L is the boundary of the surface S. Let L move towards the 'south pole' of the sphere. Then S approaches the whole spherical surface, and the length of the boundary L approaches zero. In the limit that the surface S is closed, it bounds the volume of a sphere. Let us recapitulate: The boundary of a closed surface is zero, and a closed surface is the boundary of a volume. Thus, in Wheeler's words *"the boundary of a boundary is zero"*.

Consider now our cubical coordinate volume. The edges along which the vector $\vec{A}$ is parallel transported, are parts of the boundary of the surfaces bounding the cube. When $\vec{A}$ is parallel transported around all the six surfaces of the cube all the edges are traversed twice, once in each direction, as indicated on three of the surfaces in Fig. 11.1. These displacements have signs depending upon the direction in which an edge is traversed, so all the dispacements add up to zero. The boundary (the edges with signs bounding all the surfaces) of a boundary (the surfaces bounding the cube) is zero. This implies that there will be no change at all of the vector $\vec{A}$ when it is parallel transported though the boundaries of all six surfaces, traversing each edge twice, once in each direction. Accordingly

$$\Delta A^{\mu} = 0.$$

The vector $\vec{A}$ is presupposed to be different from zero. Therefore at least one of the components A^{ν} in Eq. (11.22) are different from zero. Also the volume ΔV is different from zero. Therefore, the quantity inside the parenthesis must be equal to zero, which is just Eq. (11.20). Transforming to a covariant equation valid in an arbitrary coordinate system by replacing partial derivatives (commas) by covariant derivatives (semicolons), we arrive at the Bianchi identity, Eq. (11.21). We have thereby arrived at a geometrical interpretation of the Bianchi identity: It expresses that *the boundary of a boundary is zero*, as we saw above in the case of a sphere and a cube.

As mentioned above the conservation laws of energy and momentum are represented mathematically by putting the divergence of the (symmetrical) energy-momentum tensor (of rank 2) equal to zero. For this reason Einstein searched for a symmetrical curvature tensor of rank 2 with vanishing divergence. The Ricci tensor is a symmetrical curvature tensor of rank 2. But has it a vanishing divergence?

Applying the Bianchi identity we shall find an unambiguous answer to this question. Inserting $\alpha = \mu$ in the first term of the Bianchi identity, we get

$$R^{\mu}{}_{\nu\mu\beta;\gamma} = R_{\nu\beta;\gamma}.$$

This is the covariant derivative of the Ricci tensor. Contracting ν with γ we get $R^{\gamma}{}_{\beta;\gamma}$. This is the covariant divergence of the Ricci tensor.

From this it is seen that the value of the divergence of the Ricci tensor can be calculated by first contracting α with μ in the Bianchi identity, and then ν with γ in the resulting equation. Let us go to the task.

As we saw above, if we put $\alpha = \mu$ in the Bianchi identity, the first term is a covariant derivative of the Ricci tensor. In the second term we apply the relations

$$R^{\mu}{}_{\nu\gamma\mu} = -R^{\mu}{}_{\nu\mu\gamma} = -R_{\nu\gamma}.$$

The second term in the Bianchi identity is thereby changed to $R_{\nu\gamma;\beta}$. We then get what is called the contracted Bianchi identity,

$$R_{\nu\beta;\gamma} - R_{\nu\gamma;\beta} + R^{\mu}{}_{\nu\beta\gamma;\mu} = 0.$$

Contracting β with ν, i.e. raising the index β and putting $\beta = \mu$, gives

$$R_{;\gamma} - R^{\nu}{}_{\gamma;\nu} + R^{\mu\nu}{}_{\nu\gamma;\mu} = 0, \tag{11.23}$$

here we have defined

$$R \equiv R^{\nu}{}_{\nu}. \tag{11.24}$$

The quantity R is a curvature scalar, i.e. a tensor of rank 0. Since it is a contraction of the Ricci curvature tensor, it is called the *Ricci curvature scalar*.

The second term of Eq. (11.23) is the divergence of the Ricci tensor. As regards the third term we note that the Riemann tensor is antisymmetric in its first two indices. Consequently we have

$$R^{\mu\nu}{}_{\nu\gamma;\mu} = -R^{\nu\mu}{}_{\nu\gamma;\mu} = -R^{\mu}{}_{\gamma;\mu} = -R^{\nu}{}_{\gamma;\nu},$$

where we have replaced the summation index μ by ν in order that the second and the third terms in Eq. (11.23) shall look similar. Substituting $-R^{\nu}{}_{\gamma;\nu}$ for $R^{\mu\nu}{}_{\nu\gamma;\mu}$ in the third term of Eq. (11.23) results in

$$R_{;\gamma} - R^{\nu}{}_{\gamma;\mu} - R^{\nu}{}_{\gamma;\mu} = 0.$$

Therefore

$$R_{;\gamma} = 2\, R^{\mu}{}_{\gamma;\mu}$$

or

$$R^{\mu}{}_{\gamma;\mu} = \frac{1}{2}\, R_{;\gamma}. \tag{11.25}$$

We have now calculated the divergence of the Ricci tensor. It is equal to the covariant derivative of the Ricci curvature scalar. This vanishes only if the curvature scalar is constant. And even if this sometimes is the case (we shall later see that the Ricci curvature scalar is equal to zero everywhere in a region with vanishing energy-momentum tensor), this curvature scalar is not always constant, which means that the right-hand side of Eq. (11.25) is not equal to zero.

Thus, the Ricci tensor is not divergence free. So the Ricci tensor was not the answer to Einstein's search for a divergence-free curvature tensor of rank 2. However, Eq. (11.25) is of decisive importance as a starting point for the very last steps towards Einstein's field equations.

In order to write $R_{;\gamma}$ as the divergence of a tensor of rank 2, we use a trick. We multiply R with a *constant* tensor of rank 2, namely the Kronecker symbol, and take the divergence of the resulting tensor,

$$\left(\delta^{\mu}{}_{\gamma}\, R\right)_{;\mu} = \left(\delta^{\mu}{}_{\gamma}\right)_{;\mu} R + \delta^{\mu}{}_{\gamma}\, R_{;\mu} = 0 \cdot R + R_{;\gamma} = R_{;\gamma}.$$

Inserting this into Eq. (11.25) gives

$$\left(\delta^{\mu}{}_{\gamma}\, R\right)_{;\mu} - 2\, R^{\mu}{}_{\gamma;\mu} = \left(\delta^{\mu}{}_{\gamma}\, R - 2\, R^{\mu}{}_{\gamma}\right)_{;\mu} = 0.$$

Multiplying by $-1/2$ we get

$$\left(R^{\mu}{}_{\gamma} - \frac{1}{2}\,\delta^{\mu}{}_{\gamma}\, R\right)_{;\mu} = 0.$$

At last we have found a symmetric curvature tensor of rank 2,

$$E^{\mu}{}_{\gamma} \equiv R^{\mu}{}_{\gamma} - \frac{1}{2}\,\delta^{\mu}{}_{\gamma}\, R, \tag{11.26}$$

which is divergence free,

$$E^{\mu}{}_{\gamma;\mu} = 0. \tag{11.27}$$

The vanishing divergence of this tensor is a geometric unchangeable property which comes from the theorem that 'the boundary of a boundary is zero'. The tensor with mixed components $E^{\mu}{}_{\gamma}$ is named *Einstein's curvature tensor*. In four-dimensional spacetime Eq. (11.27) represents four identities, one for each value of the index γ, which the Einstein tensor fulfils.

The mathematical deductions have now been completed. The Einstein tensor is the adequate answer to the mathematical problem Einstein wrestled with for more than three years: which curvature tensor, if any, might be proportional to the energy-momentum tensor?

Having reached this solution late in the Autumn of 1915, Einstein promptly put his divergence free curvature tensor proportional to the energy momentum tensor, and thereby obtained the gravitational field equations of the general theory of relativity. *Einstein's field equations* are

$$E^{\mu}{}_{\nu} = \kappa\, T^{\mu}{}_{\nu} \tag{11.28}$$

where the proportionality constant κ is related to Newton's constant of gravitation (see Ch. 12).

The Einstein tensor is a symmetrical tensor of rank 2. In four-dimensional spacetime a tensor of rank 2 has sixteen components, which is often written as a 4×4 matrix. Due to the symmetry of the tensor, the components on each side of the diagonal are equal, leaving ten components that are not necessarily equal (just like the metric tensor). One might think, therefore, that Eq. (11.28) represents ten independent field equations. This is, however, not the case due to the four identities (11.27) that the Einstein tensor fulfils. Thus, there are only six independent field equations. These six second order partial differential equations are used to determine the components of the metrical tensor. However, they can determine only six of the ten independent components, leaving four of them free. This freedom,

Einstein noted, is of decisive importance. As we saw in Ch. 5, the components of the metric tensor depends upon the choice of coordinate system. In flat spacetime, for example, the metric is different in a Cartesian coordinate system and a spherical coordinate system. The freedom to choose a suitable coordinate system requires that four of the components of the metric tensor are not determined by the field equations. And this was just what Einstein found: only four of the ten field equations are independent due to the four identities (11.27) fulfilled by all Einstein tensors, independent of the geometrical properties of the spacetime they represent.

Inserting the expression (11.26) for the mixed components of the Einstein tensor, the field equations take the form

$$R^{\mu}{}_{\nu} - \frac{1}{2}\,\delta^{\mu}{}_{\nu}\,R = \kappa\,T^{\mu}{}_{\nu}. \tag{11.29}$$

Lowering the index μ in Eq. (11.29), and using that the Kronecker symbols with one superscript and one subscript, are the mixed components of the metric tensor (see Eq. (5.79)), we get a much used form of Einstein's field equations

$$R_{\mu\nu} - \frac{1}{2}\,g_{\mu\nu}\,R = \kappa\,T_{\mu\nu}. \tag{11.30}$$

Einstein's field equations may also be expressed in terms of the contravariant tensor components

$$R^{\mu\nu} - \frac{1}{2}\,g^{\mu\nu}R = \kappa\,T^{\mu\nu}. \tag{11.31}$$

This form of the field equations will be used in Ch. 12.

The field equations, together with the geodesic equation, Eq. (8.7), are the fundamental equations of the theory of relativity. In Eq. (11.30) the left-hand side represents curvature, and the right-hand side kinematical and dynamical properties of matter. The field equations are the mathematical expression of "mass is and space curves", and the geodesic equation expresses "space curves and mass moves with it".

In the next chapter we need to write Eq. (11.30) in a different way. Contraction of this equation and use of Eqs. (11.30) and (11.24) gives

$$R - \frac{1}{2}\,4\,R = \kappa\,T, \tag{11.32}$$

where

$$T \equiv T^{\nu}{}_{\nu} = T^{1}{}_{1} + T^{2}{}_{2} + T^{3}{}_{3} + T^{4}{}_{4}.$$

Equation (11.32) can be written as

$$R = -\kappa\,T. \tag{11.33}$$

Inserting Eq. (11.33) into Eq. (11.30) we find

$$R_{\mu\nu} + \frac{1}{2} g_{\mu\nu} \kappa T = \kappa T_{\mu\nu}$$

or

$$R_{\mu\nu} = \kappa \left(T_{\mu\nu} - \frac{1}{2} g_{\mu\nu} T \right). \tag{11.34}$$

One interesting property of so-called 'empty spacetime' may be noted immediately. Since empty space has vanishing energy-momentum tensor, the right-hand side of Eq. (11.34) is zero in such a region. This means that the Ricci tensor vanishes in empty spacetime. In other words the field equations of empty spacetime are

$$R_{\mu\nu} = 0. \tag{11.35}$$

It follows immediately that also the Ricci curvature scalar vanishes in empty spacetime. This does not mean, however, that space is flat in empty spacetime. The curvature of spacetime is represented by the Riemann tensor of spacetime, not the Ricci tensor.

Chapter 12
Einstein's theory of spacetime and gravitation

We have now completed our intended introduction to the mathematics used in the general theory of relativity. It remains to explain the central physical contents of the theory. Let us first offer a brief summary of the fundamental concept of Newton's theory of gravitation.

12.1 Newtonian kinematics

Particles move in a three-dimensional Euclidean space. Position vectors with finite length can be placed anywhere in this space. The position of a particle can be measured with respect to an arbitrarily chosen point of reference. Also time is universal and absolute. In particular it is possible for all observers to agree as to which events are simultaneous.

The velocity of a particle is the derivative of its position vector with respect to time

$$\vec{v} = \frac{d\vec{r}}{dt}.$$

The velocity describes how fast and in which direction a particle moves, relative to a given reference point, which may be the position of an observer.

Acceleration is the derivative of the velocity with respect to time

$$\vec{a} = \frac{d\vec{v}}{dt}.$$

The acceleration describes how fast the velocity changes, not only in magnitude (speed), but also in direction.

Ø. Grøn and A. Næss, *Einstein's Theory: A Rigorous Introduction for the Mathematically Untrained*, DOI 10.1007/978-1-4614-0706-5_12,

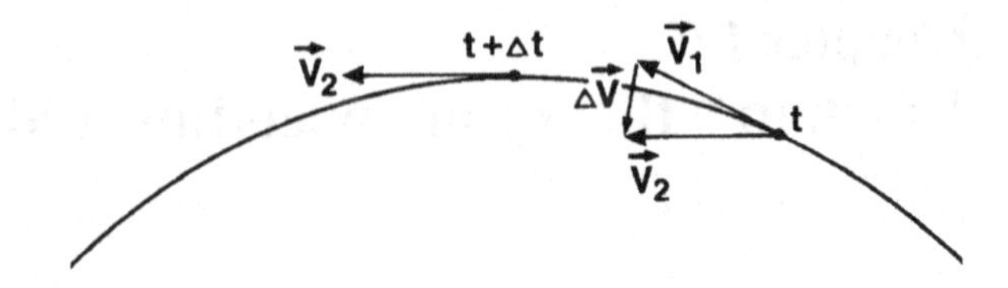

Fig. 12.1 Worldline of an accelerating body

Consider a body that moves along a circular path, for example the Moon as it moves around the Earth. Since the direction of the velocity changes all the time, the motion is accelerated all the time, even if the speed is constant.

In Fig. 12.1 $\Delta\vec{v}$ is the change of velocity from P_1 to P_2. The acceleration of the body is

$$\vec{a} = \frac{d\vec{v}}{dt} = \lim_{\Delta t \to 0} \frac{\Delta\vec{v}}{\Delta t} = \lim_{\Delta t \to 0} \frac{\vec{v}_2 - \vec{v}_1}{\Delta t}.$$

From this expression and Fig. 12.1 it is seen that when $\Delta t \to 0$, the acceleration is directed towards the centre of the circular path.

12.2 Forces

Isaac Newton (1642–1726) proposed that no force is necessary in order to keep a body moving. If a body stops moving when it is left to itself, it is because it meets friction. One is usually not able to remove completely the force of friction. A body is perhaps never quite 'left to itself'!

One speaks of a force when a body acts on another so that it is either deformed or its velocity changes. Today we recognize four *fundamental forces* in this sense.

1. The electromagnetic force
   This is a force that acts between electrically charged bodies. Also if two electrically neutral objects, with inhomogeneous charge distributions, for example two atoms, are near each other, they can act on each other by means of electromagnetic forces. Molecular bonds, for example, are due to such forces. Friction forces are also of electromagnetic nature.
2. The gravitational force
   This force acts between all masses. The gravitational force is always attractive (see Fig. 12.2). The reason that the Moon does not move along a straight path, but along a circle with an acceleration directed towards the centre of the circle (the Earth), is that the Earth acts upon the Moon with a gravitational force (see Fig. 12.2).
3. The weak nuclear force
   This is a force which is responsible for a certain radioactive emission from atomic nuclei. Its range is extremely small, shorter than the radius of a neutron.

Fig. 12.2 Earth and trajectories of falling objects

4. The so-called 'colour force'
   The name 'colour force' is fanciful. The word 'colour' has here nothing to do with human perceptions. It is used as designation for a new kind of charge, called 'colour-charge'. Particles with colour-charge act on each other with a very strong force called 'colour force'. Each proton and neutron consist of three 'quarks' (named so by the american physicist Murray Gellmann, after a line, 'three quarks for Muster Mark' in James Joyce's *Finnegan's Wake*). The quarks have colour charge. It is the colour force that binds them together in the protons and the neutrons.
   The range of this force is very small, about the radius of an atomic nucleus. The protons and neutrons are colour-neutral. Yet, the strong nuclear force, which binds protons and neutrons together in atomic nuclei, are due to the colour force, in a similar manner that forces between electrically neutral atoms, which make up molecules, are due to the electric force.

For systems as small as atomic nuclei, ordinary Newtonian theory of force does not apply. One has to use quantum mechanics. All forces of ordinary experience such as friction, pressure, the forces in a rope, the force that makes things fall etc. are due to either the electromagnetic or the gravitational force. They are the only long-range forces.

12.2.1 Newton's three laws

The fundamental laws of classical mechanics, his doctrine about forces, was formulated by Newton as follows.

Newton's first law: A body which is not acted upon by any forces (or by forces with a sum which is equal to zero) is either at rest, or it moves along a straight path with constant speed.

Some comments are in order. Firstly, this law is valid only if the velocity of the body is measured relative to an *inertial reference frame*, i.e. a non-accelerated and non-rotating frame not influenced by forces. Such a frame is traditionally called 'freely moving'.

Secondly, one may wonder if 'not influenced by forces' implies 'non-rotating'. In order to remove any doubt, think of a rotating neutron star (a pulsar). If it is not acted upon by any forces it will rotate forever. Just as a particle will proceed moving with constant velocity when no force acts upon it, an extended system will proceed to rotate with constant angular velocity when no force acts upon it, like a friction-free spinning top.

Thirdly, what is the logical status of Newton's first law? Is it a physical law, or a *definition* of the concept 'inertial reference frame'?

Are there laws of nature? Something behind what goes on and steers everything? Johann Wolfgang Goethe seems to deny it: *"Die Natur hat weder kern noch Schale, alles is auf einem Male"* (Nature has neither core nor shell, everything is there at once). One requirement of a physical law is that it is conceivable (non-contradictory?) that it could be broken by the actual behaviour of a physical object. Analyzing Newton's first law it is advisable to introduce the concept of *implicit definition*. Newton's first law can be said to have double contents, separated as follows:

1. There exist reference frames in which free particles move with constant speed along straight lines.
2. These reference frames are called inertial frames.

Newton's second law: The acceleration of a body is proportional to the force acting on it, and inversely proportional to its mass.

The proportionality constant can be chosen equal to one, so that the law can be written

$$\vec{a} = \frac{\sum \vec{F}}{m} \quad \text{or} \quad \sum \vec{F} = m\vec{a} \tag{12.1}$$

where $\sum \vec{F}$ is the vector sum of the forces acting on the body.

Newton's second law involves three terms: 'acceleration', 'mass', and 'force', and we have not yet given precise definitions of the two latter ones. A thorough discussion of these terms is outside the main topic of this text. Note, however, that the term 'mass' appears in two fundamentally different contexts in Newtonian dynamics. In connection with Newton's law of gravitation the 'mass' of a body expresses a quantity, namely the strength of the gravitational force towards it. But in connection with Newton's second law 'mass' expresses the resistance of the body to change is motion. In order to make clear that these quantities are different, we should talk about *gravitational* mass, and *inertial* mass, respectively.

Force is sometimes *defined* as mass times acceleration. Then 'Newton's second law' is a definition, (force $\equiv$ mass $\times$ acceleration) and it should not be possible to confront it with experimental results. However such 'confrontation' may be said to be a most useful part of Newtonian dynamics. One way out of this dilemma, is again to accept that what is called a 'law' has in reality a double meaning. In the present case it has an empirical content: that the acceleration of an arbitrary body is proportional to a certain quantity. And it defines a term by means of this empirical relation. The mentioned quantity is called the *force* that acts upon the body.

Newton's third law: If a body A acts on another body B with a force $\vec{F}_B$, then B acts back on A with a force $\vec{F}_A$ in the opposite direction and of the same magnitude,

$$\vec{F}_A = -\vec{F}_B.$$

According to Newton's third law forces will always appear in pairs. It is suitable, then, to think of forces as interactions between pairs of bodies.

One may wonder if it is possible to understand this law as a convention. Does it have empirical content? Could the world be such that this 'law' is not obeyed? Suppose we find that A acts upon B. Can we possibly find that nevertheless B does not react upon A? If every force is an *interaction* between two bodies, and if this term implies a sort of 'democracy' in the sense that it is only a matter of the physicists' choice whether one says that A acts upon B, or B acts upon A, then Newton's third law would always be obeyed. In this case the question above would be answered by 'no'. However, in a rotating reference frame Newton's third law is *not* obeyed. In such a system a centrifugal force acts upon a body. But this body does not act back on another body with an equal and oppositely directed force. So the question above should be answered by 'yes'. Conclusion: Newton's third law *has* 'an empirical content. It is not a terminological convention.

12.3 Newton's theory of gravitation

Newton realized that gravitation governs the motion of the planets (as illustrated in Fig. 12.2). He also formulated the law giving the gravitational force between two spherical bodies in general.

Consider two bodies with masses m_1 and m_2. Let $\vec{r} = r\,\vec{e}_r$ be a vector representing the position of m_2 relative to m_1 (see Fig. 12.3).

Newton made the hypothesis that the gravitational force acting from m_1 upon m_2 is

$$\vec{F} = -G\,\frac{m_1 m_2}{r^2}\,\vec{e}_r, \qquad (12.2)$$

where $\vec{e}_r$ is a unit vector directed away from m_1 along the connecting line through m_1 and m_2, and $G = 6.6 \times 10^{-33}\mathrm{m}^3\mathrm{kg}^{-1}\mathrm{s}^{-2}$ is Newton's gravitational constant.

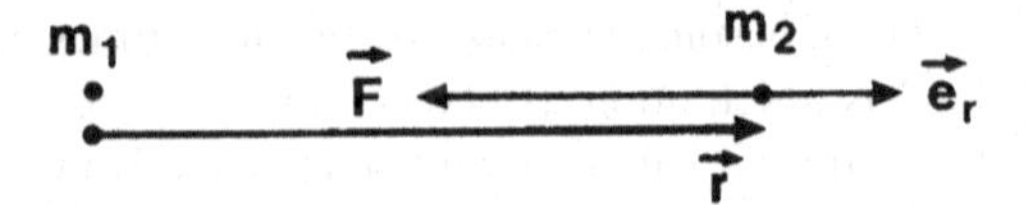

Fig. 12.3 Two bodies with masses m_1 and m_2

Equation (12.2) expresses *Newton's law of gravitation*. Incidentally the body m_2 acts upon m_1 with an oppositely directed gravitational force of the same magnitude, according to Newton's third law. This vector-equation tells that the force is directed along the line connecting the centres of the bodies, and acts in the negative $\vec{e}_r$ direction towards the mass m_1 that generates the force. It is an attractive force. If one is interested in the magnitude of the force only, Eq. (12.2) is reduced to the more familiar form

$$F = G\,\frac{m_1 m_2}{r^2}.$$

Essentially, Newton's theory of gravitation has the following ingredients. Universal time, Euclidean space, the three dynamical laws, and the gravitational law. The force of gravitation is implicitly assumed to act instantaneously (actio in distantia). If God suddenly removed some of the Sun's mass, so that it instantly disappeared, the gravitational force from the Sun upon the Earth would change instantly.

12.4 Consequences of special relativity for a theory of gravitation

Einstein constructed the special theory of relativity in 1905. It is said to be based on two 'postulates' or rather 'principles'. They have the status of physical hypotheses of a rather basic kind.

The special principle of relativity. Let F_1 and F_2 be two inertial reference frames in a region free of gravitational fields. To every physical process Q_1 there exists a physical process Q_2 such that Q_2 as observed in F_2 is identical to Q_1 as observed in F_1. The reason for the word 'special' is that the identity refers only to a special case: identity between *inertial* reference frames. According to this principle no experiment gives information about the velocity of one's own laboratory. One is always permitted to consider an inertial laboratory, i.e. a non-rotating laboratory not acted upon by any forces, as at rest. This means that velocity is always velocity in relation to something. There exists no absolute velocity. Since the expressions of the laws of nature are formalized descriptions of physical processes, it follows that these laws must be formulated without reference to the velocity of any reference frame. Only velocities of objects *relative to a chosen reference* is of physical significance: 'velocity is relative'. In the theory of relativity the statement that 'particle P_1 has velocity $\vec{v}_1$' has no sense unless a reference frame is introduced in relation to which the velocity is said to be $\vec{v}_1$.

The constancy of the velocity of light. The velocity of light is the same in every direction independently of the velocity of the observer or emitter. This implies that it is exactly the same relative to any freely moving reference frame whatsoever.

These two principles, contradictory according to Newton, furnish the basis of the special theory of relativity. We shall here be particularly concerned with one consequence of that theory, namely the increase of mass with (relative) velocity. This is expressed by the formula

$$m = m_0 \left(1 - \frac{v^2}{c^2}\right)^{-1/2} . \tag{12.3}$$

Here m is the mass of a body moving with a velocity v relative to an observer, and m_0 is its mass when it is at rest. From Eq. (12.3) it is seen that $\lim_{v \to c} m = \infty$. The mass of a body increases towards infinity when its velocity approaches c. Due to the relation between energy, E, and mass, m, expressed by $E = mc^2$, this means that one has to supply an infinite amount of energy in order to propel a particle all the way to the velocity of light. This has often been interpreted to imply that particles with a velocity greater than c cannot exist.

The Indian physicist C. G. S. Sudarshan has pointed out that the reasoning behind this conclusion is similar to the reasoning of people in India of old days, who concluded that there could not exist people North of Himalaya, for it was impossible to pass through these mountains. In fact, the theory of relativity permits the existence of particles travelling faster than light. They are called *tachyons*, and their physical existence is doubtful. From the formula (12.3) it can only be deduced that such travel cannot be realized by *increasing* the velocity from below that of light. Tachyons must be *created* on the 'upper side of the light barrier'! They remain there. They cannot be slowed down to c without being supplied with an infinite amount of energy.

However, due to the relativity of simultaneity, certain causality paradoxes would appear if tachyons were able to carry information. Think of an 'Einstein train' moving past a station. In Ch. 5 we showed that if two events, one at the front end of the train and one at the hindmost end, happened simultaneously as observed on the train, then the hindmost event would happen first as observed on the station. Imagine that we construct a 'tachyon telephone' with tachyons moving infinitely fast. We send a message from the station that arrives at the front end of the train immediately. The message is reflected to the emitter via a tachyon line attached to the train, so that it arrives at the hindmost end of the train just as it passes the emitter at the station. The signal arrives at that point practically simultaneously with the reflection event as observed on the train. Thus, as observed on the station the signal arrives before it was reflected. Since it was reflected at practically the same moment it was emitted, the signal arrives back at the emitter before it was emitted. The message could be: destroy the telephone as soon as this message is received!

Since all sorts of waves can transfer information, gravitational waves are able to carry information. The velocity of gravitational waves is the 'spreading velocity' of the gravitational force. This, together with the possibility of a 'causality paradox' implies that the gravitational force cannot act instantaneously according to the special theory of relativity. This means that Newton's theory of gravitation contradicts the special theory of relativity. There is a relation of logical incompatibility between the two.

The anti-relativistic character of Newton's theory of gravity could not be discovered empirically in the case of matter moving very much slower than light, since the relativistic deviations from the Newtonian results depend upon the factor v^2/c^2, where v is the velocity of the, say, a planet. This is the reason why Newton's theory of gravitation has functioned so well for several centuries and still does so. In fact the calculations of the rocket adjustments needed in order for example to hit Saturn with a spacecraft, are performed by means of Newton's theory of gravitation. The rocket don't move fast enough to make it necessary to warrant the introduction of the theory of relativity.

Einstein found it compelling, as a matter of principle, to construct a gravitational theory in accordance with the special theory of relativity. He knew that the theory of electromagnetic forces and fields formulated by Maxwell in about 1860 was consistent with special relativity. And the law of the electrical force between two charges, known as Coulomb's law, has just the same form as Newton's gravitational law. He also knew that it was Faraday's introduction of the *field* concept that paved the way for Maxwell's 'relativistic theory of electromagnetism'.

A first step towards a satisfactory theory, is to formulate Newton's theory of gravitation as a *field theory*, in which the force on a particle is due to a gravitational field acting on a particle, *locally*, at the position of the particle. In this way one can get rid of the non-relativisic notion of 'action at a distance' (which corresponds to the unphysical limit $c \to \infty$ in the relativistic theory of gravitation, because the force of gravity is spreading with the velocity of light in this theory, just like the electromagnetic force).

In a field theory the changes of the field are described locally by means of differential equations. For this purpose it will be convenient to introduce something called the *gravitational potential* ϕ. This is a function defining the potential energy of a particle with unit mass in a gravitational field. The potential energy of a body at a position P is equal to the work needed to move the body from a position where it has no potential energy, to P. The position where the potential energy is equal to zero, can be chosen freely. In the gravitational field at the surface of the Earth, for example, it is often chosen at sea level. The value of the potential increases in the upward direction in a gravitational field.

A simple illustration of the concept gravitational potential is provided by a uniform gravitational field in a room, say. Let us choose the zero energy level at the floor. The acceleration of gravity g is assumed to be constant. The weight of a particle with mass m is mg. The particle is lifted with constant velocity upwards from the floor to a height h above the floor. The force needed to lift the particle is mg, and the distance is h. The work, W, is the force times the distance, i.e. $W = mgh$. The potential is the work per unit mass, thus, the potential, ϕ, at a height h above the zero energy level in a uniform gravitational field is $\phi = gh$.

The potential at every point with this height has the value gh. This means that in a uniform gravitational field a horizontal surface is an equipotential surface.

Outside a spherical mass distribution the equipotential surfaces are spherical surfaces. The potential increases with increasing distance from the centre. In this case one often defines the zero energy level on a spherical surface with infinitely large radius. Then the potential is increasingly negative the closer one is to the centre.

In a field (i.e. not 'action at a distance') theory of gravitation space is imagined to be filled with a gravitational field. The field vector is the acceleration of gravity, $\vec{a}$, which is defined to equal 'minus the gradient of the gravitational potential', ϕ. According to Eq. (10.14) the gradient of a scalar field ϕ is given as

$$\nabla\phi = \frac{\partial\phi}{\partial x}\vec{e}_x + \frac{\partial\phi}{\partial y}\vec{e}_y + \frac{\partial\phi}{\partial z}\vec{e}_z$$

in a Cartesian coordinate system. In terms of the gradient operator ∇ (see Eq. (10.15) or Appendix A) the acceleration of gravity $\vec{g}$ is

$$\vec{g} = -\nabla\phi. \tag{12.4}$$

The gravitational potential is determined locally by the mass distribution. The differential equation that tells how the mass distribution determines the potential is called *Poisson's equation*. It is a second order differential equation involving the so-called *Laplacian* (see Appendix A).

In Newton's theory of gravitation, mass is the source—or rather the sink (due to the attractive character of gravity)—of the gravitational field. In the field formulation of the theory, this means mathematically that the divergence of the gravitational field is proportional to the mass density. Thus

$$\operatorname{div}\vec{g} = -4\pi G\rho.$$

Inserting the right-hand side of Eq. (12.4) leads to

$$\operatorname{div}\operatorname{grad}\phi = 4\pi G\rho.$$

Introducing the Laplacian defined in Eq. (A.1) this equation can be written

$$\nabla^2\phi = 4\pi G\rho. \tag{12.5}$$

This is Poisson's equation. It is the analogy of Einstein's field equations in the field theory formulation of Newton's theory of gravity. Equations (12.4) and (12.5) constitute what may be called 'Newton's theory of gravitation formulated as a field theory'. In the 'action at a distance version of Newton's theory of gravity, one had to assume the form of the force law. In the field theory formulation this assumption is replaced by the more general assumption that mass is the source of the gravitational field, and the force law can be deduced as a solution of this equation.

At a point where the mass density vanishes, the equation is reduced to

$$\nabla^2 \phi = 0, \tag{12.6}$$

which is called *Laplace's equation.*

We shall now show—for the special case of spherical symmetry—that this equation implies Newton's law of gravitation

$$\vec{g} = -\frac{GM}{r^2}\vec{e}_r, \tag{12.7}$$

where M is the mass inside a sphere with centre at the origin and radius r, and $\vec{g}$ is the acceleration of gravity outside the spherical mass distribution. Thus we shall solve Eq. (12.6) at an arbitrary point with vanishing mass density in a spherically symmetric region.

In the present case the Laplace operator shall act upon a gravitational potential ϕ that is a function of r only. Then the last two terms of Eq. (A.15) vanish because $\partial\phi/\partial\theta = \partial\phi/\partial\varphi = 0$. In the remaining term we can replace the partial derivatives with ordinary derivatives. Laplace's equation then takes the form

$$\frac{1}{r^2}\frac{d}{dr}\left(r^2\frac{d\phi}{dr}\right) = 0.$$

Since $1/r^2$ is different from zero, this equation reduces to

$$\frac{d}{dr}\left(r^2\frac{d\phi}{dr}\right) = 0. \tag{12.8}$$

From this equation, we shall find the acceleration of gravity, which is given in Eq. (12.4). In the present case the acceleration points in the radial direction, and the only non-vanishing component is

$$g^r = -\frac{d\phi}{dr}.$$

Since integration is the same as antiderivation, integration of the left-hand side of Eq. (12.8) gives just the expression inside the parenthesis. And since the derivative of a constant is zero, integration of the right-hand side gives a constant. Thus, we get

$$r^2\frac{d\phi}{dr} = K,$$

where K is an arbitrary constant. Dividing both sides of this equation by r^2 leads to

$$\frac{d\phi}{dr} = \frac{K}{r^2}.$$

Choosing $K = GM$ we get Newton's law of gravitation, Eq. (12.7).

So far Newton's theory formulated as a field theory reminds very much of the theory of an electrostatic field. There is, however, a fundamental difference between the theory of the electric field and gravitation. The electric field is uncharged. It does not contribute to the charge of the source. The electric field does not act by electric forces upon itself. This has the mathematical consequence that the field equations of the electric field are *linear* differential equations.

The quantity that replaces electrical charge in the case of gravitation, is mass. In other words, the 'gravitational charge' is mass. A gravitational field has energy. According to the relativistic equivalence between energy and mass ($E = mc^2$), a gravitational field has mass. Consequently a gravitational *field* is itself a source of gravitational attraction. A gravitational field acts upon itself. The differential equation (12.5) is not able to describe the action of a gravitational field upon itself. According to Newtonian theory the gravitational field has no mass and does not act upon itself. In this theory Poisson's equation is a correct field equation for gravitation. But it cannot be a relativistically correct equation.

We should mention that even if the concept of a gravitational *force* is absent in general relativity, the non-linear character of gravitation must be present in a relativistic theory of gravitation. But in general relativity it is interpreted in terms of geometry, not in terms of forces.

12.5 The general theory of relativity

The following three principles make up the fundamental physical assumptions of the general theory of relativity.

1. The special principle of relativity.
2. The principle of the constancy of the velocity of light.
3. The weak principle of equivalence (see below).

The first two principles are the building blocks of the special theory of relativity, which may be considered as a decisive step in the conceptual development leading to the general theory of relativity. Of special importance in this connection is that special relativity demands of us that we give up the old idea of absolute (non-relative) time and space, and think instead of our existence in a four-dimensional spacetime.

The principle of relativity is concerned with physical phenomena. It motivates the introduction of a formal requirement called *the covariance principle*. This principle may be formulated in the following way: The general laws of nature are to be expressed by equations which hold good for all systems of coordinates, that is, they are co-variant with respect to any change of coordinate system. This requirement is not directly concerned with physical phenomena, but with the way of talking about them. As we know from the preceding chapters, tensor equations have a coordinate independent form: they are said to be form-invariant, also called 'covariant'.

One should note, however, that a covariant equation, when given physical interpretation, does not necessarily obey the principle of relativity. This is due to the following circumstance. Because it is a physical principle, the principle of relativity is concerned with relations that can be observed. If you are going to investigate the physical consequences of a tensor equation, you have to establish certain relations between tensor components and observable physical quantities. The relations have to be separately defined. From the tensor equations that are covariant, *and* the defined relations between the tensor components and the observable physical quantities, one can deduce equations between observable physical quantities. The special principle of relativity demands that the laws which are expressed by these equations, can be stated in the same way in every inertial reference frame, and that the laws do not refer to any universally privileged reference frame, PRF. If we term the velocity of an arbitrary frame relative to PRF the *absolute* velocity of the frame, the principle of relativity requires that the laws of nature do not refer to any absolute velocity, i.e. the equations relating physical quantities should not contain any absolute velocity.

If a physical theory is expressed by means of tensor equations, the theory is said to be written in a *manifestly covariant* form. As written in this way, the theory will automatically fulfil the covariance principle, but it need not fulfil the principle of relativity.

We shall now go on and introduce the famous *principle of equivalence*. In the literature one finds two versions of this principle: the weak and the strong principle of equivalence.

In order to be able to understand the significance of the weak principle of equivalence, one should remember the fact that the term *mass* appears in two kinds of contexts in Newton's theory of gravitation. On the one hand it appears as the *gravitational mass*, m_{g}, of a body. On the other hand, mass appears in Newton's second law as a measure of how strongly a body resists acceleration. This is the *inertial mass*, m_{i}, of a body.

If Newton's second law, Eq. (12.1), is combined with Newton's law of gravitation, Eq. (12.2), we may write

$$-G\,\frac{m_{\mathrm{g}}\,M_{\mathrm{g}}}{r^2}\,\vec{e}_r = m_{\mathrm{i}}\,\vec{a},$$

for a particle with gravitational mass m_{g} and inertial mass m_{i} at a distance $\vec{r} = r\vec{e}_r$ from the centre of a spherical body with gravitational mass M_{g}. The acceleration of the particle is

$$\vec{a} = -\frac{GM_{\mathrm{g}}}{r^2}\,\frac{m_{\mathrm{g}}}{m_{\mathrm{i}}}\,\vec{e}_r. \tag{12.9}$$

Note that the acceleration depends upon the quotient between the gravitational and the inertial mass of the particle.

Experiments have without exception confirmed the hypothesis that bodies made of different materials fall with equal acceleration under the same carefully described circumstances. From this hypothesis and Eq. (12.2) we derive the well attested hypothesis: m_g/m_i has the same numerical value for all bodies whatsoever. This sameness of numerical value is the contents of the so called weak principle of equivalence.

In such important cases of empirical confirmations (so far) without exception, what a physical equation expresses is usually endowed with a honorific title 'physical law' or 'law of nature'. We have used the term 'hypothesis' because future instances of observational disconfirmation are possible. According to the philosophy of possibilism, we have no guarantee whatsoever that the future in the relevant area will resemble the past. There might be changes so deep and universal that nobody could meaningfully declare 'this time the conditions are such that a crucial test is possible, and we shall find out whether our previous series of confirmations will continue indefinitely or not'. Also if there were too many abrubt changes, we may not be able to observe them. The process of observation seems only to be possible if we can rely, for instance, on at least a moderate steadiness of the constitution of the material world and how it behaves. Fundamentally chaotic and erratic behaviour encompassing the observing subject seems to be incompatible with the act of obtaining knowledge about observable properties of the material world.

It was tempting for Einstein (and only Einstein?), *not* to rest with a law stating that two different entities, m_g and m_i, always are quantitatively the same. Einstein asked: why not think of m_g and m_i as one and the same physical entity, like the 'Morning Star' and 'Evening Star' being names for the same planet Venus? The validity of the weak principle of equivalence could then be proved because measures of m_g and m_i would then be measures of one and the *same* entity. An entity that was simply called 'mass'. It would have to be a very special kind of entity showing an extreme generality, characterising every physical kind of system, including, for example electromagnetic fields, and the particles of microphysics. Also the naming could be changed. By the weak principle of equivalence one might mean what is expressed by the six words *gravitational and inertial mass are identical*. This in turn makes the principle not something proved. It is, however, a physical hypothesis of immense generality. A stroke of Einstein's genius was to recognize this hypothesis and make it a part of the foundations of his remarkable general theory of relativity.

According to this principle Eq. (12.9) is reduced to Eq. (12.7). Note that *no* property whatsoever of the accelerated particle is referred to in the equation. Neither the mass of the particle nor what it consists of means anything for its acceleration, *only* its *position* in the gravitational field.

This made Einstein realize that the acceleration of a particle at a certain position in a gravitational field might be a property of space itself at this position. This has *nothing* to do with any property of the particle. But even if it is a property of space itself, we do not necessarily leave physics and enter mathematics. We enter the area of geometrical properties of something physical. So Einstein conceived the revolutionary project of a *geometrical* conception of gravity: the essence of the general theory of relativity.

But how is this possible some may ask. Is not geometry a part of mathematics and gravity a part of physics? The question is serious. Let us look at an example.

We may offer a purely geometrical description of a perfect sugar cube. All angles are ninety degrees, all sides flat and equal. We leave out chemical properties, taste and everything else, confining ourselves to purely geometrical properties. By calling the description of gravity geometrical, we refer to geometrical properties of physical spacetime. But in contrast to the description of the sugar cube we may add: there *are*[1] no other properties, chemical, electrical or otherwise.

Even if Newton declared that he made no hypotheses as to the physical cause of the gravitational force, the conceptual structure of his theory invites to search for such a cause. In the second half of the eighteenth century, one tried for example to explain the gravitational force as a result of hydrodynamic phenomena in the so-called 'ether'. Einstein's theory of gravitation is different. According to this theory there is no gravitational force.

An important consequence of the weak principle of equivalence is that the 'influence' (see below) of a field of *weight* upon all 'local' (the meaning of this word in the present context is explained below) *mechanical* experiments can be simulated by an *accelerated frame of reference*, say an Einstein lift, in a region free of gravitating masses. Consequently, every 'influence' of a field of weight upon a local mechanical experiment can be 'transformed away', by going into a freely falling frame of reference. The so-called *strong principle of equivalence* generalizes these statements to encompass experiments involving physical processes of arbitrary nature, not just mechanical ones.

The strong principle of equivalence makes the conception of a causal *influence* of a field of weight upon for example light superfluous and misleading. We should stick to just kinematics (science of movement), as far as gravitation is concerned, totally leaving out dynamics (science of force).

Consider a ray of light directed horizontally at the surface of the Earth. According to the Newtonian (dynamical) conception of gravitation, a gravitational field acts upon light if it has mass, and forces the light ray to bend downwards. If light is mass-less there is no gravitational force, and the light ray is perfectly straight. The understanding of gravitation obtained in the general theory of relativity is free of any dynamical relationships, and leads in fact to a prediction which is different from the Newtonian one in the present example. The situation is described as follows. If the observer moves non-inertially, in other words, if he is not in free fall, but is acted upon by some kind of force (non-gravitational, since in general relativity no gravitational force exists), then he shall find that a ray of light is bended. This bending is understood not to be an observer-independent feature of the motion of light. Nothing acts upon the light and makes it bend. If the observer moves inertially, then he shall observe that, locally, the ray of light is straight. General relativity thus predicts that if light moves with a finite velocity, then a non-inertial observer shall register a bending of a light ray, *whether light has mass or not.*

---

[1]We should here warn the reader: spacetime may not be as simple as that. Quantum field theory indicates that what we call vacuum is tremendously complicated and badly understood. But that is another story.

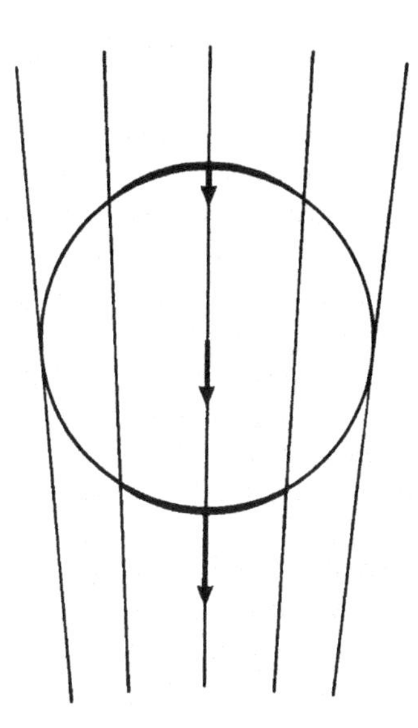

Fig. 12.4 Inhomogeneous gravitational field

As mentioned above, what is, in Newton's theory of gravitation, understood as the influence of a field of weight upon the results of a local experiment, can be transformed away, by going into a freely falling frame of reference.

The word 'local' means local both in space and time. The reason for this restriction is the existence of what in Newton's theory of gravitation is called tidal forces. These are the forces that cause the ebb and flood of tides on the Earth.

Figure 12.4 depicts the Earth in the gravitational field of the Sun. The Sun is in the 'downward' direction in the plane of the paper. The field lines approach each other in the direction of the Sun, indicating that the field is stronger closer to the Sun. The lengths of the arrows indicate that acceleration of gravity, in the field of the Sun, is larger at the side of the Earth which turns towards the Sun, than at the opposite side. Compared to the centre of the Earth, i.e. to the Earth as a whole, the water at the sunny side is drawn more strongly towards the Sun, and the water at the opposite side more weakly. This, together with the similar properties of the Moon's gravitational field at the surface of the Earth, causes the tides. The difference of gravitational pull is fortunately small. Otherwise we might have horrible tidal waves hundreds of metres high.

A gravitational field in which the acceleration of gravity is the same everywhere is called *homogeneous*. Then the gravitational field lines are parallel. A gravitational field in which the acceleration of gravity is position dependent, has field lines that are not parallel, as those of Fig. 12.4. Such a gravitational field is said to be *inhomogeneous*. An inhomogeneous gravitational field cannot be transformed away by going into a freely falling reference frame. But measurements of tidal forces demand a certain duration and spatial extension. 'Local' has here the following meaning: The measurements are to be restricted in duration and spatial extension, so that tidal forces cannot be measured within the accuracy of measurement that can be obtained with the equipment of the observer. The extension of the area that can be considered 'local', depends upon the character and accuracy of this equipment.

In this connection it is important to note the difference between inertial reference frames F_{sr} in special relativity and inertial reference frames F_{gr} in general relativity. The special relativistic inertial frames are *unaccelerated* reference frames of *arbitrary extension in a space free of gravitational fields*, i.e. *flat* spacetime. The general relativistic inertial reference frames are *local* systems falling freely in *curved* spacetime.

The *strong principle of equivalence* may now be stated as follows. Given a certain measuring accuracy, to every physical process Q_1 in F_{gr} there exists a physical process Q_2 in an inertial frame F_{sr} in Minkowski spacetime, so that Q_2 is found by observation in F_{sr} to be identical to Q_1 observed in F_{gr}.

The strings of a music instrument, for example, vibrate in the same manner whether you play in a flat region far from any masses, or in a satellite just above the Earth. You can play with it equally successfully at either place.

The strong principle of equivalence announces the physical equivalence of all inertial reference frames.

Since the process of transforming away a field of weight by going into a freely falling reference frame may be reversed, there is also a corresponding equivalence between non-inertial reference frames. The acceleration field in such a frame has the same physical effects, whether it is caused by acceleration of the reference frame, or it is due to a nearby mass.

However, if we compare observations in a non-inertial and an inertial reference frame, differences will appear. Light, for example, is deflected in a non-inertial reference frame, and moves along a straight path in an inertial reference frame. One may in several ways, both by optical and by mechanical means, find out by local experiments, whether one is in a non-inertial or an inertial reference frame.

In spite of this, Einstein generalized the special principle of relativity to a general principle of relativity, encompassing accelerated motion. Due to the locally observable difference between inertial and non-inertial reference frames, the generalization of the special principle of relativity cannot be just to replace 'inertial reference frames' with 'arbitrary reference frames' in its formulation.

Even if one restricts oneself to using inertial reference frames, a physical process is determined not only by the laws of nature, but also by the initial conditions. If one refers to an arbitrary coordinate system, the process will in addition depend upon the metric of the coordinate system. However, if one formulates the natural laws so that they are valid without any changes for every metric $g_{\mu\nu}$, one may formulate a *general principle of relativity* as follows: Natural laws are independent of coordinate systems and even reference frames, and so are their expressions. In other words: It should not be necessary to change the verbal formulation of any natural law when changing the reference frame.

In the special case of the propagation of light, the law is: light follows null geodesic curves, that is geodesic curves in spacetime such that the spacetime intervals are zero along the curve (see Sect. 5.14). This sentence is perfectly adequate whether the reference frame is accelerated or not. The shape of a geodesic curve depends upon the first derivative of the metric due to the presence of the Christoffel symbols in the geodesic equation, and it follows from this law that light is deflected in a non-inertial reference frame, but not in an inertial frame. The fact that light is deflected in a non-inertial reference frame, implies that *the velocity of light is constant only in local inertial reference frames*.

As mentioned above, an important consequence of the principles of equivalence is that gravitation may be described geometrically. In the general theory of relativity the gravitational attraction between two bodies is *not* perceived as the result of an interaction between the bodies. Instead it is described as the curvature of spacetime near the bodies. There is no longer any gravitational dynamics (of particles in a gravitational field), only gravitational kinematics—theory of movement.

Newton's law of gravitation is in general relativity replaced by equations that tell how the presence of mass or energy *corresponds* to curved spacetime and vice versa. These equations—the field equations of Einstein—do not follow logically from the principles we have talked about above. In order to arrive at these equations Einstein demanded that conservation of energy and momentum shall follow from the field equations. He also wanted to make the equations as simple as possible. Noting that higher order derivatives would make the equations inordinately complicated, he limited himself to a curvature tensor involving only first and second derivatives of the metric tensor.

Einstein was well aware of the fact that natural phenomena of a certain class, for example those having to do with gravitation, can be represented conceptually and mathematically by a manifold of theories. For Einstein a most important criterium for selecting one theory above the others is that of simplicity. In Einstein's Herbert Spencer lecture, delivered at Oxford, June 10, 1933, he said:

> Our experience hitherto justifies us in believing that nature is the realization of the simplest conceivable mathematical ideas.

And in *Forum Phil.* **1**, 173 (1930) he wrote:

> I do not consider the main significance of the general theory of relativity to be the prediction of some tiny observable effects, but rather the simplicity of its foundations and its consistency.

Note that 'simple' here does not mean 'easy'. A simple theory is above all conceptually and mathematically *economic*. It introduces a smallest possible number of principles and concepts representing a largest possible set of physical phenomena in a mathematically consistent way. In order to realize such a theory one may need to introduce advanced mathematics, but a guiding principle is: Construct the theory with maximum breath of field of valid application, with maximal conceptual economy, and represent its conceptual structure in as few and simple mathematical terms as possible, expressed with an appropriate and suggestive notation.

12.6 The Newtonian limit of general relativity

We shall describe a static and very weak gravitational field, like the gravitational field in the Solar system. We start by giving a relativistic representation of the acceleration of gravity.

The motion of a free particle is expressed by the geodesic equation, Eq. (8.7). Making use of the proper time τ of the particle as parameter, this equation takes the form

$$\frac{d^2x^\mu}{d\tau^2} + \Gamma^\mu{}_{\alpha\beta}\frac{dx^\alpha}{d\tau}\frac{dx^\beta}{d\tau} = 0, \qquad (12.10)$$

where $x^\mu = (x^1, x^2, x^3, x^4)$ and $x^4 = ct$.

Consider a free particle instantaneously at rest. Then the spatial components, $u^j = dx^j/d\tau$, of the particle's four-velocity vanish, $dx^j/d\tau = 0$. The time component is $u^4 = dx^4/d\tau = d(ct)/d\tau = c\,dt/d\tau$. In the summation over α and β in Eq. (12.10) only the term with $\alpha = \beta = 4$ is non-vanishing. Equation (12.10) represents a set of four equations, one for each value of μ. Putting $\mu = j$ one then finds the equation of motion in the j direction

$$\begin{aligned}\frac{d^2x^j}{d\tau^2} &= -\Gamma^j{}_{44}\frac{dx^4}{d\tau}\frac{dx^4}{d\tau} = -\Gamma^j{}_{44}\frac{d(ct)}{d\tau}\frac{d(ct)}{d\tau} \\ &= -\Gamma^j{}_{44}\,c\frac{dt}{d\tau}\,c\frac{dt}{d\tau} = -\Gamma^j{}_{44}\,c^2\left(\frac{dt}{d\tau}\right)^2. \end{aligned} \qquad (12.11)$$

The coordinate clocks of an observer at the position of the particle at the instant it is at rest, are chosen to be synchronized with the standard clock of the particle. Since we want to calculate the acceleration $\vec{a} = a^j\vec{e}_j$ of the particle at this particular instant, we have $d\tau = dt$, and $a^j = d^2x^j/dt^2 = d^2x^j/d\tau^2$.

The *acceleration of gravity* is defined as the acceleration of a freely falling particle instantaneously at rest. Because $d\tau/dt = 1$ we get from Eq. (12.11)

$$a^j = -c^2\Gamma^j{}_{44}. \qquad (12.12)$$

We have thereby obtained a physical, not merely geometrical, interpretation of the Christoffel symbols $\Gamma^j{}_{44}$. They represent the components of the acceleration of gravity.

We now assume that the gravitational field is very weak at the position of the particle. Relativistically this implies that the curvature of spacetime is small. The metric tensor deviates very little from the Minkowski metric $\eta_{\mu\nu}$.

The line element of flat spacetime as expressed by Cartesian coordinates and the usual time coordinate, has the form

$$ds^2 = dx^2 + dy^2 + dz^2 - c^2\,dt^2. \qquad (12.13)$$

The components of the metric tensor are

$$\eta_{xx} = \eta_{yy} = \eta_{zz} = 1 \quad \text{and} \quad \eta_{tt} = -c^2. \tag{12.14}$$

It is practical to introduce a fourth coordinate $x^4 = c\,t$, which may be called 'light distance' since it represents the distance that light passes during a time t. Inserting this coordinate into the last term of Eq. (12.13), the line element takes the form

$$ds^2 = dx^2 + dy^2 + dz^2 - (dx^4)^2. \tag{12.15}$$

We read from this that the components of the metric tensor are

$$\eta_{xx} = \eta_{yy} = \eta_{zz} = 1 \quad \text{and} \quad \eta_{44} = -1. \tag{12.16}$$

Both Eqs. (12.14) and (12.16) are called the Minkowski metric. We shall use the form (12.16).

We now introduce a tensor with components $h_{\mu\nu}$ which represents the deviation from the Minkowski metric. We may write

$$g_{\mu\nu} = \eta_{\mu\nu} + h_{\mu\nu}. \tag{12.17}$$

Since the gravitational field is assumed to be very weak, the deviation from the Minkowski metric is very small. Thus, the components $h_{\mu\nu}$ have magnitudes much less than 1, i.e. $|h_{\mu\nu}| \ll 1$. We also assume that the tensor $h_{\mu\nu}$ is diagonal. Then the contravariant components of the metric tensor are given by Eq. (5.75), that is

$$g^{\mu\mu} = 1/g_{\mu\mu}. \tag{12.18}$$

We need to calculate the Christoffel symbol $\Gamma^j{}_{44}$ which is present in Eq. (12.12). Inserting $\tau = j$, $\nu = 4$, and $\lambda = 4$ in Eq. (7.30), we get

$$\Gamma^j{}_{44} = \frac{1}{2} g^{j\mu} \left(\frac{\partial g_{\mu 4}}{\partial x^4} + \frac{\partial g_{\mu 4}}{\partial x^4} - \frac{\partial g_{44}}{\partial x^\mu} \right). \tag{12.19}$$

Since we describe a static gravitational field, the metric tensor is time-independent. Then the first two terms on the right-hand side of Eq. (12.19) vanish, and the equation is reduced to

$$\Gamma^j{}_{44} = -\frac{1}{2} g^{j\mu} \frac{\partial g_{44}}{\partial x^\mu}. \tag{12.20}$$

Since the metric is diagonal, meaning that $g^{j\mu} = 0$ for $\mu \neq j$, only $\mu = j$ contributes in the summation over μ. Equation (12.20) then further reduces to

$$\Gamma^j_{44} = -\frac{1}{2} g^{jj} \frac{\partial g_{44}}{\partial x^j}.$$

Using Eq. (12.18) we get

$$\Gamma^j_{44} = -\frac{1}{2g_{jj}}\frac{\partial g_{44}}{\partial x^j}.$$

Substituting for g_{jj} and g_{44} from Eq. (12.17) follows

$$\Gamma^j{}_{44} = -\frac{1}{2}\frac{1}{(\eta_{jj} + h_{jj})}\frac{\partial(\eta_{44} + h_{44})}{\partial x^j}.$$

Since η_{44} is constant (equal to -1) the derivative of η_{44} vanishes, so that

$$\Gamma^j{}_{44} = -\frac{1}{2}\frac{1}{(\eta_{jj} + h_{jj})}\frac{\partial h_{44}}{\partial x^j}.$$

Now $\eta_{jj} = 1$ for all j and $h_{jj} \ll 1$, so we can neglect h_{jj} and obtain

$$\Gamma^j{}_{44} = -\frac{1}{2}\frac{\partial h_{44}}{\partial x^j}.$$

Inserting this into Eq. (12.12) we have

$$a^j = \frac{c^2}{2}\frac{\partial h_{44}}{\partial x^j}. \tag{12.21}$$

Since $g_{44} = -1 + h_{44}$, this equation indicates how, in the Newtonian limit, the time–time (i.e. four–four) component of the metric tensor determines the acceleration of gravity.

According to Eq. (12.4) the j component of the acceleration of gravity is expressed in terms of the gravitational potential ϕ, by

$$a^j = -\frac{\partial \phi}{\partial x^j}.$$

Comparing with Eq. (12.21) we get

$$-\frac{\partial \phi}{\partial x^j} = \frac{c^2}{2}\frac{\partial h_{44}}{\partial x^j}.$$

From which follows

$$\phi = -\frac{c^2}{2}h_{44}.$$

Multiplying each side by -2 and dividing by c^2, we find the following expression for h_{44} in terms of the (Newtonian) gravitational potential

$$h_{44} = -\frac{2\phi}{c^2}.$$

The above equations contain concepts completely foreign to Newtonian thinking. This means that it is misleading to think of them as a generalization of Newtonian thinking. Nevertheless Newton's equations are logically derivable from Einstein's theory in the case of very weak gravitational fields and as applied to objects moving with velocities very much smaller than that of light.

Since, according to Eq. (12.21), only the component h_{44} of the tensor $h_{\mu\nu}$ contributes to the acceleration of gravity in the Newtonian limit, we are free to choose a coordinate system so that all the other components of $h_{\mu\nu}$ vanish. Then the line element of spacetime can be written

$$ds^2 = dx^2 + dy^2 + dz^2 - (1 - h_{44})\, c^2\, dt^2,$$

or

$$ds^2 = dx^2 + dy^2 + dz^2 - \left(1 + \frac{2\phi}{c^2}\right) c^2\, dt^2. \tag{12.22}$$

There is only one function ϕ to be determined by the field equations. In this special case there is only one independent field equation, which can be taken as the 44 component of Eq. (11.34)

$$R_{44} = \kappa\left(T_{44} - \frac{1}{2}\, g_{44}\, T\right). \tag{12.23}$$

Because ϕ and its first derivatives are much less than 1, we may use the expression (11.11) of the Riemann curvature tensor, valid in a local Cartesian coordinate system

$$R_{\mu\nu\alpha\beta} = \frac{1}{2}\left(g_{\mu\beta,\nu\alpha} - g_{\mu\alpha,\nu\beta} - g_{\nu\beta,\mu\alpha} + g_{\nu\alpha,\mu\beta}\right).$$

In particular, putting $\nu = 4$ and $\beta = 4$,

$$R_{\mu 4\alpha 4} = \frac{1}{2}\left(g_{\mu 4,4\alpha} - g_{\mu\alpha,44} - g_{44,\mu\alpha} + g_{4\alpha,\mu 4}\right).$$

Considering a static field, all terms with time-derivatives (terms with a number 4 somewhere after a comma) are equal to zero. In this case we get

$$R_{\mu 4\alpha 4} = -\frac{1}{2}\, g_{44,\mu\alpha} = -\frac{1}{2}\, \frac{\partial^2 g_{44}}{\partial x^\mu \partial x^\alpha}. \tag{12.24}$$

Since $g_{44} = -1 - 2\phi/c^2$, the derivatives of g_{44} are $-2/c^2$ times the derivatives of ϕ. Equation (12.24) then becomes

$$R_{\mu 4\alpha 4} = \frac{1}{c^2}\, \frac{\partial^2 \phi}{\partial x^\mu \partial x^\alpha}.$$

Contracting μ with α leads to

$$R_{44} = R^{\alpha}{}_{4\alpha 4} = \frac{1}{c^2}\frac{\partial^2 \phi}{\partial x_\alpha \partial x^\alpha}. \tag{12.25}$$

We now make use of

$$\begin{aligned}
\frac{\partial^2}{\partial x_\alpha \partial x^\alpha} &= \frac{\partial^2}{\partial x_4 \partial x^4} + \frac{\partial^2}{\partial x_j \partial x^j} \\
&= \frac{\partial^2}{\eta_{44}\partial x^4 \partial x^4} + \frac{\partial^2}{\partial x_j \partial x^j} \\
&= \frac{\partial^2}{(-1)\partial(ct)\partial(ct)} + \frac{\partial^2}{\partial x_j \partial x^j} \\
&= -\frac{\partial^2}{c^2\partial t^2} + \frac{\partial^2}{\partial x_j \partial x^j}.
\end{aligned} \tag{12.26}$$

Substituting the last expression in (12.26) for $\partial^2\phi/\partial x_\alpha \partial x^\alpha$, Eq. (12.25) takes the form

$$R_{44} = -\frac{\partial^2 \phi}{c^4 \partial t^2} + \frac{1}{c^2}\frac{\partial^2 \phi}{\partial x_j \partial x^j}.$$

Since, according to our assumptions, derivatives with respect to time are equal to zero, we have

$$R_{44} = \frac{1}{c^2}\frac{\partial^2 \phi}{\partial x_j \partial x^j}.$$

Using Eq. (A.13) in Appendix A, this equation can be written

$$R_{44} = \frac{1}{c^2}\nabla^2 \phi. \tag{12.27}$$

Considering the components of the energy-momentum tensor of a perfect fluid, as given in Eq. (10.27), we see that in the Newtonian limit the term $T_{44} = \rho c^2$ is dominating. All the other terms can be neglected compared to T_{44}. From the rule that indices can be lowered (see Eq. (5.76)) by means of the Minkowski metric in the Newtonian limit, we find, making use of $\eta^{44} = -1$

$$T \equiv T^4{}_4 = \eta^{4\alpha} T_{\alpha 4} = \eta^{44} T_{44} = -T_{44}. \tag{12.28}$$

Neglecting h_{44} in the first step and using $\eta_{44} = -1$ and Eq. (12.28) in the second step below leads to

$$T_{44} - \frac{1}{2} g_{44} T \approx T_{44} - \frac{1}{2}\eta_{44} T = T_{44} - \frac{1}{2} T_{44} = \frac{1}{2} T_{44}.$$

The 44 component of Einstein's field equations, Eq. (12.23), can in the present approximation be written

$$R_{44} = \frac{1}{2}\kappa\, T_{44} = \frac{1}{2}\kappa\,\rho\, c^2. \tag{12.29}$$

From Eqs. (12.27) and (12.29) we derive

$$\nabla^2\phi = \frac{1}{2}\kappa\,\rho\, c^4. \tag{12.30}$$

This represents the Newtonian limit of Einstein's gravitational field equations. The equation is approximately valid only for very weak gravitational fields.

According to the definition (A.1) of the Laplacian Eq. (12.30) can be written as

$$\text{div grad}\ \phi = \frac{1}{2}\kappa\rho c^4.$$

From Eq. (12.4) the acceleration of gravity is given by

$$\vec{a} = -\,\text{grad}\,\phi.$$

From the last two equations it is seen that Eq. (12.30) can be given the form

$$\text{div}\,\vec{a} = -\frac{1}{2}\kappa c^4\rho.$$

According to the interpretation of the divergence, as discussed in Sect. 10.2, this equation expresses that the acceleration of gravity is a field that converges (due to the minus sign) towards regions with a non-vanishing mass-density. This negative sign is the mathematical expression of the fact that gravitation is an attractive force.

Comparing Eqs. (12.30) and (12.5) we see that the relativistic equations are compatible with the 'Newtonian' gravitational field equation if $(1/2)\kappa\rho c^4 = 4\pi G\rho$. Solving this equation with respect to κ we get

$$\kappa = 8\pi\, G/c^4. \tag{12.31}$$

This quantity has been termed 'Einstein's gravitational constant', because κ is the proportionality constant between Einstein's curvature tensor and the energy-momentum tensor in Einstein's field equations.

12.7 Repulsive gravitation

In this section we shall still consider very weak gravitational fields, i.e. nearly flat spacetimes, that are described by the line element (12.22), but we shall now investigate the gravitational properties of perfect fluids with non-negligible stresses.

Raising one index in the field equation (12.29) we get

$$R^4{}_4 = \kappa \left(T^4{}_4 - \frac{1}{2} \delta^4{}_4 T \right).$$

Inserting $\delta^4{}_4 = 1$ and $T = T^x{}_x + T^y{}_y + T^z{}_z + T^4{}_4$ leads to

$$\begin{aligned} R^4{}_4 &= \kappa \left[T^4{}_4 - \frac{1}{2} \left(T^x{}_x + T^y{}_y + T^z{}_z + T^4{}_4 \right) \right] \\ &= \frac{\kappa}{2} \left(T^4{}_4 - T^x{}_x - T^y{}_y - T^z{}_z \right). \end{aligned} \tag{12.32}$$

Raising one index in Eq. (12.27) by means of the Minkowski metric, only changes the sign, so we have

$$R^4{}_4 = -(1/c^2)\, \nabla^2 \phi.$$

Substituting the expression (12.31) for κ and Eq. (12.32) for $R^4{}_4$, it follows that

$$\nabla^2 \phi = -\frac{4\pi G}{c^2} \left(T^4{}_4 - T^x{}_x - T^y{}_y - T^z{}_z \right). \tag{12.33}$$

Consider a perfect fluid at rest. In the weak field approximation the non-vanishing contravariant components of the energy-momentum tensor of the fluid are then given by Eq. (10.27),

$$T^{ij} = p\, \delta^{ij} \quad \text{and} \quad T^{44} = \rho c^2.$$

The mixed components are

$$T^i{}_j = p\, \delta^i{}_j \quad \text{and} \quad T^4{}_4 = -\rho c^2.$$

Inserting these components into Eq. (12.33) gives

$$\nabla^2 \phi = 4\pi G \left(\rho + 3p/c^2 \right). \tag{12.34}$$

Comparing this equation with the corresponding Newtonian equation (12.5), we define a relativistic gravitational mass density ρ_g by

$$\nabla^2 \phi = 4\pi G \rho_g. \tag{12.35}$$

Equations (12.34) and (12.35) imply

$$\rho_g = \rho + 3p/c^2. \tag{12.36}$$

This is the relativistic mass density that generates the gravitational field which gives free particles an acceleration of gravity. Here $p > 0$ represents pressure, and $p < 0$

represents tension. Thus the pressure in a star, for example, increases the attractive gravitational field. This is the reason that according to general relativity, sufficiently massive stars will collapse to black holes (see chapter 13). According to Newton's theory of gravitation a large pressure can prevent such a collapse, but relativistically the pressure is itself a source of increased gravitational attraction. If there exists a medium for which the tension is so large that $p < -(1/3)\rho c^2$, then such a medium will be a source of repulsive gravitation. In chapter 14 on cosmology we shall see that vacuum energy may be such a medium. However, proceeding towards the limit $c \to \infty$, representing Newton's theory with instantaneous action at a distance, the possibility of repulsive gravitation vanishes.

12.8 The 'geodesic postulate' derived from the field equations

The principle that free particles follow geodesic curves is often called 'the geodesic postulate'. Early in the history of general relativity this was considered an independent assumption, which could not be logically derived from the other principles of the theory. It is an empirical proposition, part of physics, not of geometry, and in principle open to testing and revision by observations. In order to exhibit the extremely important conceptual economy of Einstein's theory of spacetime and gravitation, we feel we have an obligation to show, to those readers who are still willing to do some mathematics, that 'the geodesic postulate' clearly follows as a logical consequence of the field equations. The derivation necessitates many steps, but each is rather elementary.

Consider a system of free particles in curved spacetime. This system can be regarded as a pressure-free gas. Such a gas is called *dust*. From Eq. (10.35) follows that it is described by an energy-momentum tensor

$$T^{\mu\nu} = \rho u^\mu u^\nu,$$

where ρ is the rest density of the dust as measured by an observer at rest in the dust, and u^μ the components of the four-velocity of the dust-particles. All particles of the dust have the same velocity, so that the dust moves like a rigid system.

Einstein's field equations, Eq. (11.31), as applied to spacetime filled with dust, take the form

$$R^{\mu\nu} - \frac{1}{2}\, g^{\mu\nu}\, R = \kappa\, \rho\, u^\mu u^\nu.$$

Because the divergence of the left-hand side is zero (see Eq. (11.27)), the divergence of the right-hand side must be zero, too

$$(\rho\, u^\mu u^\nu)_{;\nu} = 0$$

or

$$(\rho u^\nu u^\mu)_{,\nu} = 0.$$

The quantity in the parenthesis we now regard as a product of ρu^ν and u^μ. By the rule for differentiating a product we get

$$(\rho u^\nu)_{;\nu} u^\mu + \rho u^\nu u^\mu{}_{;\nu} = 0. \tag{12.37}$$

Since the four-velocity of any object in spacetime has a magnitude equal to the velocity of light (see Eq. (10.23)), we have

$$u_\mu u^\mu = -c^2. \tag{12.38}$$

Differentiation gives

$$(u_\mu u^\mu)_{;\nu} = 0.$$

Using, again, the rule for differentiating a product, we get

$$u_{\mu;\nu} u^\mu + u_\mu u^\mu{}_{;\nu} = 0. \tag{12.39}$$

Applying the rule (5.77) for raising an index (second and fourth equality below), and the freedom of changing a summation index from α to μ, say, (last equality) we get

$$\begin{aligned} u_{\mu;\nu} u^\mu &= u^\mu u_{\mu;\nu} = g^{\mu\alpha} u_\alpha u_{\mu;\nu} = u_\alpha g^{\mu\alpha} u_{\mu;\nu} \\ &= u_\alpha u^\alpha{}_{;\nu} = u_\mu u^\mu{}_{;\nu}. \end{aligned}$$

Thus the second term of Eq. (12.39) is equal to the first one. Accordingly Eq. (12.39) says that the sum of two equal terms are equal to zero. It follows that each of them are equal to zero. So we have

$$u_\mu u^\mu{}_{;\nu} = 0. \tag{12.40}$$

We now multiply each term of Eq. (12.37) by u_μ. From this follows

$$(\rho u^\nu)_{;\nu} u_\mu u^\mu + \rho u^\nu u_\mu u^\mu{}_{;\nu} = 0.$$

Using Eq. (12.38) in the first term, and Eq. (12.40) in the last term, which then vanishes, we get

$$(\rho u^\nu)_{;\nu} (-c^2) = 0.$$

Dividing by $-c^2$,

$$(\rho u^\nu)_{;\nu} = 0.$$

Putting this into Eq. (12.37) we find that the first term vanishes,

$$\rho u^\nu u^\mu{}_{;\nu} = 0.$$

Since $\rho \neq 0$ we must have

$$u^{\nu} u^{\mu}{}_{;\nu} = 0. \tag{12.41}$$

This is just the geodesic equation (8.5). Conclusion: *It follows from Einstein's field equations that free particles move along geodesic curves of space time.*

12.9 Constants of motion

When we are going to find the motion of particles moving in gravitational fields by solving the geodesic equation, it is very useful to find quantities that have conserved values during the motion of the particle. Such quantities are called *constants of motion*. They can be found from the geodesic equation. In this connection it is practical to lower the index μ in Eq. (12.41). Hence we write the geodesic equation in the form

$$u^{\nu} u_{\mu;\nu} = 0.$$

The covariant derivatives of the covariant components are given by Eq. (7.25)

$$u^{\nu} \left(u_{\mu,\nu} - u_{\alpha} \Gamma^{\alpha}{}_{\mu\nu}\right) = 0.$$

Consequently

$$u_{\mu,\nu} u^{\nu} = \Gamma^{\alpha}{}_{\mu\nu} u_{\alpha} u^{\nu}.$$

Making use of Eq. (7.8) on the left-hand side we get

$$\frac{d u_{\mu}}{d\tau} = \Gamma^{\alpha}{}_{\mu\nu} u_{\alpha} u^{\nu}.$$

Substituting the expression (7.30) for the Christoffel symbols gives

$$\begin{aligned} \frac{d u_{\mu}}{d\tau} &= \frac{1}{2} g^{\alpha\beta} \left(g_{\beta\mu,\nu} + g_{\beta\nu,\mu} - g_{\mu\nu,\beta}\right) u_{\alpha} u^{\nu} \\ &= \frac{1}{2} \left(g_{\beta\mu,\nu} + g_{\beta\nu,\mu} - g_{\mu\nu,\beta}\right) g^{\alpha\beta} u_{\alpha} u^{\nu}. \end{aligned} \tag{12.42}$$

According to Eq. (5.77), $g^{\alpha\beta} u_{\alpha} = u^{\beta}$, so Eq. (12.42) takes the form

$$\frac{d u_{\mu}}{d\tau} = \frac{1}{2} \left(g_{\beta\mu,\nu} + g_{\beta\nu,\mu} - g_{\nu\mu,\beta}\right) u^{\beta} u^{\nu}. \tag{12.43}$$

Note that $g_{\beta\mu,\nu} - g_{\nu\mu,\beta}$ is antisymmetric in β and ν, meaning that this quantity changes sign if β and ν are exchanged, $g_{\beta\mu,\nu} - g_{\nu\mu,\beta} = -(g_{\nu\mu,\beta} - g_{\beta\mu,\nu})$.

Furthermore, $u^\beta u^\nu$ is symmetric under exchange of β and ν, i.e. $u^\beta u^\nu = u^\nu u^\beta$. Multiplying the first and the third term inside the parenthesis in Eq. (12.43) by the factor $u^\beta u^\nu$ outside the parenthesis, we find

$$\left(g_{\beta\mu,\nu} - g_{\nu\mu,\beta}\right) u^\beta u^\nu = g_{\beta\mu,\nu} u^\beta u^\nu - g_{\nu\mu,\beta} u^\beta u^\nu .$$

Exchanging the summation indices β and ν in the first term, i.e. letting $\beta \rightarrow \nu$ and $\nu \rightarrow \beta$, we get

$$\begin{aligned}\left(g_{\beta\mu,\nu} - g_{\nu\mu,\beta}\right) u^\beta u^\nu &= g_{\nu\mu,\beta} u^\nu u^\beta - g_{\nu\mu,\beta} u^\beta u^\nu \\ &= g_{\nu\mu,\beta}\left(u^\nu u^\beta - u^\beta u^\nu\right).\end{aligned}$$

Due to the symmetry of $u^\beta u^\nu$ in β and ν the terms in the parenthesis cancel each other, and we obtain

$$\left(g_{\beta\mu,\nu} - g_{\nu\mu,\beta}\right) u^\beta u^\nu = 0.$$

The only property we have used in deducing this result is symmetry properties of the factors $g_{\beta\mu,\nu} - g_{\nu\mu,\beta}$ and $u^\beta u^\nu$. Thus we have proved the following useful result: *Summation over the indices of the factors in a product between an antisymmetric and a symmetric quantity always gives zero.*

This result implies that only the second term in the parenthesis of Eq. (12.43) contributes to the summation over β and ν, ultimately providing

$$\frac{du_\mu}{d\tau} = \frac{1}{2} g_{\beta\nu,\mu} u^\beta u^\nu . \tag{12.44}$$

This is the form of the geodesic equation we have sought for. It shows that if $g_{\beta\nu,\mu} = 0$, then $du_\mu/d\tau = 0$, which implies that u_μ is then a constant of motion. Thus we have proved the following result: *If all the components of the metric tensor are independent of a coordinate x^μ, then the covariant component u_μ of the four velocity is a constant along the trajectory of any freely moving particle.*

12.10 The conceptual structure of the general theory of relativity

We are now, finally, able to draw conceptual lines leading to the general theory of relativity. The diagram, shown in Fig. 12.5, does not show logical derivations. It may be said to suggest decisive lines of motivation and development.

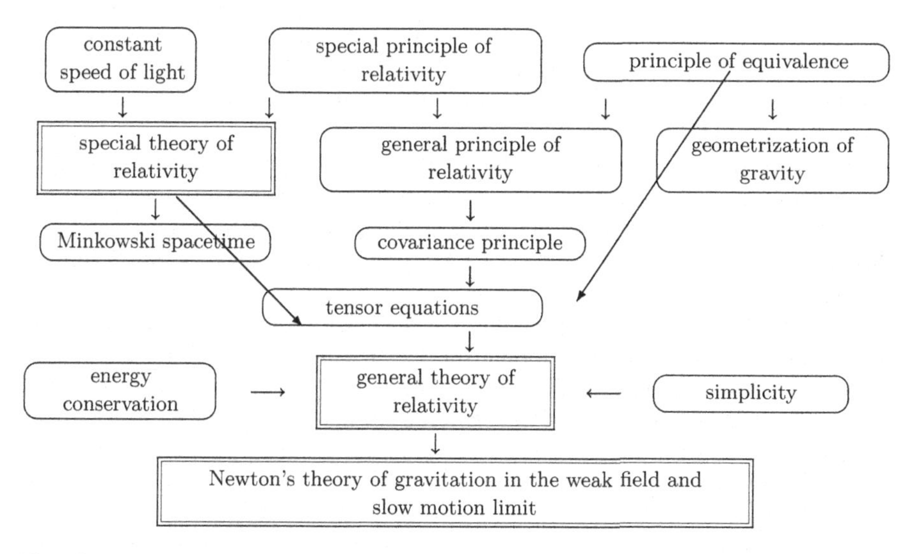

Fig. 12.5 Conceptual structure

12.11 General relativity versus Newton's theory of gravitation

You may have heard sentences such as 'Einstein has proved that Newton was wrong', implying that Newton's theory of gravity was proved to be false when the general theory of relativity was accepted as correct. One also sometimes hears that 'the general theory of relativity generalized Newton's theory of gravity'.

None of these sentences can withstand criticism based on fundamental insights reached at in contemporary philosophy of science.

Consider the conceptual structure of Newton's theory of gravity on the one hand, and the general theory of relativity on the other hand. The essential concept of Newton's theory is that of force. And this concept does not even exist in Einstein's theory understood as a theory of space, time and gravitation. The general theory of relativity is, right from its fundamental principles to its mathematical formulation, a totally new conception and invention. It does not generalize Newton's theory. It replaces it!

But clearly there *is* a connection between Newtons's theory of gravitation and Einstein's theory of relativity. They are both physical theories, and the region of applicability of Einstein's theory encompasses the region of applicability of Newton's theory. Furthermore, the fundamental principles of general relativity is formulated in terms of concepts existing in Newton's theory, such as space, time, mass, motion and soforth. However, the new concept of *curved spacetime* was introduced in the general theory. This made it possible to construct a theory in which concepts and priciples that are independent of each other in Newton's theory, are intimately related in Einstein's theory. For example space and time are united into a four-dimensional spacetime, energy and mass are equivalent, the equation of continuity (energy conservation) and the equation of motion of neutral matter (momentum conservation) follows as a consequence of the field equations in Einstein's theory. This means that Einstein's theory is conceptually more economical than Newton's theory.

When we say that Einstein's theory has a larger region of applicability than Newton's theory, we mean that some phenomena that are correctly described by Einstein's theory, are either not described at all by Newton's theory, or Newton's theory predicts wrong results for them. One example is that time goes slower further down in a gravitational field.

Conceptual simplicity and generality are obtained at the cost of introducing a more complicated mathematical formalism. The distance, as regards the amount of calculations, from the formulation of the general principles, to the predictions of physical phenomena, is greater in Einstein's theory than in Newton's.

Also the level of abstraction is greater in Einstein's theory than in Newton's. Einstein's theory is expressed in terms of a formalism—the tensor formulation of differential geometry—that was developed for geometrical, rather than physical purposes. The physical meaning of the formalism within Einstein's theory is established through our *physical interpretation* of the theory. For example, four-dimensional wace, which is a basic geometrical concept in general relativity, is interpreted as the spacetime of the universe we live in. Time-like geodesic curves are interpreted as paths of free particles. Interpreting the formalism physically, we can extract observational predictions from the equations. The predictions of the theories can then be compared.

Comparing Galilean kinematics and Newton's theory of gravity with Einstein's general theory of relativity, one finds:

1. Einstein's theory is richer (more general) than Newton's, i.e. it has a wider range of applicability than Newton's theory.
2. In the limit of very weak fields and low velocities, the predictions of Newton's and Einstein's theories are *practically* identical.

In those cases where the measurable predictions of the two theories are different, observations have agreed with general relativity. This does not mean that Newton's theory is wrong. It sets, however, a limit for the range of applicability of Newton's theory.

Today the general theory of relativity is established as a theory of great beauty, and a wide field of applications. During the latest twenty years, thousands of research articles have been published, exploring the properties and consequences of general relativity. Yet, at the very frontier of today's physics research is the effort to construct a new unified theory, encompassing both gravitational phenomena, quantum phenomena, and the fundamental forces. We must expect that a new conceptual framework has to be constructed. And there will probably continue to be an increasingly long way, mathematically, from the basic principles of the theory to its physical predictions. The bonus will be great. We shall reach new depths in our comprehension of our universe. Still, the insights Einstein gave us will never fade.

12.12 Epistemological comment

Is general relativity verified? Is it falsified? Neither. Is it verifiable? Is it falsifiable? Logically (not pragmatically) speaking: Neither.

From a terminological point of view it must immediately be added that 'verified' is taken to be synonymous with 'shown or proven to be true', and if a *precisely formulated* proposition p is true, then, according to our concept of truth, it cannot possibly be false, and further research is therefore pointless. A corresponding terminology is presupposed in relation to the term 'falsified'. If something is false, not only unanimously considered to be false, no *possibility* of truth is present.

It is often held that theories in mathematical physics, for instance general relativity as a set of propositions, p, are falsifiable, but not verifiable. If p in these cases were falsifiable through research, 'p is false' would have to be a conclusion, formally correctly derived from true premises. If one or more premises are conjectural, 'p is false' is only hypothetical.

Theories in mathematical physics are very different from generalizations from observations such as 'all ravens are black'. Experimental setups are in modern mathematical physics immensely complex. The relevance of a concrete, dateable experiment depends upon the adequacy of the experimental design. The assertion that a particular setup is adequate may be spelled out in a series of propositions; $q, r, s, \ldots$, connected with 'and'. The conclusion that p is false depends upon the truth of a series of general propositions, many of them called 'laws of nature', for instance mechanical and optical laws presupposed valid when using a certain set of machinery and apparatus in general. There is in short no definite end of the series $q, r, s, \ldots$, required as true premises, and even if there were, many members of the series clearly are propositions which are unverifiable. They are not all shown or proven to be true. A whole paradigm of practice is involved.

The conclusion 'p is false' is therefore in the case of p being a theory of mathematical physics, unfalsifiable. Or, we do not see any surplus of good reasons or good consequences from a decision to declare p to have been falsified.

There is nothing regrettable in this lack of falsification. Research involves cases of confimation and disconfirmation, all more or less open to revision. Sometimes the series of kinds of disconfirmations are of such a considerable weight that it is absurd to continue experimenting. The theory is with good reason *abandoned*, unfalsified.

Einstein admits of wonder. The theories of modern mathematical physics are full of wonder. Nothing is verified, nothing is falsified. Everything is in a process of change and improvisation. Lucky are those who are in the middle of the turmoil!

Chapter 13
Some applications of the general theory of relativity

The first eleven chapters of our text were devoted to the development of the mathematical structure of Einstein's theory of relativity. In chapter 12 we discussed the physical principles of the theory. But there is a third region of inquiry which the reader may want to enter: the multitude of applications of the general theory. Applied to the world of stars and galaxies, to our universe at large, new insights are obtained, and new possibilities of phenomena and objects that we may possibly discover observationally, are revealed. It is, for instance, a consequence of the theory that there may exist black holes somewhere.

Unfortunately, the mathematical derivations of the many astonishing and extremely interesting consequences of the theory, are no less difficult and complex than what the reader has been through so far. However, by means of a sufficient number of small steps we can reach what we want. To give you a choice in how to read this chapter (and the next), we have chosen to place the detailed calculations of the components of the Ricci curvature tensor for the present applications, in Appendices B and C.

Einstein devised, however, a method to extract some essential consequences of the general theory without having to solve the field equations. One can start by analyzing physical phenomena in the ficticious gravitational fields that appear in accelerated and rotating reference frames, and then by an application of the principle of equivalence, deduce that the same effects that one found would also take place in permanent gravitational fields associated with massive bodies. The calculations are simpler in such applications of the theory than in those that involve the field equations. In the next two sections we shall demonstrate the power of this method, by deducing an expression for a relativistic effect called the gravitational time dilation.

Of course this effect can also be deduced more rigorously for permanent gravitational fields. In Sect. (13.4) the expression for the gravitational time dilation shall be deduced directly from the solution of Einstein's field equations outside a massive, spherical body.

Ø. Grøn and A. Næss, *Einstein's Theory: A Rigorous Introduction for the Mathematically Untrained*, DOI 10.1007/978-1-4614-0706-5_13,

13.1 Rotating reference frame

Mathematically, the simplest and most interesting application of Einstein's 'equivalence principle method' is based upon an analysis of clocks in a rotating reference frame. In the present section we shall become familiar with such a frame, which may be thought of as a merry-go-round equipped with measuring rods and clocks.

We start by introducing an inertial reference frame I in flat spacetime, and consider clocks on a plane, circular disk. We introduce plane polar coordinates r' and θ' in I, and a time coordinate t'. The time t' is measured on clocks at rest in I. With these coordinates the infinitesimal radial distance is simply equal to the radial coordinate differential dr', and the distance along a circle with radius r' about the axis is $r'd\theta'$ (from the definition of θ' as measured in radians, see Sect. 4.1). Thus the line element takes the form

$$ds^2 = dr'^2 + r'^2 d\theta'^2 - c^2 dt'^2. \tag{13.1}$$

Then we introduce a second reference frame, R, which rotates steadily. It may be thought of as a merry-go-round with the axis of rotation at rest in I. A useful mental exercise is to imagine that you are an observer at rest in R. Then you can think of the experience of being positioned on a rotating merry-go-round of glass just above the inertial disc I, so that you can compare the readings made with the measuring equipment in R and that in I. The measuring equipment in R is represented by a comoving system of plane polar coordinates r and θ in R, and time coordinate t. We choose a coordinate time t that is per definition measured by *coordinate clocks* that are adjusted and synchronized so that they show the same time as the non-rotating (inertial) clocks, i.e. $t = t'$.

We shall now deduce a coordinate transformation between comoving coordinates in I and R. Imagine a radial line representing $\theta = \theta_0$ engraved on the merry-go-round, and a similar line representing $\theta' = \theta'_0$ engraved on the non-rotating plane disc in I. Let P be a point on the line engraved on the rotating disc R of glass. In R this point has a constant angular coordinate $\theta = \theta_0$. In I the radial line on the glass rotates so that its angular coordinate increases as measured in I. The line in I coinsides with that in R at a point of time $t = 0$.

We need to know a quantity called *angular velocity*. Ordinary velocity is a measure of how fast a particle moves along a curve. Angular velocity, on the other hand, is a measure of how fast a body rotates. The angular velocity of R relative to I, ω, is defined as the rate of change of angle of the radial reference line in R as measured in I,

$$\omega \equiv \frac{d\theta'}{dt'}. \tag{13.2}$$

Making use of $t' = t$ and multiplying each side of the equation by dt we get

$$d\theta' = \omega \, dt.$$

Steady rotation means that the angular velocity ω is constant. Integration results in

$$\theta' = \theta'_0 + \omega t,$$

where θ'_0 is a constant of integration. The angular coordinate of the radial reference line in R is constant,

$$\theta = \theta_0.$$

Since $\theta'_0 = \theta_0$ we get

$$\theta' = \theta + \omega t.$$

This is the coordinate transformation of the angular coordinate between R and I.

Using the same radial coordinate in R and I, we can write the coordinate transformation between R and I as follows

$$r' = r, \quad \theta' = \theta + \omega t, \quad \text{and} \quad t' = t.$$

Differentiation gives

$$dr' = dr, \quad d\theta' = d\theta + \omega\, dt, \quad \text{and} \quad dt' = dt.$$

Inserting this into Eq. (13.1), we find

$$\begin{aligned} ds^2 &= dr^2 + r^2 (d\theta + \omega\, dt)^2 - c^2\, dt^2 \\ &= dr^2 + r^2\, d\theta^2 + r^2\omega^2\, dt^2 - c^2\, dt^2 \\ &\quad + 2\, r^2\, \omega\, d\theta\, dt \\ &= dr^2 + r^2\, d\theta^2 - \left(1 - r^2\omega^2/c^2\right) c^2\, dt^2 \\ &\quad + 2\, r^2\, \omega\, d\theta dt. \end{aligned} \tag{13.3}$$

In the following section we shall apply this line element to show that the theory of relativity implies the existence of, and provides a formula for, a sort of time dilation that is different from the velocity dependent time dilation of special relativity.

13.2 The gravitational time dilation

In order to discuss a possible position dependence of the rate of time in the rotating reference frame, we need a position independent reference for the rate of time, i.e. a set of clocks going equally fast irrespective of their position. This is here represented by the rate of time as measured on the coordinate clocks, because they go at the same rate as the non-rotating clocks, and we know that their rate is position independent.

The rate of time in any reference frame is represented by the proper time τ. According to the definition (5.115) a proper time interval $d\tau$ is given in terms of the line element by

$$d\tau^2 = -(1/c^2)\,ds^2. \tag{13.4}$$

Clocks that measure $d\tau$ are called *standard clocks*. They are 'natural clocks' that are not adjusted in any way as they are moved to different positions. A standard clock at a distance r from the axis, has at first been adjusted to go at a correct rate and to show the correct time (that of the non-rotating clocks) while at the axis. Then it has been moved *slowly* (so as not to be slowed down by the special relativistic, velocity dependent time dilation) to its position. When the clock is at rest in the rotating reference frame, its path through spacetime has $dr = d\theta = 0$. In this special case Eqs. (13.3) and (13.4) result in

$$d\tau = \left(1 - r^2\omega^2/c^2\right)^{1/2} dt. \tag{13.5}$$

This formula shows that for a given value of dt the value of $d\tau$ gets less with increasing value of r. Since the coordinate clocks have a position independent rate, that of the clock at the axis, this means that the standard clocks are slower the further they are from the axis. The position dependence of the rate of the standard clocks at rest in R can be measured both by an observer at rest in I and one at rest in R.

As observed from the inertial frame I, a clock at rest in R moves along a circular path. According to the definition (13.2) of angular velocity it passes over an angle $d\theta' = \omega\,dt'$ during a time dt'. The angle is measured in radians (see Sect. 4.1), i.e. arclength divided by radius. Thus, the distance covered by the clock during a time dt' is $d\ell' = r\,d\theta' = r\omega\,dt'$. The velocity of the clock is

$$v = d\ell'/dt' = r\omega. \tag{13.6}$$

This equation says that the velocity of a particle moving along a circular path is equal to the radius of the circle times the angular velocity of the particle.

Using Eq. (13.6) we see that the special relativistic formula for the velocity dependent time dilation, Eq. (5.100) gives just Eq. (13.5). However, as observed from the rotating frame R the clock is at rest. So the explanations of the fact that the clocks further away from the axis are slower, are different in the inertial frame and in the rotating frame.

Einstein's explanation is as follows. An observer in the rotating reference frame experiences an acceleration of gravity directed away from the axis. In this field 'up' is towards the axis, and 'down' away from the axis, since a stone which we drop in the merry-go-round falls away from the axis. And, said Einstein, a gravitational field due to the acceleration or rotation of the reference frame is equivalent, in its action on both material systems, and on the rate of time, to a gravitational field due to masses. This is a consequence of the principle of equivalence. Thus, the *position dependent* rate of standard clocks in a rotating reference frame is interpreted as a gravitational effect. The conclusion is that *the rate of time is slower farther down in a gravitational field*. But as seen from below time goes fast up there.

Note that in the theory of relativity the concepts 'up' and 'below' are defined with respect to the field of acceleration of gravity, experienced locally. This implies, for example, that 'upwards' is oppositely directed in Norway and Australia. A free particle falls downwards, by definition. In the case of the merry-go-round 'downwards' means farther away from the axis, $r = 0$.

The movements of the orchestra-conductor will appear faster than normal to the musicians of an orchestra in an abnormally deep grave. And this, in fact, has the remarkable effect that the orchestra plays at just the correct tempo as seen by the conductor. But the musicians do not age as fast as the conductor.

In accordance with Einstein's interpretation the gravitational time dilation can be expressed in terms of the potential difference between two positions in a gravitational field. In order to calculate the potential of the gravitational field experienced in R, we have to find an expression for the acceleration of gravity in this field. To do this we have to find the acceleration of an observer at rest in R. Such an observer moves along a circular path with constant velocity, given by Eq. (13.6), i.e.

$$\vec{v} = r\,\omega\,\vec{e}_{\hat{\theta}'},$$

where $\vec{e}_{\hat{\theta}'}$ is a tangential unit vector. According to Eq. (4.45) the coordinate basis vector in this direction is $\vec{e}_{\theta'} = r\,\vec{e}_{\hat{\theta}'}$, which gives

$$\vec{v} = \omega\,\vec{e}_{\theta'}.$$

The acceleration is the rate of change of velocity, $\vec{a} = d\vec{v}/dt$. Therefore

$$\vec{a} = \omega\,\frac{d\vec{e}_{\theta'}}{dt'}. \tag{13.7}$$

The rate of change of the basis vector $\vec{e}_{\theta'}$ along a circular path is given by Eq. (6.17) with $dr = 0$. Hence

$$\frac{d\vec{e}_{\theta'}}{dt'} = -r\,\frac{d\theta'}{dt'}\,\vec{e}_r = -r\,\omega\,\vec{e}_r.$$

Inserting this into Eq. (13.7) we get

$$\vec{a} = -r\,\omega^2\,\vec{e}_r. \tag{13.8}$$

This equation shows that the acceleration of observers at rest in R is directed radially towards the centre of the disc. Therefore it is called *centripetal acceleration.*

The acceleration of gravity, $\vec{g}$, felt by an observer in R is equal to his own acceleration, but oppositely directed, i.e.

$$\vec{g} = -\vec{a} = r\,\omega^2\,\vec{e}_r. \tag{13.9}$$

In Newtonian theory it is called *centrifugal acceleration*, and one introduces a force $m\vec{g}$ which is called the *centrifugal force* and is reckoned as a fictive force, not a real force. In general relativity one does not introduce any fictive force, and the acceleration field $\vec{g}$ is reckoned as a genuine gravitational field on line with the gravitational field at the Earth.

We shall now calculate the potential in the gravitational field given by Eq. (13.9). This concept was introduced in Sect. (12.4). The position with zero potential can be chosen freely, and we choose to put the potential equal to zero at the axis of the merry-go-round. Since the gravitational field points radially outwards from the axis, all points at the same distance from the axis has the same value of the potential, equal, by definition, to the work which must be performed to move a unit mass from the axis to a distance r from the axis.

Work is defined as force times distance. Usually the force depends upon the position, so it is not possible to calculate the work just by multiplying a certain value of the force by the distance. The correct method is first to set up an expression for the work needed to move the body an infinitesimal distance, and then integrate over the whole distance.

Imagine that a body is moved with constant velocity from the axis to a point at a distance r from the axis. The force that prevents the body from falling freely outwards with increasing velocity, must act inwards. This force, per unit mass, is $-g$. Thus the change of the potential during a displacement dr is

$$d\phi = -g\,dr. \tag{13.10}$$

Inserting $g = r\,\omega^2$ from Eq. (13.9) into Eq. (13.10) gives

$$d\phi = -\omega^2\,r\,dr.$$

This is the expression for the change of the potential during an infinitesimal radial displacement. Integrating from 0 to r, noting that ω is constant and using the rule (3.33), we find

$$\phi = -\omega^2 \int_0^r r\,dr = -\frac{1}{2}\,r^2\omega^2.$$

This is the gravitational potential in R at a distance r from the axis of rotation.

Substituting 2ϕ for $-r^2\omega^2$ in Eq. (13.5), and noting that the coordinate clocks in R have a position independent rate equal to that of standard clocks at the axis, i.e. $dt = d\tau_0$, we get

$$d\tau = \left(1 + 2\phi/c^2\right)^{1/2} d\tau_0, \tag{13.11}$$

where $d\tau_0$ refers to time as measured on a standard clock at the position with zero potensial. Equation (13.11) is *the equation of the gravitational time dilation.*

There are two conceptions of physical relationships concerning gravitational phenomena, according to the general theory of relativity; the causal and the acausal. According to the causal conception the position dependent rate of time in a gravitational field is an effect due to gravity, i.e. gravity *causes* the gravitational time dilation. The acausal conception is different. Time measures in spacetime depend upon the position of the clock in a gravitational field. At each position there is a definite rate of time; slower or more rapid. This has to do purely with interrelations of time itself with no causality involved. Whether one prefers the causal or the acausal point of view is a matter of taste. The theory does not force any of these conceptions upon us.

13.3 The Schwarzschild solution

Having arrived at Einstein's field equations, we may ask: "Which are the solutions to this set of equations? What are the properties of the spacetimes implied by these equations?" The field equations in the form (11.34) are very general. They can describe all sorts of spacetimes: flat (Minkowski) spacetime, curved spacetime outside massive particles, spacetime inside stars, cosmic spacetimes in which space expands and so forth.

The solutions of the field equations are the ten functions that make up the components of the metric tensor. The field equations are a set of six second order partial differential equations, and generally the ten metric components are functions of all four spacetime coordinates. Six equations can only determine six of the metric functions—fortunately. This leaves us the freedom of choosing a coordinate system appropriate for the spacetime which is to be investigated. Still, the mathematical problem of solving the field equations is often extremely difficult, and only for spacetimes with a high degree of symmetry are we able to solve the equations in terms of elementary functions.

The symmetry of a space means that the properties of the space do not change under a certain motion. Think of spacetime outside a particle, for example. Imagine a spherical surface with centre on the particle. No properties of space change if you move from one point to another arbitrary point on the surface. The metric is independent of the position on the spherical surface. Then we say that this space is *spherically symmetric*. This is the type of spaces we shall consider in this chapter and the next.

A symmetry has important mathematical consequences. It may simplify enormously the task of solving the field equations. In the first place the symmetry suggests the type of coordinate system that one should introduce and reduces the number of unknown metric functions from six, to possibly only one or two, and secondly it implies that the components of the metric tensor are functions not of all four spacetime coordinates, but possibly only of one or two of the coordinates.

In this chapter we shall consider spacetime outside a static (i.e. time independent), spherically symmetric mass distribution. In the case of flat space one would then introduce spherical coordinates, as given in Sect. 6.3. From Eq. (6.22) we get $g_{rr} = \vec{e}_r \cdot \vec{e}_r = 1$, $g_{\theta\theta} = \vec{e}_\theta \cdot \vec{e}_\theta = r^2$, and $g_{\varphi\varphi} = \vec{e}_\varphi \cdot \vec{e}_\varphi = r^2 \sin^2\theta$. Hence, the line element of Euclidean 3-dimensional space, as expressed in spherical coordinates, is

$$d\ell^2 = dr^2 + r^2\, d\theta^2 + r^2 \sin^2\theta\, d\varphi^2. \tag{13.12}$$

The line element of flat spacetime, with these coordinates, is

$$ds^2 = d\ell^2 - c^2\, dt^2,$$

or

$$ds^2 = dr^2 + r^2\, d\theta^2 + r^2 \sin^2\theta\, d\varphi^2 - c^2 dt^2. \tag{13.13}$$

We know that there is a gravitational field outside a mass distribution. Spacetime is curved in such a region. Allowing for this, we make a generalization of the form (13.3) of the line element, writing

$$ds^2 = \mathrm{e}^{\lambda(r)} dr^2 + r^2 d\theta^2 + r^2 \sin^2\theta\, d\varphi^2 - \mathrm{e}^{\nu(r)} c^2 dt^2, \tag{13.14}$$

where $\lambda(r)$ and $\nu(r)$ are functions of r only. The letter $\mathrm{e} = 2.71828\ldots$ in Eq. (13.14) has nothing to do with basis vectors (no arrows!), but denotes the basis of the exponential function (see Ch. 3). Instead of introducing just functions $A(r)$ and $B(r)$ we use the exponential form, because this simplifies the field equations. This is just a trick which turns out to be very convenient.

The form (13.14) of the line element preserves the spherical symmetry, and the time independence. The corresponding components of the metric tensor are

$$\begin{aligned}
g_{rr} &= \mathrm{e}^{\lambda(r)}, \\
g_{\theta\theta} &= r^2, \\
g_{\varphi\varphi} &= r^2 \sin^2\theta, \\
g_{tt} &= -c^2 \mathrm{e}^{\nu(r)}.
\end{aligned} \tag{13.15}$$

In the following we shall leave out the functional dependence on r when we write the exponential functions e^{ν} and e^{λ}, but it is of course still understood that these expressions denote functions of r.

Since we are going to find the geometry of the region outside the mass distribution, we must solve the field equations (11.35) for vacuum, $R_{\mu\nu} = 0$.

Using Eqs. (B.17), (B.18) and (B.22) in Appendix B, the vacuum field equations for the static, spherically symmetric space may be written

$$R_{tt}/\mathrm{e}^{\nu-\lambda} = \frac{\nu''}{2} - \frac{1}{4}\nu'\lambda' + \frac{1}{4}\nu'^2 + \frac{\nu'}{r} = 0, \tag{13.16a}$$

$$R_{rr} = -\frac{\nu''}{2} + \frac{1}{4}\nu'\lambda' - \frac{1}{4}\nu'^2 + \frac{\lambda'}{r} = 0, \tag{13.16b}$$

and

$$R_{\theta\theta} = 1 - \left(1 + \frac{r}{2}\nu' - \frac{r}{2}\lambda'\right) \mathrm{e}^{-\lambda} = 0. \tag{13.16c}$$

These differential equations shall now be solved. Inspecting the equations we see that a simple equation is obtained if the first two of Eqs. (13.16) are added, namely

$$\frac{\nu' + \lambda'}{r} = 0.$$

Consequently

$$\nu' + \lambda' = (\nu + \lambda)' = 0.$$

Since the derivative of $\nu + \lambda$ vanishes, this quantity must be equal to a constant, $\nu + \lambda = b$, which is determined from the condition that spacetime is flat infinitely far from the mass distribution. Then the line element (13.14) is reduced to the form (13.13) as $r \to \infty$. Since $e^0 = 1$, this means that ν and λ must vanish as $r \to \infty$. Consequently $b = 0$ and

$$\lambda = -\nu. \tag{13.17}$$

Inserting this in Eq. (13.16c), we get

$$1 - \left(1 + r\,\nu'\right) \mathrm{e}^{\nu} = 0,$$

or

$$\left(1 + r\,\nu'\right) \mathrm{e}^{\nu} = 1. \tag{13.18}$$

According to the product rule for differentiation, and the chain rule,

$$(r\,\mathrm{e}^{\nu})' = \mathrm{e}^{\nu} + r\,(\mathrm{e}^{\nu})' = \mathrm{e}^{\nu} + r\,\mathrm{e}^{\nu}\,\nu' = \left(1 + r\,\nu'\right) \mathrm{e}^{\nu}.$$

Hence Eq. (13.18) can be written as

$$(r\,\mathrm{e}^{\nu})' = 1. \tag{13.19}$$

Since integration is 'antiderivation', the integral of the left-hand side is equal to $r\,\mathrm{e}^{\nu}$. Integrating the right-hand side by means of Eq. (3.33) with $p = 0$ (because $x^0 = 1$), we get the answer $r + C$, where C is a constant. Integration of Eq. (13.19) thus gives

$$r\,\mathrm{e}^{\nu} = r + C,$$

or

$$\mathrm{e}^{\nu} = 1 + \frac{C}{r}. \tag{13.20}$$

The constant C will be determined by considering the Newtonian limit of general relativity (see Sect. 12.6). In this limit the time-time component of the metric tensor is written as

$$g_{tt} = -c^2\,(1 - h_{tt})\,. \tag{13.21}$$

The acceleration of gravity is given in terms of the derivative of the metric tensor by Eq. (12.21). In the present case this equation is reduced to

$$a = \frac{c^2}{2}\,h'_{tt}. \tag{13.22}$$

According to Eq. (12.7) (i.e. Newton's law of gravitation) the acceleration of gravity at a distance r outside a spherical body with mass M, is

$$a = -\frac{GM}{r^2}. \tag{13.23}$$

The minus sign means that the acceleration points in the direction of decreasing r. Putting the right-hand sides of Eqs. (13.22) and (13.23) equal to each other, and solving with respect to h'_{tt}, gives

$$h_{tt'} = -\frac{2GM}{c^2}\frac{1}{r^2}.$$

We see that h'_{tt} is equal to a constant factor times $1/r^2$. Integrating $1/r^2$ by means of Eq. (3.33), this time with $p = -2$, (note that $1/r^2 = r^{-2}$ and $r^{-1} = 1/r$) leads to

$$h_{tt} = \frac{2GM}{c^2 r} + K. \tag{13.24}$$

The integration constant K is determined by the condition that spacetime is flat infinitely far from the mass M. This implies that $h_{tt} \rightarrow 0$ as $r \rightarrow \infty$, which is possible only if $K = 0$. Inserting Eq. (13.24) with $K = 0$ into Eq. (13.21) results in

$$g_{tt} = -c^2\left(1 - \frac{2GM}{c^2 r}\right)$$

which shows that the constant C in Eq. (13.20) is $C = -2GM/c^2$. According to Eqs. (13.15), (13.17) and (13.20)

$$g_{rr} = \mathrm{e}^{\lambda} = \mathrm{e}^{-\nu} = \frac{1}{e_\nu} = \frac{-c^2}{g_{tt}} = \frac{1}{1 - \frac{2GM}{c^2 r}}. \tag{13.25}$$

Equation (13.25) provides the solution to Einstein's vacuum field equations outside a spherical body at rest. Inserting these expressions in the line element (13.14) we get

$$\begin{aligned} ds^2 = {} & \frac{dr^2}{1 - \frac{2GM}{c^2 r}} + r^2\,d\theta^2 + r^2\sin^2\theta\,d\varphi^2 \\ & - \left(1 - \frac{2GM}{c^2 r}\right) c^2\,dt^2. \end{aligned} \tag{13.26}$$

This is the famous *Schwarzschild solution* of Einstein's field equations.

The Schwarzschild solution constitutes an expression of the metric of spacetime, namely a *relativistically valid* metric of a static, spherically symmetric spacetime, and thereby also of the gravitational field of the Sun. Since the masses of the planets are much less than the mass of the Sun, the Schwarzschild metric represents a good approximation of the spacetime near the Sun, and is adequate to describe the gravitational effects in the Solar system.

The quantity

$$r_S \equiv \frac{2GM}{c^2}, \tag{13.27}$$

is a *length* which characterizes the mass of a body. It is called the *Schwarzschild radius* of the body. Inserting values for Newton's gravitational constant, the velocity of light and the mass of the Sun, we get r_S=3 km for the Sun. Thus the Schwarzschild radius of a body with mass M is $r_S=(M/M_\odot)3$ km, where $M_\odot$ is the mass of the Sun. The Earth, for example has a Schwarzschild radius equal to 1 cm. A man with mass 100 kg has a Schwarzschild radius equal to 1.5×10^{-25} m, which is very much less that the radius of an atomic nucleus. The physical significance of the Schwarzschild radius will become clear below, in connection with our discussion of black holes.

Inserting the expression (13.27) for the Schwarzschild radius into Eq. (13.26), the Schwarzschild line element takes the form

$$ds^2 = \frac{dr^2}{1-\frac{r_S}{r}} + r^2\,d\theta^2 + r^2\,\sin^2\theta\,d\varphi^2 - \left(1-\frac{r_S}{r}\right)c^2\,dt^2. \tag{13.28}$$

At a large distance from a body compared to its Schwarzschild radius, $r \gg r_S$, the line element is approximately equal to Eq. (13.13) which represents flat spacetime as expressed in spherical coordinates. At the surface of the Earth, for example, $r_S/r \approx 10^{-8}$, and at the surface of the Sun $r_S/r \approx 10^{-5}$. This means that the gravitational field is weak in the whole of the Solar system. This is the reason for the success of Newton's theory of gravitation as applied to bodies in the Solar system.

The components of the metric tensor are

$$g_{rr} = \left(1-\frac{r_S}{r}\right)^{-1}, \tag{13.29a}$$

$$g_{\theta\theta} = r^2, \tag{13.29b}$$

$$g_{\varphi\varphi} = r^2\,\sin^2 \tag{13.29c}$$

and

$$g_{tt} = -\left(1-\frac{r_S}{r}\right)c^2. \tag{13.29d}$$

This is the Schwarzschild metric.

13.4 The Pound–Rebka experiment

It is a demonstration of the eminent ability of experimental physicists that they have managed to measure the gravitational time dilation on the surface of the Earth in an experiment with maximal extension 22.5 metres. It is remarkable that this phenomenon was predicted by Einstein 45 years before anybody was able to measure it. Let us follow Einstein's simple explanation.

Einstein argued that light waves (moving freely) can neither vanish nor be created between the emitter and the receiver. So, in a spacetime where nothing changes with time, the same number of waves per second must arrive at the receiver as are sent out from the emitter. But, said Einstein, this conclusion is correct only if a tacit assumption is accepted, namely that the clocks at the emitter and the receiver go equally fast.

Consider standard clocks spatially *at rest* in the Schwarzschild spacetime. For these clocks $dr = d\theta = d\varphi = 0$, and Eq. (13.28) is reduced to

$$ds^2 = -\left(1 - \frac{r_S}{r}\right) c^2\, dt. \tag{13.30}$$

According to Eq. (5.115) the proper time interval measured on the standard clocks is given by

$$d\tau = \frac{\sqrt{-ds^2}}{c}.$$

Combining with Eq. (13.30) leads to

$$d\tau = \sqrt{1 - \frac{r_S}{r}}\, dt. \tag{13.31}$$

Since the Schwarzschild metric is static, the coordinate clocks, showing the time t, must be synchronized and adjusted so that they go equally fast independently of their position. From Eq. (13.31) we can therefore conclude that the less r is, the slower the standard clocks go. In other words time goes slower farther down in a gravitational field.

Let us now return to Einstein's argument. Assume that light is emitted from a height $h > 0$ to the floor, at $h = 0$. Since the standard clocks at the floor are slower than those at the height h, one measures that more light waves arrive at the receiver per second than are emittet at the height h. Einstein therefore concluded that one would measure a frequency increase at the receiver; a blue shift of the light.

Pound and Rebka performed the experiment in 1960 with $h = 22.5$ metres.

If r_S and R are the Schwarzschild radius and radius of the Earth, respectively, the proper time interval $d\tau_R$ at the receiver and $d\tau_E$ at the emitter, are given by

$$d\tau_R = \sqrt{1 - \frac{r_S}{R}}\, dt$$

and

$$d\tau_E = \sqrt{1 - \frac{r_S}{R+h}}\, dt.$$

Since $r_S = 1$ cm and $R = 6300$ km we have $x \equiv r_S/R = 1.5 \times 10^{-9}$. We can therefore, with sufficient accuracy, use the approximation $\sqrt{1-x} \approx 1 - x/2$, obtained by retaining the first two terms of the MacLaurin series, Eq. (2.84), of the function $f(x) = \sqrt{1-x}$. Accordingly

$$d\tau_R \approx \left(1 - \frac{1}{2}\frac{r_S}{R}\right) dt, \tag{13.32a}$$

$$d\tau_E \approx \left(1 - \frac{1}{2}\frac{r_S}{R+h}\right) dt, \tag{13.32b}$$

where dt is the period of the light as measured with the coordinate clocks, and $d\tau_R$ and $d\tau_E$ are the periods as measured with standard clocks at the receiver and the emitter, respecively.

The 'blue shift' of the light is denoted by z and defined by

$$z \equiv \frac{d\tau_E - d\tau_R}{d\tau_R}.$$

Inserting the expressions (13.32), and approximating the denominator by dt (which is permissible since the numerator is very small, and the denominator is very close to dt), we get,

$$z \approx \frac{r_S}{2}\left(\frac{1}{R} - \frac{1}{R+h}\right) = \frac{r_S}{2}\left(\frac{R+h-R}{R(R+h)}\right).$$

Ignoring h in the denominator in the last expression, we finally arrive at

$$z \approx \frac{1}{2}\frac{r_S h}{R^2}.$$

Inserting the numerical quantities we get the following prediction $z = 2.5 \times 10^{-15}$. The measurements showed agreement with this prediction.

13.5 The Hafele–Keating experiment

The special relativistic *velocity* dependent time dilation and the general relativistic (gravitational) *position* dependent time dilation are physically different. But sometimes they occur together, for example when one travels in an airplane.

In the Hafele–Keating experiment both types of time dilations were demonstrated by travelling around the Earth with caesium clocks in Jumbo-jets, comparing the travelling time as measured with these clocks, with that measured on a similar clock positioned at the airport. We shall calculate what general relativity predicts for this experiment.

According to Eq. (5.115) the proper time, i.e. the time shown by an ordinary clock, is given in terms of the line element by

$$d\tau = \frac{\sqrt{-ds^2}}{c}.$$

Inserting the general expression for the line element from Eq. (5.120) leads to

$$d\tau = \frac{1}{c}\sqrt{-g_{\mu\nu}dx^{\mu}dx^{\nu}}.$$

Specializing to the diagonal metric of orthogonal coordinate systems, we get

$$d\tau = \frac{1}{c}\sqrt{(-g_{\mu\mu})(dx^{\mu})^2}.$$

The summation over μ goes over the four values 1, 2, 3, and 4. In this sum with four terms we write first the term with $\mu = 4$, and then use Einstein's summation convention with $\mu = i$ for the remaining three terms. The fourth coordinate is $x^4 = c\,t$ (see Sect. 12.6). Thus

$$\begin{aligned} d\tau &= \frac{1}{c}\sqrt{-g_{44}c^2dt^2 - g_{ii}(dx^i)^2/c} \\ &= \frac{1}{c}\sqrt{-g_{44}\,c^2dt^2 - g_{ii}\left(\frac{dx^i}{dt}\right)^2 dt^2}. \end{aligned}$$

The components of the velocity of a clock are

$$v^i = \frac{dx^i}{dt},$$

so

$$d\tau = \frac{1}{c}\sqrt{-g_{44}\,c^2\,dt^2 - g_{ii}\,(v^i)^2\,dt^2}.$$

Putting the common factor dt^2 outside the square root, it comes out as dt, and similarly, putting the factor of $1/c$ inside the square root, it becomes $1/c^2$. We then have

$$d\tau = \sqrt{-g_{44} - g_{ii}\left(\frac{v^i}{c}\right)^2}\,dt. \tag{13.33}$$

According to Eq. (4.7) the square of the velocity is the scalar product of the velocity vector by itself

$$v^2 = \vec{v} \cdot \vec{v}.$$

From Eq. (1.26) we get

$$v^2 = g_{ij} v^i v^j.$$

In the case of a diagonal metric this is reduced to

$$v^2 = g_{ii} (v^i)^2.$$

Inserting this into Eq. (13.33) leads to

$$d\tau = \sqrt{-g_{44} - \frac{v^2}{c^2}}\, dt. \tag{13.34}$$

We now assume that spacetime outside the Earth is adequately modelled by the Schwarzschild solution of Einstein's field equations, Eq. (13.28). Thus [compare Eqs. (12.14) and (12.16)]

$$g_{44} = g_{tt}/c^2 = -\left(1 - \frac{r_S}{r}\right), \tag{13.35}$$

where r_S is the Schwarzschild radius of the Earth, defined in Eq. (13.27) with M representing the mass of the Earth in the present case. Substituting Eq. (13.35) into Eq. (13.34) gives

$$d\tau = \sqrt{1 - \frac{r_S}{r} - \frac{v^2}{c^2}}\, dt. \tag{13.36}$$

This is an expression of the time dilation that includes both the velocity dependent time dilation of special relativity and the position dependent (gravitational) time dilation of general relativity. In the case of a vanishing gravitational field, $r_S = 0$, and Eq. (5.117) is recovered, and in the case of a clock at rest, $v = 0$, Eq. (13.31) is recovered.

In the Hafele–Keating experiment one clock was at rest at the airport and one was travelling around the Earth, once westwards and once eastwards. Let Ω be the angular velocity of the Earth due to its daily rotation, and R the radius of the Earth. A point on the Equator moves with angular velocity Ω along a circle with radius R. According to Eq. (13.2) a satellite at rest relative to the surface of the Earth at a height h above the surface, has a velocity $(R + h)\,\Omega$. Using the Galilean law of velocity addition, valid for velocities small compared to the velocity of light, an airplane flying with velocity u relative to the surface of the Earth at a height h, has a velocity

$$v = (R + h)\,\Omega + u.$$

Inserting this, and the expression (13.27), with M equal to the mass of the Earth and $r = R + h$, into Eq. (13.36), we get

$$d\tau = \sqrt{1 - \frac{GM}{(R+h)\,c^2} - \frac{[(R+h)\,\Omega + u]^2}{c^2}}\,dt.$$

The clock at the airport has $h = u = 0$, which gives

$$d\tau_0 = \sqrt{1 - \frac{G\,M}{R\,c^2} - \frac{R^2\,\Omega^2}{c^2}}\,dt.$$

In order to simplify the calculation we[1] now assume that the airplanes travel with constant velocity and height just above the Equator. Choosing positive u in the same direction that the Earth rotates in, $u > 0$ for a clock travelling eastwards and $u < 0$ for a clock travelling westwards. The difference between the travelling times as measured by the air-borne clocks, and as measured by the clock at the airport is

$$\Delta\tau - \Delta\tau_0 = \left(\frac{\Delta\tau}{\Delta\tau_0} - 1\right)\Delta\tau_0$$
$$= \left(\frac{\sqrt{1 - \frac{GM}{(R+h)\,c^2} - \frac{[(R+h)\,\Omega+u]^2}{c^2}}}{\sqrt{1 - \frac{GM}{R\,c^2} - \frac{R^2\,\Omega^2}{c^2}}} - 1\right)\Delta\tau_0.$$

The travelling time is about 24 hours, i.e. $\Delta\tau_0 = 1.2 \times 10^5$ seconds. The predicted time differences are,

$$(\Delta\tau - \Delta\tau_0)_{\mathrm{eastwards}} = -1.2 \times 10^{-7}\ \mathrm{s}$$

and

$$(\Delta\tau - \Delta\tau_0)_{\mathrm{westwards}} = 2.5 \times 10^{-7}\ \mathrm{s},$$

for travelling eastwards and westwards, respectively. These predictions were confirmed by the measurements with about 20% accuracy.

[1] Actually, Hafele and Keating integrated the equations numerically along the travelling routes, but the result we obtain by our analytical calculation is sufficiently accurate to agree with the results of the numerical calculations within the measuring uncertainty.

13.6 Mercury's perihelion precession

The planets move along elliptic orbits around the Sun. However the ellipses are not exactly closed. The point closest to the Sun gets a small displacement for each round. The closest point to the Sun is called perihelion, and the mentioned displacement is called the precession of the perihelion.

We shall consider the motion of the innermost planet Mercury. For this planet the precession is 532″per cen-tury = 532 seconds of arc per century. Of this 489″ per century could be accounted for by Newton's theory of gravitation. It is an effect of the other planets upon Mercury. However 43″ per century could not be accounted for by Newton's theory.

Mercury moves along a geodesic curve in the Schwarzschild spacetime outside the Sun. We choose the orientation of the coordinate system so that Mercury moves in the equatorial plane, $\theta = \pi/2$. The spacetime is static and spherically symmetric. Thus, the metric is independent of the coordinates t and φ. It follows from the geodesic equation in the form (12.44) that u_t and u_φ are constants of motion of Mercury.

The most efficient way of solving the geodesic equation, is to insert the constants of motion into the four velocity identity (12.38), which in the present case takes the form

$$u^r u_r + u^\varphi u_\varphi + u^t u_t = -c^2. \tag{13.37}$$

Here $u^r = dr/d\tau$. Since the metric is diagonal, we get from Eqs. (5.75) and (5.76),

$$u_r = g_{rr} u^r, \tag{13.38a}$$

$$u^\varphi = g^{\varphi\varphi} u_\varphi = u_\varphi / g_{\varphi\varphi}, \tag{13.38b}$$

and

$$u^t = u_t / g_{tt}. \tag{13.38c}$$

Inserting this into Eq. (13.37) leads to

$$g_{rr}\left(\frac{dr}{d\tau}\right)^2 + \frac{u_\varphi^2}{g_{\varphi\varphi}} + \frac{u_t^2}{g_{tt}} = -c^2.$$

Substituting the components of the Schwarzschild metric (13.29), we get

$$\frac{(dr/d\tau)^2}{1 - r_S/r} + \frac{u_\varphi^2}{r^2} - \frac{u_t^2}{1 - r_S/r} = -c^2,$$

or

$$\frac{(dr/d\tau)^2}{1 - r_S/r} = \frac{u_t^2}{1 - r_S/r} - \left(c^2 + \frac{u_\varphi^2}{r^2}\right).$$

Multiplying by $1 - r_S/r$ leads to

$$\left(\frac{dr}{d\tau}\right)^2 = u_t^2 - \left(1 - \frac{r_S}{r}\right)\left(c^2 + \frac{u_\varphi^2}{r^2}\right). \tag{13.39}$$

In this equation the radial coordinate of Mercury is to be thought of as a function of time. The rather complicated differential equation is somewhat simplified if we express it in terms of the function $y = 1/r$ instead of r. Writing $r = 1/y = y^{-1}$ and differentiating, we find

$$\frac{dr}{d\tau} = \frac{d(y^{-1})}{d\tau} = -y^{-2}\frac{dy}{d\tau} = -\frac{1}{y^2}\frac{dy}{d\tau}.$$

Inserting this into Eq. (13.39) we get

$$\frac{1}{y^4}\left(\frac{dy}{d\tau}\right)^2 = u_t^2 - (1 - r_S\, y)\left(c^2 + u_\varphi^2\, y^2\right). \tag{13.40}$$

This equation of motion can be solved to find y and thereby r as a function of time. However, what interests us here is the shape of the trajectory followed by Mercury. We would like to know r as a function of the angle φ rather than as a function of time. Therefore we introduce φ as a variable instead of τ. The connection between φ and τ is given by Eq. (13.38b). Noting that $u^\varphi = d\varphi/d\tau$ and $g_{\varphi\varphi} = r^2$ in the equatorial plane, we get

$$\frac{d\varphi}{d\tau} = g^{\varphi\varphi}u_\varphi = \frac{u_\varphi}{g_{\varphi\varphi}} = \frac{u_\varphi}{r^2} = u_\varphi\, y^2.$$

Thus, differentiation with respect to φ and differentiation with respect to τ is connected by

$$\frac{d}{d\tau} = u_\varphi\, y^2\frac{d}{d\varphi}$$

or

$$\frac{dy}{d\tau} = u_\varphi\, y^2\frac{dy}{d\varphi}.$$

Inserting this into the left-hand side of Eq. (13.40), and multiplying out the right hand side, we obtain

$$u_\varphi^2\left(\frac{dy}{d\varphi}\right)^2 = u_t^2 - c^2 + c^2 r_S\, y - u_\varphi^2\, y^2 + u_\varphi^2\, r_S\, y^3.$$

This equation can be simplified by differentiating each term,

$$2\,u_\varphi^2\,\frac{dy}{d\varphi}\frac{d^2y}{d\varphi^2} = c^2\, r_S\frac{dy}{d\varphi} - 2\,u_\varphi^2\, y\,\frac{dy}{d\varphi} + 3\,u_\varphi^2\, r_S\, y^2\,\frac{dy}{d\varphi}.$$

Dividing by $2\,u_\varphi^2\,(dy/d\varphi)$ leads to

$$\frac{d^2y}{d\varphi^2} = \frac{c^2 r_S}{2\,u_\varphi^2} - y + \frac{3r_S}{2}\,y^2.$$

Adding y to each side we obtain the usual form of the equation of the trajectory of Mercury

$$\frac{d^2y}{d\varphi^2} + y = a + by^2, \quad a = \frac{c^2 r_S}{2\,u_\varphi^2}, \quad \text{and} \quad b = \frac{3}{2} r_S. \tag{13.41}$$

From observations we know that the trajectory of Mercury is a nearly circular ellipse. Thus we are not interested in finding the general solution of Eq. (13.41). We want to find a nearly circular solution. Putting $d^2y/d\varphi^2 = 0$ in Eq. (13.41), we find that it has a solution representing a circle with inverse radius y_0 given by

$$y_0 = a + b\,y_0^2. \tag{13.42}$$

With a small deviation from circular motion, the value of y is changed from y_0 by a small amount, which we denote by y_1, i.e. $y_1 \ll y_0$. Inserting $y = y_0 + y_1$ into Eq. (13.41) we get

$$\frac{d^2y_1}{d\varphi^2} + y_0 + y_1 = a + b\,(y_0 + y_1)^2 = a + b\,y_0^2 + 2\,b\,y_0\,y_1 + b\,y_1^2.$$

Subtracting Eq. (13.42) gives

$$\frac{d^2y_1}{d\varphi^2} + y_1 = 2\,b\,y_0\,y_1 + b\,y_1^2.$$

Since $y_1 \ll y_0$, we ignore the last term, and arrive at the equation

$$\frac{d^2y_1}{d\varphi^2} + y_1 \approx 2\,b\,y_0\,y_1.$$

Subtracting $2\,b\,y_0\,y_1$ on each side of the equation we get

$$\frac{d^2y_1}{d\varphi^2} + (1 - 2\,b\,y_0)\,y_1 \approx 0.$$

Defining

$$f \equiv \sqrt{1 - 2\,b\,y_0}, \tag{13.43}$$

and reintroducing the equality sign, this equation takes the form

$$\frac{d^2 y_1}{d\varphi^2} + f^2\, y_1 = 0. \tag{13.44}$$

We shall show that the function

$$\tilde{y}_1 = \epsilon\, y_0 \cos(f\varphi), \tag{13.45}$$

where ϵ is a constant, is a solution of Eq. (13.44). Differentiation by means of Eq. (4.25) and the chain rule (2.31) we get

$$\frac{d\,\tilde{y}_1}{d\varphi} = -\epsilon\, y_0\, f\, \sin(f\,\varphi).$$

One more differentiation gives

$$\frac{d^2 \tilde{y}_1}{d\varphi^2} = -\epsilon\, b\, f^2 \cos(f\,\varphi).$$

Inserting $\tilde{y}_1$ from Eq. (13.45) leads to

$$\frac{d^2 \tilde{y}_1}{d\varphi^2} = -f^2\, \tilde{y}_1,$$

which for $y_1 = \tilde{y}_1$ is equivalent to Eq. (13.44).

We have then found the solution for the trajectory of Mercury

$$y = y_0 + y_1 = y_0\, [1 + \epsilon\, \cos(f\,\varphi)]$$

or

$$\frac{1}{r} = \frac{1}{r_0}\, [1 + \epsilon\, \cos(f\,\varphi)]\,.$$

For $f = 1$, i.e. $b = 0$, this expression describes an elliptic orbit which is fixed in space, since the minimum value of r would then always take place at the same position, when the angle has increased by $\varphi = 2\pi$. This corresponds to the Newtonian case. However, taking the relativistic effect into account, Mercury has to move an angle φ so that $f\,\varphi = 2\pi$, before it returns to the point closest to the Sun, i.e. it must move an angle $\varphi = 2\pi/f$. Thus, Mercury moves not 2π between each time it passes the point where it is closest to the Sun, but 2π plus an extra angle

$$\Delta\varphi = 2\pi\,(1/f - 1). \tag{13.46}$$

This is the precession angle per orbit. Replacing f in Eq. (13.46) by the expression (13.43) gives

$$\Delta\varphi = 2\pi\left(\frac{1}{\sqrt{1-2\,b\,y_0}} - 1\right). \tag{13.47}$$

Here $b\,y_0 \ll 1$. We can then use the two first terms of the MacLaurin series (Eq. (2.84) with $x^2 = 2\,b\,y_0$) for this expression,

$$\frac{1}{\sqrt{1-2\,b\,y_0}} \approx 1 + b\,y_0.$$

Substituting this into Eq. (13.47) we arrive at

$$\Delta\varphi = 2\pi\,b\,y_0.$$

Replacing b by $(3/2)r_S$ (see Eq. (13.41)) finally gives

$$\Delta\varphi = 3\pi\,r_S\,y_0 = 3\pi(r_S/r_0)$$

per revolution. Here r_S is the Schwarzschild radius of the Sun, and r_0 is the radius of the nearly circular orbit of Mercury. Inserting numerical values gives $\Delta\varphi = 5.03 \times 10^{-7}$ radians per revolution. This corresponds to a precession of 43″ per century!

13.7 Gravitational deflection of light

One of the first predictions Einstein made from his general theory of relativity was that light grazing the Sun should be deflected by 1.75″. In this section we shall see how this prediction can be calculated.

In Sect. 5.12 we saw that a light-like interval has $ds^2 = 0$. This means that in the case of light, Eq. (13.37) is replaced by

$$u^r u_r + u^\varphi u_\varphi + u^t u_t = 0.$$

Consequently the equation of the trajectory of photons in the Schwarzschild spacetime is obtained by putting $a = 0$ in Eq. (13.41),

$$\frac{d^2y}{d\varphi^2} + y = by^2. \tag{13.48}$$

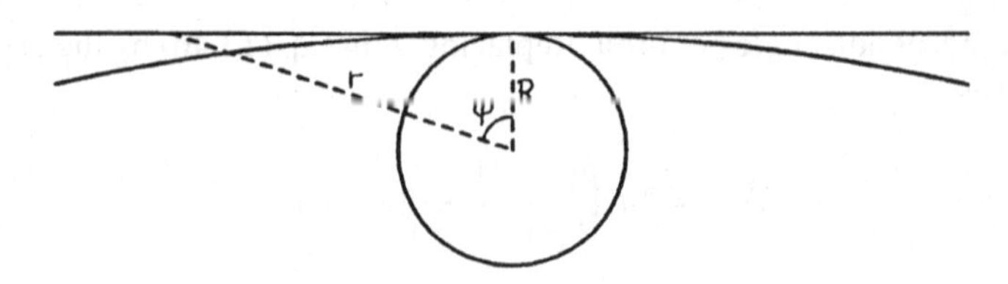

Fig. 13.1 Light ray deflected by the Sun

We are not interested in the general solution of this equation. Light grazing the Sun is deflected by a very small amount. The trajectory of this light is nearly a straight line. From Fig. 13.1 is seen that

$$\cos\varphi = R/r,$$

or, denoting the y function representing this line by y_0,

$$y_0 = \frac{1}{r} = \frac{\cos\varphi}{R}. \tag{13.49}$$

This function has the same form as $\tilde{y}_1$ of Eq. (13.45). Accordingly it fulfills an equation of the same form as Eq. (13.44),

$$\frac{d^2 y_0}{d\varphi^2} + y_0 = 0. \tag{13.50}$$

Since we seek a solution close to the straight line, we can write

$$y = y_0 + y_1, \quad y_1 \ll y_0. \tag{13.51}$$

Inserting this into Eq. (13.48) leads to

$$\frac{d^2 y_0}{d\varphi^2} + \frac{d^2 y_1}{d\varphi^2} + y_0 + y_1 = b\,(y_0 + y_1)^2.$$

Subtracting Eq. (13.50) we get

$$\frac{d^2 y_1}{d\varphi^2} + y_1 = b\,(y_0 + y_1)^2.$$

At the right-hand side we can neglect y_1 compared to y_0,

$$\frac{d^2 y_1}{d\varphi} + y_1 = b\,y_0^2.$$

Substituting the expression (13.49) we have the final form of the equation of the photon trajectory's deviation from a straight line

$$\frac{d^2 y_1}{d\varphi^2} + y_1 = \frac{b}{R^2} \cos^2 \varphi. \quad (13.52)$$

Due to the appearance of the function $\cos^2 \varphi$ at the right-hand side of this equation, we guess that a particular solution may be written

$$y_1 = A + B \cos^2 \varphi, \quad (13.53)$$

where A and B are constants, which we shall try to determine so that the expression (13.53) do satisfy Eq. (13.52). Differentiating the expression (13.53) by means of the chain rule (2.31) with $y = u^2$, $u = \cos \varphi$ and $x = \varphi$, using Eq. (2.36) with $n = 2$ and Eq. (4.25), we obtain

$$\frac{dy_1}{d\varphi} = -2B \cos \varphi \sin \varphi. \quad (13.54)$$

From the product rule (2.24) together with the rules (4.24) and (4.25) we get

$$\frac{d^2 y_1}{d\varphi^2} = -2B \left(-\sin^2 \varphi + \cos^2 \varphi \right) = 2B \left(\sin^2 \varphi - \cos^2 \varphi \right). \quad (13.55)$$

According to Eq. (4.11) $\sin^2 \varphi = 1 - \cos^2 \varphi$. Inserting this into Eq. (13.55) leads to

$$\frac{d^2 y_1}{d\varphi^2} = 2B \left(1 - 2 \cos^2 \varphi \right) = 2B - 4B \cos^2 \varphi. \quad (13.56)$$

Substituting the expressions (13.53) and (13.56) into Eq. (13.52) we find

$$2B - 4B \cos^2 \varphi + A + B \cos^2 \varphi = \frac{b}{R^2} \cos^2 \varphi$$

or

$$2B + A - 3B \cos^2 \varphi = \frac{b}{R^2} \cos^2 \varphi.$$

In order that the left-hand and right-hand sides of this equation shall indeed be identical for all values of φ, the constants A and B must obey the following equations

$$2B + A = 0 \quad \text{and} \quad -3B = \frac{b}{R^2}.$$

Thus

$$A = \frac{2b}{3R^2} \quad \text{and} \quad B = -\frac{b}{3R^2}.$$

Inserting these values of A and B into Eq. (13.53) gives

$$y_1 = \frac{b}{3R^2}\left(2 - \cos^2\varphi\right). \tag{13.57}$$

Inserting the expressions (13.49) and (13.57) into Eq. (13.51), we get the equation of the trajectory of the light grazing the Sun

$$y = y_0 + y_1 = \frac{\cos\varphi}{R} + \frac{b}{3R^2}\left(2 - \cos^2\varphi\right)$$

or

$$\frac{1}{r} = \frac{\cos\varphi}{R} + \frac{b}{3R^2}\left(2 - \cos^2\varphi\right).$$

Denoting the value of φ in the limit $r \to \infty$ for φ_∞ we get

$$0 = \frac{\cos\varphi_\infty}{R} + \frac{b}{3R^2}\left(2 - \cos^2\varphi_\infty\right). \tag{13.58}$$

Since the deflection is very small for light grazing the Sun, we know that $\varphi_\infty \approx \pi/2$. Therefore we write $\varphi_\infty = \pi/2 + \Delta\varphi$, where $\Delta\varphi \ll 1$. Using Eq. (4.19) we find

$$\begin{aligned}\cos\varphi_\infty &= \cos(\pi/2 + \Delta\varphi)\\ &= \cos(\pi/2)\,\cos\Delta\varphi - \sin(\pi/2)\,\sin\Delta\varphi.\end{aligned} \tag{13.59}$$

Since $\cos(\pi/2) = 0$ and $\sin(\pi/2) = 1$, Eq. (13.59) reduces to

$$\cos\varphi_\infty = -\Delta\varphi.$$

Equation (13.58) then takes the form

$$0 = \frac{-\Delta\varphi}{R} + \frac{b}{3R^2}\left[2 - (\Delta\varphi)^2\right].$$

Because $\Delta\varphi \ll 1$, we can neglect the term $(\Delta\varphi)^2$ inside the parenthesis. This leads to

$$\Delta\varphi = \frac{2b}{3R}.$$

From Fig. 13.1 is seen that the deflection of the light is twice this angle

$$\Delta\varphi_{\text{tot}} = 2\Delta\varphi = \frac{4b}{3R}.$$

Inserting the value $b = (3/2)r_S$ from Eq. (13.41) we finally arrive at

$$\Delta\varphi_{tot} = \frac{2r_S}{R}.$$

Here r_S is the Schwarzschild radius of the Sun, and R is the actual radius of the Sun. Inserting numerical values gives $\Delta\varphi_{tot} = 8.48 \times 10^{-6}$ radians $= 1.75''$.

13.8 Black holes

We shall now investigate the spacetime outside a massive spherical body with mass M by considering light cones in the Schwarzschild spacetime. It will be sufficient for our purposes to find the intersections of the light cones by the (r, t)-plane. In other words we shall investigate radially moving photons, such that $d\theta = d\varphi = 0$ along the photon worldlines. Furthermore, since photons follow null geodesics, their equations of motion are obtained by putting $ds^2 = 0$ in the Schwarzschild line element, Eq. (13.26). This gives

$$\frac{dr^2}{1 - r_S/r} = \left(1 - \frac{r_S}{r}\right) c^2\, dt^2.$$

Taking the square root of each side, and solving with respect to the coordinate velocity dr/dt of the (outgoing, +, and ingoing, −) photons we get

$$\frac{dr}{dt} = \pm\left(1 - \frac{r_S}{r}\right) c. \tag{13.60}$$

This is the coordinate velocity of light. The velocities of ingoing and outgoing photons are the same. As r increases the velocity approaches the special-relativistic value, c. But for $r = r_S$ we get $dr/dt = 0$. As shown by Eq. (13.60) both the ingoing and outgoing coordinate velocity of light vanish at $r = r_S$, and the light cone of a source at this position degenerates to a line. This astonishing conclusion is due to the choice of coordinates, and indicates that the chosen coordinates are not suitable for describing the propagation of light in the vicinity of the Schwarzschild radius.

We want to use coordinates such that the light cone is open for ingoing light passing the Schwarzschild radius. A suitable time coordinate for this purpose is called ‘the ingoing Eddington–Finkelstein coordinate’. It can be found by integrating the equation of motion of a radially ingoing light signal, Eq. (13.60), with the minus sign. This equation may be written as

$$\frac{1}{1 - r_S/r}\, dr = -c\, dt.$$

Multiplying the numerator and the denominator by r we get

$$\frac{r}{r - r_S}\, dr = -c\, dt. \tag{13.61}$$

In order to facilitate integration of this equation we write

$$\frac{r}{r - r_S} = 1 + \frac{r_S}{r - r_S}.$$

Both these terms can be integrated by means of the rules in chapter 3. Putting r_S in front of the integral, since it is constant, we get

$$\int \left(1 + \frac{r_S}{r - r_S}\right) dr = \int dr + r_S \int \frac{1}{r - r_S}\, dr.$$

Thus, Eq. (13.61) can be integrated term by term as follows

$$\int dr + r_S \int \frac{1}{r - r_S}\, dr = -c \int dt. \tag{13.62}$$

In the first term on the left-hand side we note that the integral of dr is r, in the second term we use the rule (3.55) with $u = r - r_S$ (note that $u' = (r - r_S)' = r' = 1$ since the derivative of the constant r_S is zero), and at the right-hand side of Eq. (13.62) we note that the integral of dt is t. Including a constant of integration K, we find

$$r + r_S \ln|r - r_S| = -c\, t + K. \tag{13.63}$$

We now introduce a new time coordinate $\bar{t}$ such that the equation of motion of a radially moving ingoing photon takes the very simple form

$$r = -c\, \bar{t}. \tag{13.64}$$

In the $(r, \bar{t})$ coordinate system the coordinate velocity of an ingoing photon is constant and equal to c. Putting $K = 0$, Eq. (13.63) can be written

$$r = -c\, t - r_S \ln|r - r_S|.$$

Comparing with Eq. (13.64) we see that the new time coordinate is related to the old one by

$$\bar{t} = t + \frac{r_S}{c} \ln|r - r_S|. \tag{13.65}$$

This is the announced 'ingoing Eddington–Finkelstein time coordinate'. Note that the clocks that measure the new time coordinate $\bar{t}$ are not synchronized with the clocks that measure the Schwarzschild time coordinate, since $\bar{t}$ depends upon the position even if t = constant. The physical significance of this new time coordinate is that the velocity of radially ingoing light in the Schwarzschild spacetime is constant and equal to c as measured with these clocks (see Eq. (13.64)).

In order to find how an outgoing photon moves in the $(r, \bar{t})$ system we first have to express the line element of the Schwarzschild solution, Eq. (13.26), by these coordinates. This means that we must calculate dt expressed by $d\bar{t}$. We therefore solve Eq. (13.65) with respect to t (and multiply, for later convenience, each term by c),

$$c\,t = c\,\bar{t} - r_S \ln|r - r_S|.$$

Differentiation gives

$$c\,dt = c\,d\bar{t} - \frac{r_S}{r - r_S}\,dr = c\,d\bar{t} - \frac{r_S/r}{1 - r_S/r}\,dr.$$

Inserting this into the Schwarzschild line element in the form (13.26), using the rule $(a+b)^2 = a^2 + 2\,a\,b + b^2$ with $a = d\bar{t}$ and $b = -\frac{r_S/cr}{1-r_S/r}\,dr$, results in

$$\begin{aligned}
ds^2 &= \frac{dr^2}{1 - r_S/r} + r^2\left(d\theta^2 + \sin^2\theta\,d\varphi^2\right) - \left(1 - \frac{r_S}{r}\right) c^2\,dt^2 \\
&= \frac{dr^2}{1 - r_S/r} + r^2\left(d\theta^2 + \sin^2\theta\,d\varphi^2\right) \\
&\quad - \left(1 - \frac{r_S}{r}\right)\left(c\,d\bar{t} - \frac{r_S/r}{1 - r_S/r}\,dr\right)^2 \\
&= \frac{dr^2}{1 - r_S/r} + r^2\left(d\theta^2 + \sin^2\theta d\,\varphi^2\right) - \left(1 - \frac{r_S}{r}\right) \\
&\quad \times \left(c^2\,d\bar{t}^2 - \frac{2\,r_S/r}{1 - r_S/r}\,c\,d\bar{t}\,dr + \frac{(r_S/r)^2}{(1 - r_S/r)^2}\,dr^2\right) \\
&= \frac{dr^2}{1 - r_S/r} + r^2\left(d\theta^2 + \sin^2\theta d\,\varphi^2\right) - \left(1 - \frac{r_S}{r}\right) c^2\,d\bar{t}^2 \\
&\quad + 2\frac{r_S}{r}\,c\,d\bar{t}\,dr + \frac{(r_S/r)^2}{(1 - r_S/r)}\,dr^2.
\end{aligned}$$

We now collect the terms with dr^2 and get (applying $1 - x^2 = (1-x)\,(1+x)$, with $x = r_S/r$)

$$\begin{aligned}
\frac{1}{1 - r_S/r}\,dr^2 - \frac{(r_S/r)^2}{1 - r_S/r}\,dr^2 &= \frac{1 - (r_S/r)^2}{1 - r_S/r}\,dr^2 \\
&= \frac{(1 - r_S/r)\,(1 + r_S/r)}{1 - r_S/r}\,dr^2 = \left(1 + \frac{r_S}{r}\right) dr^2.
\end{aligned}$$

With the new time coordinate $\bar{t}$, the line element of the Schwarschild spacetime now takes the form of

$$ds^2 = \left(1 + \frac{r_S}{r}\right) dr^2 + r^2 \left(d\theta^2 + \sin^2\theta \, d\varphi^2\right) + 2\frac{r_S}{r} c \, d\bar{t} \, dr - \left(1 - \frac{r_S}{r}\right) c^2 \, d\bar{t}^2. \tag{13.66}$$

Like Eq. (13.26) this line element represents the Schwarzschild spacetime, but now as expressed in terms of the new time coordinate $\bar{t}$. The line element reduces to Eq. (13.13) representing flat spacetime, in the limit $r \to \infty$, i.e. infinitely far from the mass M, or if $r_S = 0$, i.e. if the mass M is removed.

We shall now consider photons moving along a radial coordinate axis. Photons follow null geodesic curves (see Ch. 5). Therefore the path of a photon is found by putting $ds^2 = 0$. Since we are considering photons moving radially, $d\theta = d\varphi = 0$ in the line element (13.66). This results in

$$\left(1 + \frac{r_S}{r}\right) dr^2 + 2\frac{r_S}{r} c \, d\bar{t} \, dr - \left(1 - \frac{r_S}{r}\right) c^2 \, d\bar{t}^2 = 0.$$

Dividing by $c^2 \, d\bar{t}^2$

$$\left(1 + \frac{r_S}{r}\right) \left(\frac{dr}{c \, d\bar{t}}\right)^2 + 2\frac{r_S}{r} \frac{dr}{c \, d\bar{t}} - \left(1 - \frac{r_S}{r}\right) = 0.$$

This is a quadratic equation for $dr/(c \, d\bar{t})$, i.e. it has the form

$$0 = a \, x^2 + b \, x + d, \tag{13.67a}$$

$$x = \frac{dr}{c \, d\bar{t}}, \tag{13.67b}$$

$$a = 1 + \frac{r_S}{r}, \tag{13.67c}$$

$$b = \frac{2r_S}{r}, \tag{13.67d}$$

$$d = \left(1 - \frac{r_S}{r}\right). \tag{13.67e}$$

In order to solve this equation with respect to x we shall first show how one can construct a quadratic term from the two terms containing x. Multiplying (13.67a) by $4a$, we get

$$4 a^2 x^2 4 a \, b \, x + 4 a \, d = 0.$$

We now use a trick that sometimes leads to a simplification, that of adding and subtracting a suitably chosen term. In the present case the term b^2 is chosen in order to write the resulting first three terms as one quadratic term. Adding and subtracting b^2 leads to

$$4\,a^2\,x^2 + 4\,a\,b\,x + b^2 + 4\,a\,d - b^2 = 0$$

or

$$(2\,a\,x + b)^2 + 4\,a\,d - b^2 = 0$$

where we have used the rule $u^2 + 2uv + v^2 = (u+v)^2$. Then

$$(2\,a\,x + b)^2 = b^2 - 4\,a\,b.$$

Taking the square root

$$2\,a\,x + b = \pm\sqrt{b^2 - 4\,a\,d}$$

or

$$2\,a\,x = -b \pm \sqrt{b^2 - 4\,a\,d}.$$

Dividing each side by $2\,a$, we get the solution of Eq. (13.67a)

$$\frac{dr}{c\,d\bar{t}} = x = \frac{-b \pm \sqrt{b^2 - 4\,a\,d}}{2\,a}. \tag{13.68}$$

Now look at the term under the square root of Eq. (13.68). Inserting the expressions for a, b, and d we get

$$b^2 - 4\,a\,d = \frac{4\,r_\mathrm{S}^2}{r^2} + 4\left(1 + \frac{r_\mathrm{S}}{r}\right)\left(1 - \frac{r_\mathrm{S}}{r}\right). \tag{13.69}$$

Using the rule $(u+v)(u-v) = u^2 - v^2$ with $u = 1$ and $v = r_\mathrm{S}/r$ gives

$$\left(1 + \frac{r_\mathrm{S}}{r}\right)\left(1 - \frac{r_\mathrm{S}}{r}\right) = 1 - \frac{r_\mathrm{S}^2}{r^2}.$$

Equation (13.69) then takes the form of

$$b^2 - 4\,a\,d = \frac{4\,r_\mathrm{S}^2}{r^2} + 4\left(1 - \frac{r_\mathrm{S}^2}{r^2}\right) = 4.$$

Inserting this and the expressions for a and b into Eq. (13.68) and multiplying with c leads to

$$\frac{dr}{d\bar{t}} = \frac{-2\,r_\mathrm{S}/r \pm 2}{2\,(1 + r_\mathrm{S}/r)}\,c = -\frac{\pm 1 + r_\mathrm{S}/r}{1 + r_\mathrm{S}/r}\,c.$$

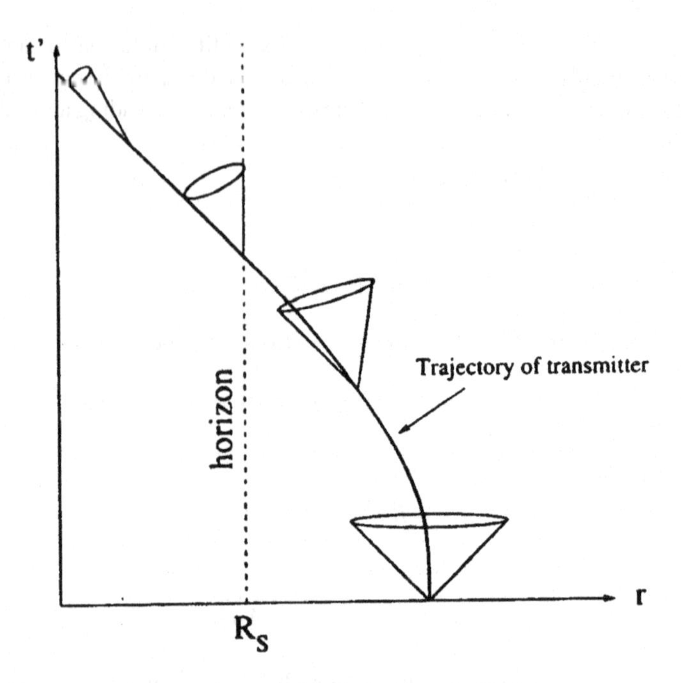

Fig. 13.2 Light cones of a transmitter crossing the horizon of a black hole

Thus, the coordinate velocities of ingoing and outgoing photons, respectively, are

$$\left(\frac{dr}{d\bar{t}}\right)_{\text{in}} = -c$$
$$\left(\frac{dr}{d\bar{t}}\right)_{\text{out}} = \frac{1 - r_S/r}{1 + r_S/r}\,c. \qquad (13.70)$$

Note that $r_S = 0$ gives $\left(\frac{dr}{d\bar{t}}\right)_{\text{out}} = c$, showing that in this case the velocity of ingoing and outgoing light is the same. This is quite natural since spacetime is flat in this case, making the special theory of relativity govern the scene. Equation (13.70) shows that the velocity of an ingoing photon is constant and equal to c, in accordance with the defining equation (13.64). On the other hand, the velocity of an outgoing photon varies with the radial position. This is different from in the special theory. The light cones corresponding to Eq. (13.70) are drawn in Fig. 13.2.

The properties of the drawing follow from the expressions for the photon velocities in Eq. (13.70). The inward cut of the light cone with the plane of the paper, which represents an ingoing photon, makes an angle equal to 45 degrees with the time axis, independently of the position of the emitter, since the inward velocity of light is equal to c at all distances in these coordinates. However, the outward velocity, given by the second expression in Eq. (13.70) is less that c. So the outward cut of the cone with the plane of the paper, makes an angle with the time axis which is less than 45 degrees. But in the limit of large r, i.e. far away from the mass distribution, the velocity of outgoing photons approach c, and the cut makes nearly 45 degrees with the time axis. At the Schwarzschild radius r_S the numerator in the second expression in Eq. (13.70) is equal to zero. This means that even a photon that is emitted radially outwards from this position, will not be able to move outwards. Thus the outward cut of the light cone is parallel to the time axis for an emitter at this position. In the case of an emitter inside the Schwarzschil radius, $r < r_S$, the velocity of the outgoing photon is negative. This means that a photon emitted outwards from such a position will nevertheless move inwards. This is shown in the figure where the whole light cone leans inwards for $r < r_S$.

From the discussion in Sect. 5.13 we know that all particles of the observed universe move so that their worldlines are inside the light cones. This means that whatever superb rockets you may possess, if your rocket ship comes inside the Schwarzschild radius there is no way out. And worse still, you cannot even send a message for help out of the region inside the Schwarzschild radius. No information can reach from this region to the outside world. Therefore the spherical surface with radius $r = r_S$ is called *the Schwarzschild horizon*. Since, as seen from the outside, no light comes from the inside, this region is called a *black hole*.

Black holes, that *may* exist according to the general theory of relativity, have probably been located in double star systems, with one invisible star seemingly disturbing the appearance of the visible star. The most prominent candidate is Cygnus X-1.

If the existence of black holes is confirmed beyond any reasonable doubt, there is ample reason to enjoy a feeling of excitement. One would then experience one of those seldom events that make vivid what Einstein meant when he said: "The most incomprehensible thing about the world is that it is comprehensible". For the man who invented the theory it would be astonishing. Nothing like black holes was thought of when the theory was constructed. It was a question of making a theory of space, time and gravitation in accordance with the most natural (for Einstein) and general principles: the principle of relativity, the principle of the local constancy of the velocity of light and the principle of equivalence. Eighty years of work by hundreds of physicists has proved that the theory has a vast manifold of interesting physical consequences. And it may be felt like a wonder: nature seems to obey the theory!

Still Einstein felt like Newton: to be playing along the shore of an ocean of ignorance. He did not rest, but went on along the shore, venturing further out into the ocean than most others. His goal was no less than a unified theory of the known fundamental forces, from which all the observable properties of the material world could be deduced. During the last thirty years of his life he searched for such a theory. Physicists are still searching for it.

Chapter 14
Relativistic universe models

Cosmology may be said to be that part of physical science that aims at giving a description of the universe at large. Such descriptions are called *universe models*, and are mathematical models interpreted physically. They are based upon observations and physical laws. These laws represent our deepest insights as to the behaviour of the material world. They are the main contents of the physical theories.

By giving the laws mathematical formulations, one may calculate from the laws how the matter behaves under given circumstances, and thus predict the behaviour of different types of models. Comparing such predictions with observations, one may obtain an idea of the validity and scope of our conceptions concerning the universe.

14.1 Observations

During the last 60 years or so one has observed processes in parts of the universe by means of large optical telescopes, radio telescopes and spaceborn observation equipment, in particular the Hubble telescope. These observations have suggested several simple properties of the universe as a whole.

Distances. The velocity of light in vacuum is approximately $c = 300,000$ km/s, which corresponds to travelling 7 times around the Earth per second. The light uses 8 minutes from the Sun to the Earth. The distance that light travels during one year is called a *light year*, and is equal to 9.46×10^{15} m. Our nearest star, Alpha Centauri, is about 4 light years away from us. This is much farther away than it sounds when we use light year as distance unit. Imagine a spacecraft travelling with a velocity 100 km/s. The spacecraft would pass the Sun after 20 days and nights, but it would take 12,000 years before the spacecraft arrives at the nearest star.

Our planetary system is positioned in an arm of a spiral galaxy, the Milky Way, about 100.000 light years from its centre. There are about 10^{11} stars in our galaxy, and there are presumably a similar number of galaxies in the universe. Our nearest galaxy is the Andromeda galaxy. It is about 2.5×10^{6} light years from the Milky

Ø. Grøn and A. Næss, *Einstein's Theory: A Rigorous Introduction for the Mathematically Untrained*, DOI 10.1007/978-1-4614-0706-5_14,

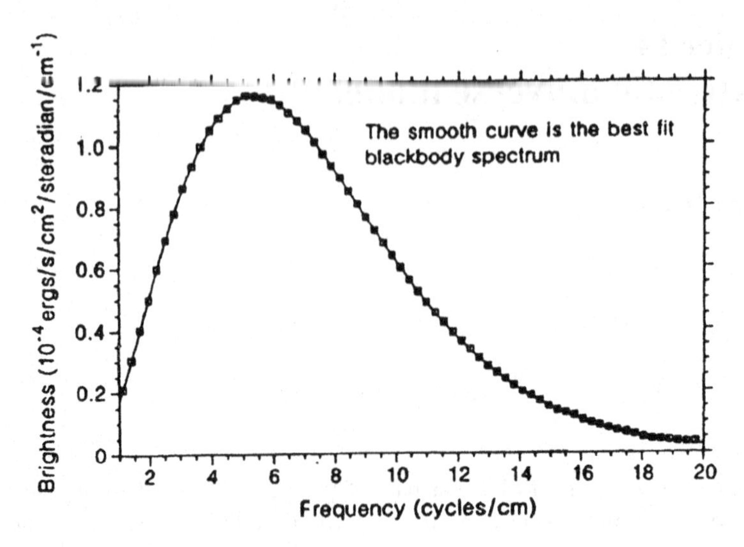

Fig. 14.1 Frequency distribution of cosmic background radiation

Way to the Andromeda galaxy. The galaxies are collected in clusters of galaxies. The mean distance between the galaxy clusters is about 10^8 light years. The farthest objects observed have a distance about 10^{10} light years from us.

Large scale homogeneity. Observations seem to indicate that over distances larger than about 10^9 light years the material of the universe is uniformly distributed. On such a scale the universe is usually assumed to be homogeneous. The validity of this assumption has, however, been discussed recently on the basis on new three-dimensional surveys of the distribution of matter in the universe.

Isotropy. The distribution of galaxies seems to be equal in all directions on a large scale, i.e. the distribution is *isotropic*. The isotropy of the universe on a large scale has been confirmed by observations of the cosmic background radiation. Its spectral distribution corresponds to black body radiation with a temperature 2.726 K (see Fig. 14.1).

Expansion. From 1925 to 1930 Edwin Hubble observed the spectral lines in the light from galaxies far away. He also estimated the distances to the galaxies by astrophysical methods. The result of these investigations is that the spectral lines corresponding to known atomic transitions are displaced towards red, i.e. towards longer wavelengths, and that this red shift is proportional to the distances of the galaxies. This is called *Hubble's law*.

The simplest explanation of this 'law' is that it is due to the Doppler effect, indicating that the galaxies are moving away from us, and faster the farther away they are (see Fig. 14.2). Thus, the universe expands.

According to Newtonian kinematics such an expansion of at least a universe of finite extension implies the existence of a centre from which everything expands.

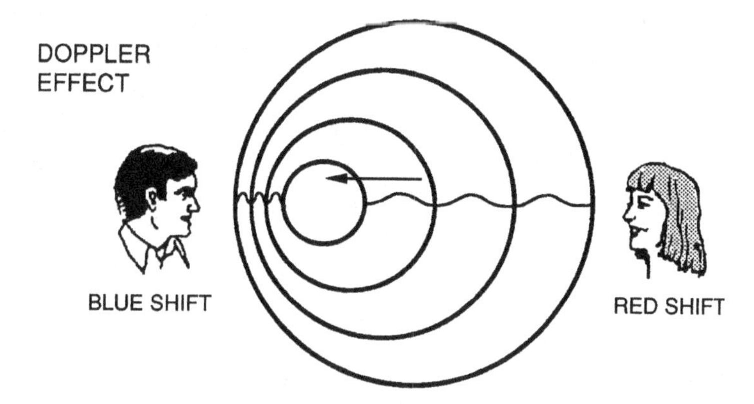

Fig. 14.2 Doppler effect

This is not so according to the conceptions of the general theory of relativity. As we have seen, space is curved according to this theory. The two-dimensional analogue of a finite isotropic universe is an expanding spherical surface. The galaxies may be imagined as dots painted on the surface. The distances between the dots increase due to the expansion. But the dots do not move on the surface. Neither is there any centre on the surface. Similarly there is no centre in the homogeneous relativistic universe models.

14.2 Homogeneous and isotropic universe models

In 1922 the Russian meteorologist A. Friedmann found a set of solutions to Einstein's field equations, describing expanding universes. The models of Friedmann are similar in all directions and at all points, i.e. they are isotropic and homogeneous. This seems to be in agreement with the observed properties of our universe on a large scale. It should be noted, however, that there is an ongoing discussion concerning a possible hierarchical structure of the distribution of matter on a large scale.

The homogeneous and isotropic universe models of Friedmann are the the simplest relativistic models. They have dominated the large scale modelling of the universe for the last seventy years. In order to become familiar with the research region termed *relativistic cosmology*, one should at first make oneself familiar with the Friedmann models. And we shall restrict ourselves to these models in the present text.

However, the models have important limitations as to their ability to explain observed properties of the universe, such as homogeneity and isotropy, because of the following circumstance: If one wants to explain a certain observed property of the universe, one has to investigate sufficiently general theoretical models, which permit the universe *not* to have this property. More general models have been constructed, but they are mathematically more complicated than the ones we shall consider.

Fig. 14.3 Spherical surface

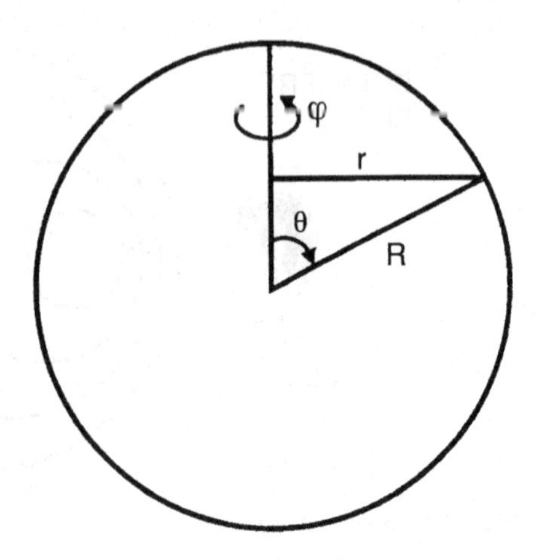

The curvature of space at a certain moment in a Friedmann model, is constant. In order to gradually develop some visual notions about spaces with constant curvature, we shall start by considering a two-dimensional space with constant positive curvature, a spherical surface (see Fig. 14.3).

The line element of flat three-dimensional space, as expressed in a spherical coordinate system, is given in Eq. (13.12). Let us consider a spherical surface with radius R. On this surface the line element is reduced to

$$d\sigma^2 = R^2\, d\theta^2 + R^2 \sin^2\theta\, d\varphi^2. \tag{14.1}$$

Here θ represents the latitude with $\theta = 0$ at the North Pole, and φ represents the longitude. The position of a point on the spherical surface is given by specifying θ and φ. We now replace the coordinate θ by a radial coordinate r representing the distance from the axis passing through the poles to a point on the spherical surface. From Fig. 14.3 and the formula (4.1a) for the sinus of an angle, follows

$$\sin\theta = r/R \tag{14.2}$$

or

$$r = R\,\sin\theta. \tag{14.3}$$

Note that r is a function of θ alone since R is constant. Differentiation by means of Eq. (4.24), then gives

$$\frac{dr}{d\theta} = R\,\cos\theta$$

or

$$\frac{dr}{\cos\theta} = R\,d\theta.$$

Squaring, and exchanging the left-hand and right-hand sides, we have

$$R^2\, d\theta^2 = \frac{dr^2}{\cos^2\theta}. \tag{14.4}$$

From Eq. (4.11) follows

$$\cos^2\theta = 1 - \sin^2\theta. \tag{14.5}$$

Substituting the expression (14.2) for $\sin\theta$ in Eq. (14.5) leads to

$$\cos^2\theta = 1 - r^2/R^2.$$

Inserting this into Eq. (14.4) we get

$$R^2\, d\theta^2 = \frac{dr^2}{1 - r^2/R^2}. \tag{14.6}$$

Inserting the right-hand side of Eq. (14.6) and the left-hand side of Eq. (14.3) into Eq. (14.1) leads to

$$d\sigma^2 = \frac{dr^2}{1 - r^2/R^2} + r^2\, d\varphi^2.$$

In chapter 9, after Eq. (9.12), we found that the internal curvature of a spherical surface with radius R is $k = 1/R^2$. Thus the line element on the spherical surface may be written as

$$d\sigma^2 = \frac{dr^2}{1 - k\, r^2} + r^2\, d\varphi^2. \tag{14.7}$$

This form of the line element is obviously valid also for a plane, which has $k = 0$, since Eq. (14.7) then reduces to the line element (5.105) of an Euclidean plane, as expressed in plane polar coordinates. It is also valid for surfaces of constant negative curvature, $k < 0$.

The line element (13.12) is a generalization of flat three-dimensional space of the line element (5.105) for a (two-dimensional) plane. Similarly, the generalization of (14.7) to a line element, which we call $d\,b^2$, for three-dimensional space with constant curvature, is

$$db^2 = \frac{dr^2}{1 - k\, r^2} + r^2\, d\theta^2 + r^2 \sin^2\theta\, d\varphi^2. \tag{14.8}$$

We shall consider an expanding universe. Our coordinate system is chosen such that the galaxies have fixed spatial coordinates. The quantity db, which appears squared in Eq. (14.8), is the coordinate distance between the origin of the coordinate system and a point with coordinates $(dr, d\theta, d\varphi)$. The physical distance is

$$dl = a(t)\, db,$$

where the function $a(t)$ is a so-called *scale factor*, which tells us how the distance to a galaxy changes with time. No direction is preferred. Assuming that the function $a(t)$ increases with time, there is an isotropic expansion.

In order to write the full line element for spacetime, we must specify what sort of coordinate clocks is to be used. Choosing a coordinate time that is equal to the time as measured by standard clocks carried by the galaxies, the line element of spacetime takes the form

$$ds^2 = a^2(t)\left(\frac{dr^2}{1-k\,r^2} + r^2\,d\theta^2 + r^2\,\sin^2\theta\,d\varphi^2\right) - c^2\,dt^2. \tag{14.9}$$

This line element describes the geometric and kinematical properties of isotropic and (spatially) homogeneous universe models. It is called the *Robertson–Walker* line element. The components of the metric tensor are

$$g_{rr} = \frac{a^2(t)}{1-k\,r^2}, \tag{14.10a}$$

$$g_{\theta\theta} = a^2(t)\,r^2, \tag{14.10b}$$

$$g_{\varphi\varphi} = a^2(t)\,r^2\,\sin^2\theta, \tag{14.10c}$$

$$g_{tt} = -c^2. \tag{14.10d}$$

14.3 Einstein's gravitational field equations for homogeneous and isotropic world models

The non-vanishing Christoffel symbols for the Robertson–Walker line element have been calculated in Appendix C. In particular $\Gamma^r{}_{tt} = 0$, and this will prove to be of physical significance. In Sect. 12.8 we showed that free particles follow geodesic curves in spacetime. Consider a free particle instantaneously at rest. From the isotropy of the models follows that the particle has no acceleration in the θ and φ directions. The r component of the geodesic equation (8.7) is

$$\frac{d^2r}{d\tau^2} + \Gamma^r{}_{\alpha\beta}\frac{dx^\alpha}{d\tau}\frac{dx^\beta}{d\tau} = 0, \tag{14.11}$$

where τ is the proper time of the particle. Since the particle is instantaneously at rest, the only non vanishing component of the four velocity (see Eq. (10.21)) is the time component $dt/d\tau$. Thus Eq. (14.11) is reduced to

$$\frac{d^2r}{d\tau^2} + \Gamma^r{}_{tt}\left(\frac{dt}{d\tau}\right)^2 = 0.$$

Since $\Gamma^r{}_{tt} = 0$ it follows that

$$\frac{d^2 r}{d\tau^2} = 0.$$

Thus the particle has no acceleration. Therefore it will remain at rest in the coordinate system. This implies that particles with constant spatial coordinates are freely moving. We can therefore identify such 'particles' with the galaxies. This means that our coordinate system is comoving with the galaxies. Thus the scale factor $a(t)$ tells how the galaxies move. According to our conceptions the movement is not imagined as a movement *through* space, but signifies the expansion of space itself.

We now come to the field equations. Inserting the expression (12.31) for Einstein's gravitational constant into Eq. (11.34), the field equations take the form

$$R_{\mu\nu} = \frac{8\pi G}{c^4}\left(T_{\mu\nu} - \frac{1}{2}\, g_{\mu\nu}\, T\right).$$

The left-hand side, i.e. the needed components of the Ricci tensor have been calculated in Appendix C. The right-hand side of the field equations involves the energy-momentum tensor of the matter. We assume that the cosmic matter is homogeneous, and may be represented as a perfect fluid. This means that the only physical properties of the matter that we take account of, is its motion, the density and the pressure or tension. Viscosity, for example, is neglected. The energy-momentum tensor of a perfect fluid is given in Eq. (10.35). Lowering the indices we obtain

$$T_{\mu\nu} = \left(\rho + p/c^2\right) u_\mu u_\nu + p\, g_{\mu\nu}.$$

According to the definition (10.21) the contravariant components of the four velocity are

$$u^\mu = \frac{dx^\mu}{d\tau}.$$

Since the galaxies are at rest in the coordinate system, only the time component is non-vanishing

$$u^t = \frac{dt}{d\tau}.$$

According to our choice of coordinate time, $dt = d\tau$, which leads to

$$u^t = 1.$$

Lowering the index, we get

$$u_t = g_{tt}\, u^t = -c^2.$$

We need the following components of the energy-momentum tensor

$$T_{tt} = (\rho + p/c^2)\,(u_t)^2 + p\,g_{tt} = (\rho + p/c^2)\,c^4 - p\,c^2$$
$$= \rho\,c^4 + p\,c^2 - p\,c^2 = \rho\,c^4$$
$$T_{\theta\theta} = p\,g_{\theta\theta} = p\,a^2\,r^2.$$

The sum of the mixed components with equal indices of the energy-momentum tensor is

$$T \equiv T^{\mu}{}_{\mu} = (\rho + p/c^2)\,u^{\mu}u_{\mu} + p\,\delta^{\mu}{}_{\mu},$$

where we have used Eq. (5.79) in the last term. From Eqs. (10.21) and (10.22) follow that the square of the four velocity is

$$u^{\mu}u_{\mu} = -c^2.$$

According to Eq. (11.2) $\delta^{\mu}{}_{\mu} = 4$. Therefore

$$T = (\rho + p/c^2)\,(-c^2) + 4\,p = -\rho\,c^2 - p + 4\,p = -\rho\,c^2 + 3\,p.$$

We can now calculate the quantities at the right-hand side of the field equations

$$T_{tt} - \frac{1}{2}\,T\,g_{tt} = \rho\,c^4 - \frac{1}{2}\,(-\rho\,c^2 + 3\,p)\,(-c^2)$$
$$= \rho\,c^4 - \frac{1}{2}\,\rho\,c^4 + \frac{3}{2}\,p\,c^2 = \frac{1}{2}\,\rho\,c^4 + \frac{3}{2}\,p\,c^2,$$

giving

$$T_{tt} - \frac{1}{2}\,T\,g_{tt} = \frac{1}{2}\,(\rho\,c^2 + 3\,p)\,c^2$$
$$= \frac{1}{2}\,(\rho + 3\,p/c^2)\,c^4. \tag{14.12}$$

Furthermore

$$T_{\theta\theta} - \frac{1}{2}\,T\,g_{\theta\theta} = p\,a^2\,r^2 - \frac{1}{2}\,(-\rho\,c^2 + 3\,p)\,a^2\,r^2$$
$$= p\,a^2\,r^2 + \frac{1}{2}\rho\,c^2a^2r^2 - \frac{3}{2}p\,a^2r^2,$$

which leads to

$$T_{\theta\theta} - \frac{1}{2}\,T\,g_{\theta\theta} = \frac{1}{2}\,(\rho\,c^2 - p)\,a^2r^2$$
$$= \frac{1}{2}\,(\rho - p/c^2)\,a^2\,r^2\,c^2. \tag{14.13}$$

From Eqs. (C.15) and (14.12) we deduce the (t, t) component of the field equations

$$-3\,\frac{\ddot{a}}{a} = \frac{8\pi G}{c^4}\,\frac{1}{2}\left(\rho + 3\,p/c^2\right)c^4,$$

or

$$a\,\ddot{a} = -\frac{4\,\pi\,G}{3}\left(\rho + 3\,p/c^2\right)a^2 \tag{14.14}$$

and from Eqs. (C.20) and (14.13) we find the (θ, θ) component of the field equations

$$\left(a\,\ddot{a} + 2\,\dot{a}^2 + 2\,k\,c^2\right)\frac{r^2}{c^2} = \frac{8\,\pi\,G}{c^4}\,\frac{1}{2}\left(\rho - p/c^2\right)a^2\,r^2\,c^2,$$

or

$$a\,\ddot{a} + 2\,\dot{a}^2 + 2\,k\,c^2 = 4\,\pi\,G\left(\rho - p/c^2\right)a^2. \tag{14.15}$$

Equations (14.14) and (14.15) are the famous Friedmann equations for isotropic and homogeneous universe models with perfect fluids.

14.4 Physical properties of the Friedmann models

Let us first find which type of expansion these models have. For this purpose we consider models with Euclidean spatial geometry, $k = 0$. From the line element (14.9) follows that for such models the physical distance from the origin to a galaxy with radial coordinate r is

$$\ell = a(t)\,r.$$

The expansion velocity of the galaxy is

$$v = \dot{l} = \dot{a}\,r = \frac{\dot{a}}{a}\,a\,r = \frac{\dot{a}}{a}\,\ell.$$

From this follows that the velocity of the galaxy is proportional to the distance. As mentioned above, this is called *Hubble's law*. His many observations of the spectral lines in the light from the galaxies suggested the hypothesis that there exists such a law. More recent observations support this.

Defining the so-called *Hubble factor*,

$$H \equiv \frac{\dot{a}}{a}, \tag{14.16}$$

the Hubble law can be written as

$$v = H\,\ell.$$

Note that for most universe models the Hubble factor is a function of the time.

The present value H_0 of the Hubble factor, as determined from observations, is

$$H_0 = h\,\frac{30\,\mathrm{km/s}}{10^6\ \mathrm{l.y.}}, \tag{14.17}$$

where h is a factor with a value limited by observations to the interval $0.5 < h < 1$. The magnitude of the allowed values for h represents the uncertainty in the value of the Hubble factor. The most recent measurements (1997) point at $h = 0,6$ as the most probable value of h. The expression (14.17) means that if the distance is increased by one million light years, then the expansion velocity increases by h times 30 km/s.

Our neighbouring galaxy, the Andromeda galaxy, is about 2.5×10^6 l.y. away from us. At this distance the expansion velocity (with $h = 1$) is about 75 km/s. This does not mean that the Andromeda galaxy moves with this velocity away from us. The galaxies are like the particles of a gigantic gas. The velocity due to the expansion is superposed upon a velocity due to a motion of the galaxies in arbitrary directions. The average velocity of this motion is about 500 km/s. Thus, only for galaxies at distances greater than about 17 million light years, will the expansion dominate over the arbitrary motions.

We shall now consider some consequences of the field equations. Equation (14.14) confirms the results (12.34) and (12.36) that we found earlier in the weak field approximation, namely that the relativistic gravitational mass density is $\rho_g = \rho + 3\,p/c^2$, implying that pressure or tension contributes to the gravitational field. For ordinary matter $p/c^2 \ll \rho$, which gives $\ddot{a} < 0$. This means that the expansion is slowed down, due to attractive gravity. The expansion was faster at earlier times.

The observed fact that the universe expands means that the particles in the cosmic matter were closer before than they are now. Going far enough back in time we come to a point of time where the cosmic matter was very closely packed. The density was extremely large. This is a state that cannot be properly described by means of the general theory of relativity. At a sufficiently early time we reach the limit of applicability of the general theory of relativity, and we need a quantum theory of gravity to describe the first moment. If we call the point of time with infinite density, as predicted by general relativity, for $t = 0$, then the region of applicability of general relativity is later than the so-called Planck time, $t_{\mathrm{Planck}} = \sqrt{G\hbar/c^5} = 10^{-43}$ s, where $\hbar$ is Planck's constant. This initial moment in the history of the universe is called the *Big Bang*.

The time from the Big Bang to now is called the age of the universe. Due mainly to the uncertainty in the value of the Hubble factor, it is uncertain by about a factor of two. If the expansion velocity was constant, the present age of the universe would be equal to the inverse of the present value of the Hubble factor. This age is called the *Hubble age*, and is denoted by t_{H_0}. Due to attractive gravity the expansion is slowed down. The universe expanded faster before. Thus the age of the universe is less that the Hubble age, which has a value

$$t_{H_0} = 1/H_0 = (10/h)\ 10^9\,\mathrm{years}. \tag{14.18}$$

Inserting $h = 0.6$ we find the most probable value $t_{H_0} = 16.7 \times 10^9$ years.

Subtracting Eq. (14.14) from Eq. (14.15) we get

$$2\dot{a}^2 + 2k\,c^2 = \frac{4\pi G}{3}\left(3\rho + \rho - \frac{3p}{c^2} + \frac{3p}{c^2}\right)a^2,$$

which leads to

$$\dot{a}^2 + k\,c^2 = \frac{8\pi G}{3}\rho a^2. \tag{14.19}$$

Introducing the Hubble factor from Eq. (14.16), and dividing each side of Eq. (14.19) by a^2, the latter equation takes the form of

$$H^2 + \frac{k\,c^2}{a^2} = \frac{8\pi G}{3}\rho. \tag{14.20}$$

A universe with Euclidean spatial geometry, $k = 0$, is said to have *critical* mass density ρ_{crit}. Hence

$$H^2 = \frac{8\pi G}{3}\rho_{\text{crit}}$$

or

$$\rho_{\text{crit}} = \frac{3H^2}{8\pi G}. \tag{14.21}$$

Inserting the present value, H_0, of the Hubble factor, we get the present value of the critical mass density, $(\rho_{\text{crit}})_0 = 2\,h^2\,10^{-26}$ kg/m^3. Inserting $h = 0.6$ gives $(\rho_{\text{crit}})_0 = 7,2 \times 10^{-27}$ kg/m^3, which corresponds to about 4 hydrogen atoms per cubic metre.

One often introduces a dimensionless parameter representing the cosmic mass density, namely the ratio of the actual cosmic mass density and the critical mass density,

$$\Omega \equiv \rho/\rho_{\text{crit}}. \tag{14.22}$$

Dividing Eq. (14.20) by H^2 and using Eq. (14.22) and that $\dot{a} = Ha$, we find

$$1 + \frac{k\,c^2}{\dot{a}^2} = \Omega$$

or

$$\Omega - 1 = \frac{k\,c^2}{\dot{a}^2}. \tag{14.23}$$

When we know a as a function of the time, this equation will tell whether the mass density approaches the critical density or not.

An equation for the rate of change of the cosmic mass density can be deduced from the field equations as follows. Differentiating Eq. (14.19) we get

$$2\,\dot{a}\,\ddot{a} = \frac{8\pi G}{3}\left(\dot{\rho}\,a^2 + 2\,\rho\,a\,\dot{a}\right).$$

From this equation and Eq. (14.14) follows

$$-\left(\rho + \frac{3\,p}{c^2}\right) a\,\dot{a} = \rho\, a^2 + 2\,\rho\, a\,\dot{a}.$$

Subtracting $2\rho a\dot{a}$, dividing by a^2 and then exchanging the left and right-hand sides we get

$$\dot{\rho} = -3\left(\rho + \frac{p}{c^2}\right)\frac{\dot{a}}{a}. \tag{14.24}$$

This is the equation for the rate of change of density of the cosmic fluid.

There are different sorts of perfect fluid characterized by different relationships between pressure and density. Such a relationship is usually called the equation of state of the cosmic fluid. In all the fluids we shall consider the pressure is proportional to the density, i.e.

$$p = (\beta - 1)\,\rho, \tag{14.25}$$

where β is a constant characterizing the fluid. Inserting Eq. (14.25) into Eq. (14.24) we get

$$\dot{\rho} = -3\left[\rho + (\beta - 1)\,\rho\right]\frac{\dot{a}}{a} = -3\left(\rho + \beta\,\rho - \rho\right)\frac{\dot{a}}{a}$$
$$= -3\,\beta\,\rho\,\frac{\dot{a}}{a}.$$

Dividing each side of this equation by ρ we have

$$\frac{\dot{\rho}}{\rho} = -3\,\beta\,\frac{\dot{a}}{a}.$$

Integrating each side of this equation by means of the rule (3.55) we find

$$\ln\rho = -3\,\beta\,\ln a + K,$$

where K is a constant. Using the rule (3.51) and adding $3\,\beta\,\ln a = \ln(a^{3\,\beta})$ on each side, we can write

$$\ln\rho + \ln(a^{3\,\beta}) = K.$$

Using the rule (3.49) we get

$$\ln(\rho a^{3\beta}) = K = \ln K_\beta$$

where K_β is a new constant. Hence

$$\rho\, a^{3\,\beta} = K_\beta. \tag{14.26}$$

This equation shows how the density of different sorts of cosmic fluids depends upon the scale factor.

We can now calculate how the the scale factor of a universe with critical mass density depends upon the time. Inserting Eq. (14.21), with $H = \dot{a}/a$, into Eq. (14.26) we find

$$\frac{3}{8\pi G}\frac{\dot{a}^2}{a}a^{3\beta} = K_\beta.$$

Multiplying each side by $8\pi G/3$, taking the square root of each side of the resulting equation, and making use of

$$\sqrt{\frac{a^{3\beta}}{a^2}} = \frac{a^{3\beta/2}}{a} = a^{\frac{3}{2}\beta-1},$$

we get

$$a^{\frac{3}{2}\beta-1}\dot{a} = \sqrt{\frac{8\pi G}{3}K_\beta}.$$

Inserting $\dot{a} = da/dt$ and multiplying by dt

$$a^{\frac{3}{2}\beta-1}\,da = \sqrt{\frac{8\pi G}{3}}\,dt.$$

Integrating, with the initial condition $a(0) = 0$, we have

$$\int_0^a a^{\frac{3}{2}\beta-1}\,da = \sqrt{\frac{8\pi G}{3}}\int_0^t dt.$$

Using the rule (3.34) we get

$$\frac{1}{(3/2)\,\beta}a^{\frac{3}{2}\beta} = \sqrt{\frac{8\pi G}{3}K_\beta}\,t.$$

Multiplying by $(3/2)\,\beta$ leads to

$$a^{\frac{3}{2}\beta} = \frac{3}{2}\beta\sqrt{\frac{8\pi G}{3}K_\beta}\,t.$$

Squaring, and using that $(a^{(3/2)\beta})^2 = a^{3\beta}$, we find

$$a^{3\beta} = \frac{9}{4}\beta^2\frac{8\pi G}{3}K_\beta\,t^2 = 6\pi\beta^2\,G\,K_\beta\,t^2. \tag{14.27}$$

Dividing the exponent by 3β we finally conclude

$$a = (6\pi\beta^2 G K_\beta)^{\frac{1}{3\beta}}\,t^{\frac{2}{3\beta}}. \tag{14.28}$$

This equation shows how the scale factor varies with time for a universe with critical mass density.

The expression (14.28) can be used to calculate the actual age of the universe in terms of its Hubble age. Differentiation of Eq. (14.28) gives

$$\dot{a} = (2\pi\beta^2 G K_\beta)^{\frac{1}{3\beta}} \frac{2}{3\,\beta} t^{\frac{2}{3\beta}-1} = \frac{2\,a}{3\,\beta\,t}.$$

Therefore

$$H = \frac{\dot{a}}{a} = \frac{2}{3\,\beta\,t}.$$

Multiplying by t and dividing by H gives

$$t = \frac{2}{3\beta}\frac{1}{H}.$$

Inserting values for the present time

$$t_0 = \frac{2}{3\beta}\frac{1}{H_0}$$

or

$$t_0 = \frac{2}{3\beta} t_{H_0}.$$

This is the sought expression. For a dust dominated universe, $\beta = 1$, and

$$t_0 = \frac{2}{3} t_{H_0}. \tag{14.29}$$

Inserting the value of the Hubble age from Eq. (14.18) into Eq. (14.29) gives the age of a dust dominated universe—more commonly known as 'matter dominated'—with critical mass density

$$t_0 = (6.7/h) \times 10^9 \text{ years}.$$

For $h = 0.6$ we find the most probable age for this universe model

$$t_0 = 11.2 \times 10^9 \text{ years}. \tag{14.30}$$

14.5 The matter and radiation dominated periods

Observations reveal that the visible matter in the universe consists mainly of very cold and thin hydrogen and helium gas for which the pressure divided by c^2 is negligibly small compared to the density. Thus we can, as a good approximation, put $p = 0$, or $\beta = 1$ in the equation of state of the cosmic matter during such a matter dominated period. Matter with vanishing pressure is called dust. Putting $\beta = 1$ in Eq. (14.26) we get

$$\rho_{\text{dust}}\, a^3 = (\rho_{\text{dust}})_0\, a_0^3, \tag{14.31}$$

where the index 0 denotes 'value at the present age of the universe'. The mass density estimated from visible matter in the universe is between 0.5% and 5% of the critical density. However, the dynamics of the galaxies and clusters of galaxies indicates that there may be large amounts of dark matter in the universe. We therefore assume that the density of the cosmic matter is equal to the critical density.

In 1965 it was discovered that the universe is filled by black body radiation. Recent measurements with the COBE satellite has shown that the radiation has a temperature of $T = 2.726K$. According to the Stefan–Boltzmann radiation law, black body radiation with this temperature has a mass density $(\rho_{\text{rad}})_0 = 4{,}5 \times 10^{-31}\ \text{kg/m}^3$.

The equation of state of black body radiation is

$$p = \frac{1}{3}\frac{\rho_{\text{rad}}}{c^2},$$

which corresponds to $\beta = 4/3$. Inserting this in Eq. (14.26) we get

$$\rho_{\text{rad}}\, a^4 = (\rho_{\text{rad}})_0\, a_0^4. \tag{14.32}$$

Thinking of earlier periods in the history of the universe, the value of the scale factor was less, and the densities of the dust and the radiation were greater. Since $\rho_{\text{rad}}\, a^4 = \text{constant}$ for radiation, and $\rho_{\text{dust}}\, a^3 = \text{constant}$ for dust, the density of the radiation increases faster when we go backwards in time than the density of the dust. Thus at a certain point of time, t_{eq} the mass density of the radiation was equal to that of the dust. Before this point of time the mass density of the radiation was greater than that of the dust, i.e. the universe was *radiation dominated*. In order to find the point of time for the transition from the radiation dominated era to the dust dominated era, we divide Eq. (14.32) by (14.31), which leads to

$$\frac{\rho_{\text{rad}}}{\rho_{\text{dust}}}\, a = \frac{(\rho_{\text{rad}})_0}{(\rho_{\text{dust}})_0}\, a_0.$$

At the point of time t_{eq} the densities ρ_{rad} and ρ_{dust} were equal, from which follows

$$a(t_{\text{eq}}) = \frac{(\rho_{\text{rad}})_0}{(\rho_{\text{dust}})_0}\, a_0.$$

Inserting the values for $(\rho_{\text{rad}})_0$ and $(\rho_{\text{dust}})_0$ we find

$$a(t_{\text{eq}}) \approx 10^{-4}\, a_0. \tag{14.33}$$

In order to proceed in the calculation of the point of time t_{rd} we must find the scale factor as function of time in a dust dominated universe with critical mass density. This is given in Eq. (14.28) with $\beta = 1$,

$$a = (6\pi G K_1)^{1/3}\, t^{2/3}. \tag{14.34}$$

The value of the scale factor at the present point of time is

$$a_0 = (6\pi G K_1)^{1/3}\, t_0^{2/3}. \tag{14.35}$$

Dividing Eq. (14.34) by Eq. (14.35) gives

$$a(t) = a_0 \left(\frac{t}{t_0}\right)^{2/3}.$$

The value of the scale factor at the transition from the radiation dominated to the dust dominated era, is

$$a(t_{eq}) = a_0 \left(\frac{t_{eq}}{t_0}\right)^{2/3}. \tag{14.36}$$

From Eqs. (14.33 and (14.36) follow

$$\left(\frac{t_{eq}}{t_0}\right)^{2/3} = 10^{-4}.$$

Multiplying the exponent by $3/2$ on each side of this equation we get

$$\frac{t_{eq}}{t_0} = (10^{-4})^{3/2} = 10^{-6}$$

or

$$t_{eq} = 10^{-6}\, t_0.$$

Inserting the value (14.30) gives $t_{eq} = 1.12 \times 10^4$ years.

Before this time the universe was radiation dominated. Inserting $\beta = 4/3$ in Eq. (14.28) results in

$$a = \sqrt{(6\pi(4/3)^2 G\, K_{4/3}}\,\sqrt{t} = \sqrt{(32/3)\pi G\, K_{4/3}}\,\sqrt{t}. \tag{14.37}$$

At the end of the radiation dominated era the scale factor was

$$a_{eq} = \sqrt{(32/3)\,\pi G\, K_{4/3}}\,\sqrt{t_{eq}}. \tag{14.38}$$

Dividing Eq. (14.37) by Eq. (14.38) we get

$$a = a_{eq}\sqrt{t/t_{eq}}.$$

Differentiating we find the expansion velocity during the radiation dominated era

$$\dot{a} = \frac{1}{2} \frac{a_{eq}}{\sqrt{t_{eq}}} \frac{1}{\sqrt{t}}. \tag{14.39}$$

Thus the expansion velocity was extremely large at early times.

The model we have considered up to now is called the *standard model* of the universe. This model is not without problems!

14.6 Problems faced by the standard model of the universe

There are problems of two different types. The first type concerns properties of the models that are in accordance with observations, but which are not explained as a consequence of the dynamics of the model. They are put into the model in an ad hoc way as initial conditions. Among such properties are the isotropy, the homogeneity, the amount of matter in the universe and the expansion of the model. The second type of problems are those where the properties of the model are in conflict with observations. The most prominent problem of this type concerns the age of the universe.

As an illustration we shall consider the problem concerned with the amount of matter in the universe. Let us introduce a parameter

$$\epsilon \equiv \Omega - \frac{1}{\Omega} \tag{14.40}$$

as a measure of the quantity of matter in the universe, where Ω is defined in Eq. (14.22). $\epsilon < 0$ if $\rho < \rho_{crit}$, $\epsilon = 0$ if $\rho = \rho_{crit}$ and $\epsilon > 0$ if $\rho > \rho_{crit}$. If the actual density is very much less than the critical density, ϵ has a large negative value, and if the density is very much larger than the critical density, ϵ has a large positive value. If the absolute value of ϵ is not very large, the density of the cosmic matter is close to the critical density. Measurements indicate that $\epsilon \approx -100$, which is a small number in this connection.

We shall calculate an initial value for ϵ, i.e. the value of ϵ at the Planck time, $t_{Planck} = 10^{-43}$ s, according to the standard model. The time variation of $\Omega - 1$ is given in Eq. (14.23). The scale factor during the matter dominated period is given by Eq. (14.34). Differentiation gives

$$\dot{a} \propto t^{-1/3},$$

where $\propto$ denotes 'proportional to'. Inserting this into Eq. (14.23) we find the time variation of $\Omega - 1$ during the matter dominated period,

$$\Omega - 1 \propto t^{2/3}.$$

The time variation of $\Omega - 1$ during the radiation dominated period follows by inserting Eq. (14.39) into Eq. (14.23), which results in

$$\Omega - 1 \propto t. \tag{14.41}$$

These expressions show that the value of Ω is departing from 1 with increasing time. Thus, the value of Ω was closer to 1 at earlier times. In our calculation of the value of ϵ at the Planck time we can therefore put $\Omega \approx 1$. Hence

$$\begin{aligned}\epsilon &= \Omega - \frac{1}{\Omega} = \frac{\Omega^2 - 1}{\Omega} \approx \Omega^2 - 1 \\ &= (\Omega + 1)(\Omega - 1) \approx 2(\Omega - 1).\end{aligned} \tag{14.42}$$

Since $\Omega - 1$ is proportional to t during the radiation dominated period, then ϵ is so too, according to Eqs. (14.41) and (14.42). We can then write

$$\epsilon_{\text{Planck}} \approx \epsilon_{\text{eq}}(t_{\text{Planck}}/t_{\text{eq}}), \tag{14.43}$$

where the index "Planck" denotes 'value at the Planck time' and "eq" 'value at the transition time from radiation dominated to matter dominated'. Similarly, since $\Omega - 1$ is proportional to $t^{2/3}$ during the matter dominated period, we can write

$$\epsilon_{\text{eq}} \approx \epsilon_0 \, (t_{\text{eq}}/t_0)^{2/3}.$$

Inserting this expression into Eq. (14.43) leads to

$$\epsilon_{\text{Planck}} \approx \frac{t_{\text{Planck}}}{t_{\text{eq}}} \frac{t_{\text{eq}}^{2/3}}{t_0^{2/3}} \epsilon_0 = \frac{t_{\text{Planck}}}{t_{\text{eq}}^{1-2/3} t_0^{2/3}} \epsilon_0 = \frac{t_{\text{Planck}}}{t_{\text{eq}}^{1/3} t_0^{2/3}} \epsilon_0.$$

Being here only interested in an order of magnitude estimate, we insert $t_{\text{Planck}} = 10^{-43}$ s, $t_{\text{eq}} = 10^4$ years $\approx 10^{11}$ s, $t_0 = 10^{10}$ years $\approx 10^{17}$ s, $\epsilon_0 = -100$ and get $\epsilon_{\text{Planck}} = -10^{-56}$. Thus, at the Planck time the mass density of the universe was extremely close to the critical density. Such an extremely accurate adjustment of an initial condition of the universe cannot be explained within the frame of the standard model of the universe.

In Sect. 14.7 of this book we shall see how new ideas that appeared at about 1980 offer a solution to this problem, and also offer an explanation of why the universe expands.

14.7 Inflationary cosmology

According to quantum field theory, it is impossible to remove all the energy from a region. Vacuum is not a space without energy, but with the least possible energy.

Some of the great advances in our understanding of the behaviour of the material universe are associated with advents of new unified theories of different phenomena.

For example, at the beginning of the eighteenth century, electrical, magnetic and optical phenomena were not related to each other. However, between 1850 and 1860 J. C. Maxwell managed to develop a unified theory of these phenomena, and an understanding that light is electromagnetic waves. This type of theoretical advance has proceeded. More than a hundred years later a Nobel prize was given for the detection of particles that were predicted by a unified theory for the weak nuclear force and the electromagnetic force. One has also constructed so-called Grand Unified Theories (GUTs) for the 'electroweak' force and the strong nuclear force. Even if these theories lack experimental confirmation, we shall explore one consequence of them, that is of great significance for the evolution of the universe.

Let us first calculate the time dependence of the critical cosmic energy density. Inserting (14.27) into (14.26) gives

$$\rho_{\text{crit}}\, 6\pi G\beta^2 K_\beta\, t^2 = K_\beta$$

or

$$6\pi G\beta^2\, \rho_{\text{crit}}\, t^2 = 1.$$

This means that

$$\rho_{\text{crit}} \propto 1/t^2, \tag{14.44}$$

which shows that the critical density was large at earlier times. Since the actual density was then very close to the critical density the actual density of the cosmic energy was very large in the beginning.

According to the Stefan–Boltzmann radiation law

$$\rho_{\text{rad}} \propto T^4. \tag{14.45}$$

Equations (14.44) and (14.45) imply that

$$T^4 \propto 1/t^2$$

or

$$T \propto 1/\sqrt{t}.$$

This equation shows that the cosmic background radiation had a very high temperature in the beginning. Also it shows that the temperature of the cosmic background radiation can be used as a cosmic clock.

The GUTs combined with the model of a universe that was extremely hot in the beginning, has as a consequence that the evolution of the universe may have been dominated by vacuum energy at its first moment, before the universe was 10^{-33} s old.

Several experiments have been performed to measure the velocity of the Earth through the vacuum, the most wellknown one being the Michelson–Morley experiment of 1887. None of these experiments managed to measure such a velocity. It is consistent with the results of these experiments to assume that the vacuum energy is of such a nature that it is impossible to measure velocity relative to the vacuum.

We shall now deduce some very interesting consequences of this assumption within the context of homogeneous and isotropic universe models. The only physical properties attributed to the vacuum energy are density and pressure or tension. This means that we can describe the vacuum as a perfect fluid. Since the vacuum has no velocity, the components of its energy-momentum tensor in an arbitrarily moving local inertial frame are those given in Eq. (10.27),

$$T^{ij} = p\,\delta^{ij} \quad \text{and} \quad T^{tt} = \rho. \tag{14.46}$$

Because it is impossible to measure velocity relative to the vacuum fluid, all components of the energy-momentum tensor must be invariant, i.e. unchanged, under arbitrary Lorentz transformations. We shall apply this requirement to the components T^{xx} and T^{tt} under a Lorentz transformation in the x direction. Denoting the transformed components by $T^{x'x'}$ and $T^{t't'}$ we then have

$$T^{x'x'} = T^{xx} \quad \text{and} \quad T^{t't'} = T^{tt}. \tag{14.47}$$

Using the second of Eqs. (5.80) the transformed components are

$$T^{x'x'} = \frac{\partial x'}{\partial x^\mu}\frac{\partial x'}{\partial x^\nu}T^{\mu\nu} \quad \text{and} \quad T^{t't'} = \frac{\partial t'}{\partial x^\mu}\frac{\partial t'}{\partial x^\nu}T^{\mu\nu}. \tag{14.48}$$

The elements of the transformation matrix is obtained by differentiating the coordinates x' and t' in Eq. (5.98),

$$\frac{\partial x'}{\partial x} = \gamma, \quad \frac{\partial x'}{\partial t} = -\gamma v, \quad \frac{\partial t'}{\partial x} = -\frac{\gamma v}{c^2}, \quad \text{and} \quad \frac{\partial t'}{\partial t} = \gamma. \tag{14.49}$$

Performing the summations over μ and ν in Eq. (14.48), and keeping only non-vanishing terms, we get

$$\begin{aligned} T^{x'x'} &= \frac{\partial x'}{\partial x}\frac{\partial x'}{\partial x}T^{xx} + \frac{\partial x'}{\partial t}\frac{\partial x'}{\partial t}T^{tt}, \\ T^{t't'} &= \frac{\partial t'}{\partial x}\frac{\partial t'}{\partial x}T^{xx} + \frac{\partial t'}{\partial t}\frac{\partial t'}{\partial t}T^{tt}. \end{aligned}$$

Inserting the expressions (14.49) leads to

$$\begin{aligned} T^{x'x'} &= \gamma^2\,T^{xx} + \gamma^2\,v^2\,T^{tt}, \\ T^{t't'} &= (\gamma^2 v^2/c^4)\,T^{xx} + \gamma^2\,T^{tt}. \end{aligned} \tag{14.50}$$

From Eqs. (14.47) and (14.50), we have

$$\begin{aligned} T^{xx} &= \gamma^2\,T^{xx} + \gamma^2\,v^2\,T^{tt}, \\ T^{tt} &= (\gamma^2 v^2/c^4)\,T^{xx} + \gamma^2\,T^{tt} \end{aligned}$$

or

$$(\gamma^2 - 1)\, T^{xx} = -\gamma^2 v^2\, T^{tt},$$
$$(\gamma^2 - 1)\, T^{tt} = -(\gamma^2 v^2/c^4)\, T^{xx}. \tag{14.51}$$

Dividing these equations by each other, we get

$$\frac{T^{xx}}{T^{tt}} = \frac{T^{tt}}{T^{xx}} c^4.$$

Hence

$$(T^{xx})^2 = (T^{tt})^2\, c^4.$$

From Eq. (14.51) follows that T^{xx} and T^{tt} have opposite signs. Thus, we take the negative root, which results in

$$T^{xx} = -T^{tt}\, c^2.$$

Inserting $T^{xx} = p$ and $T^{tt} = \rho$ from Eq. (14.46) we obtain

$$p = -\rho\, c^2. \tag{14.52}$$

This equation shows that the 'vacuum fluid' is in a state of tension. Inserting Eq. (14.52) in Eq. (12.36) for the gravitational mass density, we get

$$\rho_g = -2\rho.$$

We are lead to the conclusion that the gravitational mass density of vacuum is negative, which means that the gravity of vacuum is repulsive.

We here see the remarkable strength of the general theory of relativity. From the assumption that we cannot measure velocity relative to vacuum, the theory implies that vacuum acts upon itself with repulsive gravitation. A vacuum dominated region will have a tendency to explode!

The equation of state (14.52) is just Eq. (14.25) with $\beta = 0$. Inserting $\beta = 0$ into Eq. (14.26), and putting $a^0 = 1$, we get $\rho = K_0$, showing that the vacuum energy has constant energy density during the expansion of the universe. There will be more and more vacuum energy during the expansion.

From Eq. (14.23) follows that since the expansion velocity $\dot{a}$ increases during the accelerated expansion in a vacuum dominated period, the value of Ω approaches 1, i.e. the density of the cosmic vacuum fluid approaches the critical density. The time dependence of the scale factor during the early vacuum dominated period, which is called *the inflationary era*, may thus be found by integrating Eq. (14.19) with $k = 0$ and $\rho = \rho_{\text{vac}} = \text{constant}$,

$$\dot{a}^2 = \frac{8\pi G}{3}\, \rho_{\text{vac}}\, a^2.$$

Taking the square root and dividing by a we get

$$\frac{\dot{a}}{a} = \sqrt{\frac{8\pi G}{3}\rho_{\text{vac}}}. \tag{14.53}$$

Note that the left-hand side is the Hubble factor of this universe model. Since the right-hand side is constant, it follows that the Hubble factor of a vacuum dominated universe model with critical density, is constant. We denote this constant by H_{vac}, that is

$$H_{\text{vac}} = \sqrt{\frac{8\pi G}{3}\rho_{\text{vac}}}. \tag{14.54}$$

The transition into a vacuum dominated period happened at a point of time when the energy density of the radiation became less than a certain energy density ρ_{GUT} determined by the GUTs. Thus, according to Eq. (14.54) the Hubble factor during the inflationary era is

$$H_{\text{vac}} = \sqrt{\frac{8\pi G}{3}\rho_{\text{GUT}}}.$$

The point of time t_1 when the universe entered the inflationary era, is taken to be the Hubble age associated with the GUT-energy density, i.e.

$$t_1 = 1/H_{\text{vac}}.$$

The GUTs give $t_1 = 10^{-35}$ s.

Equation (14.53) can now be written

$$\frac{\dot{a}}{a} = H_{\text{vac}} = \frac{1}{t_1}.$$

Using that $\dot{a} = da/dt$, and multiplying by dt we thus get

$$\frac{da}{a} = \frac{1}{t_1}\,dt.$$

Integrating this equation by means of the rule (3.54) with the initial condition $a(t_1) = a_1$, we find

$$\ln a - \ln a_1 = \frac{1}{t_1}(t - t_1) = \frac{t}{t_1} - 1.$$

Applying the rule (3.50) this equation can be written as

$$\ln\frac{a}{a_1} = \frac{t}{t_1} - 1. \tag{14.55}$$

Using that $y = \ln x \Leftrightarrow x = e^y$, where e is the base of the natural logarithm, ans substituting a for x and $H_{vac}\, t$ for y, we see that Eq. (14.55) is equivalent to the equation

$$a/a_1 = e^{(t/t_1)-1}$$

or

$$a = a_1\, e^{(t/t_1)-1}.$$

Thus the universe has exponential expansion during the inflationary era.

Differentiation gives

$$\dot{a} = (a_1/t_1)\, e^{(t/t_1)-1}.$$

Inserting this into Eq. (14.23) leads to

$$\Omega - 1 = k\,(c\,t_1/a_1)^2 e^{-2[(t/t_1)-1]}.$$

This equation shows that Ω approaches 1 exponentially fast during the inflationary era.

At the beginning of the inflationary era the d ensity of the cosmic matter may have differed very much from the critical density. If, for example, $a_1 \ll c\,t_1$, then

$$\Omega(t_1) = 1 + k\,(c\,t_1/a_1)^2 \approx k\,(c\,t_1/a_1)^2 \gg 1,$$

which means that the cosmic mass density was much larger than the critical density at the initial moment of the inflationary era. The value of the parameter ϵ introduced in Eq. (14.40), was then, with good accuracy

$$\epsilon(t_1) \approx \Omega\,(t_1) \approx k\,(c\,t_1/a_1)^2. \tag{14.56}$$

However, as noted above, the density approached the critical density very fast during the inflationary era, and at the end of this era we can use the approximation (14.42). According to the GUTs the inflationary era lasted until $t_2 = 10^{-33}$ s, and the value of ϵ at the end of the inflationary era was

$$\epsilon(t_2) \approx 2\,[\Omega(t_2) - 1] = 2\,k\,(c\,t_1/a_1)\,e^{-2\,[(t_2/t_1)-1]}.$$

Using Eq. (14.56) and inserting the values of t_1 and t_2 we get

$$\epsilon(t_2) = 2\,\epsilon(t_1)\,e^{-[(t_2/t_1)-1]} = 2 \times e^{-199}\,\epsilon(t_1) \approx 10^{-87}\epsilon(t_1).$$

This shows that during the inflationary era the density became extremely close to the critical density. Hence one of the predictions of the inflationary universe models, is that the universe should still have a cosmic mass density close to the critical density. Thus the inflationary cosmology solves the problem, which was unsolved in standard model, of explaining the observed fact that the mass density is close to the critical density. In fact it seems to solve it too well. Observations indicate a cosmic mass density which is close to, but less than the critical mass density, $0.2 < \Omega < 0.4$. This is less than the mass density predicted by the inflationary models.

One final triumph of the inflationary models should be mentioned, though. We now seem to know the answer to the question: Why does the universe expand? The inflationary cosmological models offer the following answer: Once upon a time vacuum energy filled the universe, it had negative mass, and repulsive gravitation forced the universe to expand.

Appendix A
The Laplacian in a spherical coordinate system

In order to be able to deduce the most important physical consequences from the Poisson equation (12.5), which represents the Newtonian limit of Einstein's field equations, we must know the form of the Laplacian in a spherical coordinate system. This is because most important applications of any theory of gravitation involve physical systems, for example planets, stars, black holes and the whole universe, that are spherically symmetric. Of course it is possible to find the wanted form of the Laplacian in collections of mathematical formulae. However, in the spirit of the rest of this book, we here offer a detailed deduction.

The Laplace operator, or more commonly called the Laplacian, is defined as the divergence of the gradient, i.e.

$$\nabla^2 \equiv \operatorname{div} \operatorname{grad}. \tag{A.1}$$

We shall find an expression of the Laplacian valid in an arbitrary orthogonal coordinate system, and then specialize to a spherical coordinate system. In order to calculate the Laplacian in an orthogonal coordinate system, we must first find expressions for the gradient and the divergence in such coordinate systems. We start with the gradient.

An operator is something which acts upon a function and changes it in a prescribed way. The gradient operator acts upon a scalar function by differentiating it, and gives out a vector called the gradient of the function. In an arbitrary coordinate system the gradient operator is defined as a vector with covariant components

$$\nabla_i \equiv \frac{\partial}{\partial x^i}. \tag{A.2}$$

According to the rule (5.77) for raising an index, the contravariant components are

$$\nabla^i = g^{ij}\,\nabla_j = g^{ij}\,\frac{\partial}{\partial x^j}. \tag{A.3}$$

Ø. Grøn and A. Næss, *Einstein's Theory: A Rigorous Introduction for the Mathematically Untrained*, DOI 10.1007/978-1-4614-0706-5,

We shall need to know the expression for the gradient in the class of coordinate systems with orthogonal coordinate axes. Such coordinate systems have diagonal metric tensors, i.e. $g_{ij} = 0$ for $i \neq j$. In this case the only non-vanishing contravariant components of the metric tensor are given in terms of the covariant components by Eq. (5.75).

$$g^{ii} = 1/g_{ii}.$$

Inserting this into Eq. (A.3) gives

$$\nabla^i = \frac{1}{g_{ii}} \frac{\partial}{\partial x^i}.$$

Next we shall find an expression for the divergence in terms of ordinary partial derivatives and the components of the metric tensor in an orthogonal coordinate system. The divergence is given as a covariant derivative in Eq. (10.3). Inserting the expression (7.15) for the covariant derivative into Eq. (10.3), leads to

$$\text{div}\,\vec{F} = \frac{\partial F^i}{\partial x^i} + F^k \Gamma^i{}_{ki}. \tag{A.4}$$

Hence, we must find an expression for the Christoffel symbols $\Gamma^i{}_{ki}$ in an orthogonal coordinate system with a diagonal metric tensor. For this purpose we use Eq. (7.30) with $\tau = i$, $\nu = k$, $\lambda = i$, which for a diagonal metric tensor reduces to

$$\Gamma^i{}_{ki} = \frac{1}{2\,g_{ii}} \left(\frac{\partial g_{ii}}{\partial x^k} + \frac{\partial g_{ik}}{\partial x^i} - \frac{\partial g_{ik}}{\partial x^i} \right).$$

The last two terms cancel each other, so we are left with

$$\Gamma^i{}_{ki} = \frac{1}{2\,g_{ii}} \frac{\partial g_{ii}}{\partial x^k}. \tag{A.5}$$

Inserting this into Eq. (A.4) we get

$$\text{div}\,\vec{F} = \frac{\partial F^i}{\partial x^i} + F^k \frac{1}{2\,g_{ii}} \frac{\partial g_{ii}}{\partial x^i}. \tag{A.6}$$

We now define

$$g \equiv g_{11}\, g_{22}\, g_{33}. \tag{A.7}$$

Differentiation g by means of the product rule (2.22) we first put $u = g_{11}\, g_{22}$ and $v = g_{33}$ and get

$$\frac{\partial g}{\partial x^k} = \frac{\partial (g_{11}\, g_{22})}{\partial x^k} g_{33} + g_{11}\, g_{22} \frac{\partial g_{33}}{\partial x^k}. \tag{A.8}$$

Using the product rule once more, we have

$$\frac{\partial(g_{11}\, g_{22})}{\partial x^k} = \frac{\partial g_{11}}{\partial x^k}\, g_{22} + g_{11}\, \frac{\partial g_{22}}{\partial x^k}.$$

Inserting this into Eq. (A.8) gives

$$\frac{\partial g}{\partial x^k} = \frac{\partial g_{11}}{\partial x^k}\, g_{22}\, g_{33} + g_{11}\, \frac{\partial g_{22}}{\partial x^k}\, g_{33} + g_{11}\, g_{22}\, \frac{\partial g_{33}}{\partial x^k}.$$

Multiplying the numerators and the denominators of the terms at the right-hand side by g_{11}, g_{22}, and g_{33}, respectively, reordering, and using Eq. (A.7), we get

$$\frac{\partial g}{\partial x^k} = \frac{g}{g_{11}}\, \frac{\partial g_{11}}{\partial x^k} + \frac{g}{g_{22}}\, \frac{\partial g_{22}}{\partial x^k} + \frac{g}{g_{33}}\, \frac{\partial g_{33}}{\partial x^k}.$$

Using Einstein's summation convention this may be written as

$$\frac{\partial g}{\partial x^k} = \frac{g}{g_{ii}}\, \frac{\partial g_{ii}}{\partial x^k}.$$

Dividing by g and exhanging the left-hand and right-hand sides, we get

$$\frac{1}{g_{ii}}\, \frac{\partial g_{ii}}{\partial x^k} = \frac{1}{g}\, \frac{\partial g}{\partial x^k}.$$

Inserting this into Eq. (A.6) we get

$$\operatorname{div} \vec{F} = \frac{\partial F^i}{\partial x^k} + F^k\, \frac{1}{2g}\, \frac{\partial g}{\partial x^k}. \tag{A.9}$$

In order to simplify this expression, we differentiate $\sqrt{g}$. Using the rule (2.36) with $n = 1/2$ and the chain rule (2.31), we obtain

$$\frac{\partial \sqrt{g}}{\partial x^k} = \frac{\partial g^{1/2}}{\partial x^k} = \frac{1}{2}\, g^{-1/2}\, \frac{\partial g}{\partial x^k} = \frac{1}{2\, \sqrt{g}}\, \frac{\partial g}{\partial x^k}.$$

Dividing by $\sqrt{g}$ gives

$$\frac{1}{\sqrt{g}}\, \frac{\partial \sqrt{g}}{\partial x^k} = \frac{1}{2\, g}\, \frac{\partial g}{\partial x^k}.$$

Thus, Eq. (A.9) can be written as

$$\operatorname{div} \vec{F} = \frac{\partial F^i}{\partial x^i} + F^i\, \frac{1}{\sqrt{g}}\, \frac{\partial \sqrt{g}}{\partial x^i}. \tag{A.10}$$

Using, once more, the product rule for differentiation, we get

$$\frac{1}{\sqrt{g}}\frac{\partial(\sqrt{g}\,F^i)}{\partial x^i} = \frac{1}{\sqrt{g}}\left(\sqrt{g}\,\frac{\partial F^i}{\partial x^i} + F^i\,\frac{\partial\sqrt{g}}{\partial x^i}\right)$$

$$= \frac{\partial F^i}{\partial x^i} + F^i\,\frac{1}{\sqrt{g}}\,\frac{\partial\sqrt{g}}{\partial x^i}.$$

Comparing with Eq. (A.10) we see that this equation can be written as

$$\operatorname{div}\vec{F} = \frac{1}{\sqrt{g}}\,\frac{\partial}{\partial x^i}\left(\sqrt{g}\,F^i\right). \tag{A.11}$$

Inserting $F^i = \nabla^i$ from Eq. (A.3) and the expression (A.11) for the divergence into Eq. (A.1), we ultimately arrive at the expression for the Laplacian in an arbitrary orthogonal coordinate system

$$\nabla^2 \equiv \frac{1}{\sqrt{g}}\,\frac{\partial}{\partial x_i}\left(\frac{\sqrt{g}}{g_{ii}}\,\frac{\partial}{\partial x^i}\right). \tag{A.12}$$

In the case of a Cartesian coordinate system, the only non-vanishing components of the metric tensor are $g_{xx} = g_{yy} = g_{zz} = 1$. Then $\sqrt{g} = 1$, and performing the summation over j in Eq. (A.12), we get

$$\nabla^2 = \frac{\partial^2}{\partial x^2} + \frac{\partial^2}{\partial y^2} + \frac{\partial^2}{\partial z^2}.$$

Using Einstein's summation convention this may be written as

$$\nabla^2 = \frac{\partial^2}{\partial x_i\,\partial x^i} \tag{A.13}$$

valid in Cartesian coordinates.

The expression (A.12) shall now be used to calculate the Laplace operator in a spherical coordinate system. Then we need the line element of Euclidean 3-dimensional space as expressed in a spherical coordinate system. The basis vectors of this coordinate system are given in terms of the basis vectors of a Cartesian coordinate system in Eq. (6.22). The components of the metric tensor in this coordinate system are given by the scalar products of the basis vectors in Eq. (6.22). Since the vectors are orthogonal to each other only the products of each vector with itself are different from zero. Using Eq. (4.11) we find

$$g_{rr} = \vec{e}_r\cdot\vec{e}_r = \sin^2\theta\,\cos^2\varphi + \sin^2\theta\,\sin^2\varphi + \cos^2\theta$$

$$= \sin^2\theta\left(\cos^2\varphi + \sin^2\varphi\right) + \cos^2\theta$$

$$= \sin^2\theta + \cos^2\theta = 1,$$

$$
\begin{aligned}
g_{\theta\theta} &= \vec{e}_\theta \cdot \vec{e}_\theta = r^2 \left(\cos^2\theta \cos^2\varphi + \cos^2\theta \sin^2\varphi + \sin^2\theta\right) \\
&= r^2 \left[\cos^2\theta \left(\cos^2\varphi + \sin^2\varphi\right) + \sin^2\theta\right] \\
&= r^2 \left(\cos^2\theta + \sin^2\theta\right) = r^2, \\
g_{\varphi\varphi} &= \vec{e}_\varphi \cdot \vec{e}_\varphi = r^2 \left(\sin^2\theta \sin^2\varphi + \sin^2\theta \cos^2\varphi\right) \\
&= r^2 \sin^2\theta \left(\sin^2\varphi + \cos^2\varphi\right) = r^2 \sin^2\theta.
\end{aligned}
$$

Thus the line element of flat space in spherical coordinates has the form

$$
d\ell^2 = dr^2 + r^2\, d\theta^2 + r^2 \sin^2\theta\, d\varphi^2.
$$

The components of the metric tensor in a spherical coordinate system are therefore

$$
g_{rr} = 1, \quad g_{\theta\theta} = r^2, \quad \text{and} \quad g_{\varphi\varphi} = r^2 \sin^2\theta. \tag{A.14}
$$

Performing the summation over i in Eq. (A.12), with $x^1 = r$, $x^2 = \theta$, and $x^3 = \varphi$, we have

$$
\nabla^2 = \frac{1}{\sqrt{g}} \frac{\partial}{\partial r}\left(\frac{\sqrt{g}}{g_{rr}} \frac{\partial}{\partial r}\right) + \frac{1}{\sqrt{g}} \frac{\partial}{\partial \theta}\left(\frac{\sqrt{g}}{g_{\theta\theta}} \frac{\partial}{\partial \theta}\right) + \frac{1}{\sqrt{g}} \frac{\partial}{\partial \varphi}\left(\frac{\sqrt{g}}{g_{\varphi\varphi}} \frac{\partial}{\partial \varphi}\right).
$$

Inserting the expressions (A.14) and using that $g = r^4 \sin^2\theta$, we get

$$
\begin{aligned}
\nabla^2 = &\frac{1}{r^2 \sin\theta} \frac{\partial}{\partial r}\left(r^2 \sin\theta \frac{\partial}{\partial r}\right) \\
&+ \frac{1}{r^2 \sin\theta} \frac{\partial}{\partial \theta}\left(\frac{r^2 \sin\theta}{r^2} \frac{\partial}{\partial \theta}\right) \\
&+ \frac{1}{r^2 \sin\theta} \frac{\partial}{\partial \varphi}\left(\frac{r^2 \sin\theta}{r^2 \sin^2\theta} \frac{\partial}{\partial \varphi}\right).
\end{aligned}
$$

In the first term $\sin\theta$ is constant during the differentiation with respect to r, and can be moved to the numerator in front of $\partial/\partial r$. The $\sin\theta$ in the numerator and the denominator cancel each other. In the second term r^2 in the numerator and the denominator inside the parenthesis cancel each other. Finally, in the last term the $\sin\theta$ in the numerator cancels one of the $\sin\theta$ factors in the denominator inside the parenthesis. The remaining $\sin\theta$ can be put in front of $\partial/\partial\varphi$, since $\sin\theta$ is constant during differentiation with respect to φ. This finally gives the expression for the Laplacian in a spherical coordinate system

$$\nabla^2 = \frac{1}{r^2}\frac{\partial}{\partial r}\left(r^2\frac{\partial}{\partial r}\right) + \frac{1}{r^2\sin\theta}\frac{\partial}{\partial\theta}\left(\sin\theta\frac{\partial}{\partial\theta}\right) + \frac{1}{r^2\sin^2\theta}\frac{\partial^2}{\partial\varphi^2}. \tag{A.15}$$

Appendix B
The Ricci tensor of a static and spherically symmetric spacetime

The most interesting applications of Einstein's theory are concerned with systems having spherical symmetry. In Ch. 13 we consider spacetime outside static systems. Then we have to solve Einstein's vacuum field equations in a static, spherically symmetric spacetime. Since the vacuum field equations amount to putting the Ricci tensor equal to zero, we need to calculate the components of the Ricci tensor for this case.

We first have to calculate the Christoffel symbols for the metric (13.15). Then we use Eq. (7.30). We need the contravariant components $g^{\mu\nu}$ of the metric tensor. Since the metric is diagonal, they are given by Eq. (5.75), i.e. $g^{\mu\mu} = 1/g_{\mu\mu}$. Then Eq. (7.30) simplifies to

$$\Gamma^{\mu}{}_{\nu\lambda} = \frac{1}{2\,g_{\mu\mu}} \left(g_{\mu\nu,\lambda} + g_{\mu\lambda,\nu} - g_{\nu\lambda,\mu} \right). \tag{B.1}$$

In this equation and the next ones, there is no summation over the values of the index ν.

In order to calculate the Christoffel symbols, we divide the Christoffel symbols into three groups. First we consider the case with $\mu = \nu$. Inserting this into Eq. (B.1) gives

$$\Gamma^{\nu}{}_{\nu\lambda} = \frac{1}{2\,g_{\nu\nu}} \left(g_{\nu\nu,\lambda} + g_{\nu\lambda,\nu} - g_{\nu\lambda,\nu} \right).$$

The last two terms cancel each other, so we are left with

$$\Gamma^{\nu}{}_{\nu\lambda} = \frac{1}{2 g_{\nu\nu}}\, g_{\nu\nu,\lambda}. \tag{B.2}$$

Next we find an expression for the Christoffel symbols with $\lambda = \nu$ and μ different from both. Inserting $\lambda = \nu$ in Eq. (B.1) gives

$$\Gamma^{\mu}{}_{\nu\nu} = \frac{1}{2\,g_{\mu\mu}} \left(g_{\mu\nu,\nu} + g_{\mu\nu,\nu} - g_{\nu\nu,\mu} \right).$$

Ø. Grøn and A. Næss, *Einstein's Theory: A Rigorous Introduction for the Mathematically Untrained*, DOI 10.1007/978-1-4614-0706-5,

Since $\mu \neq \nu$ in this case, and the metric of the line element (13.14) has $g_{\mu\nu} = 0$ for $\mu \neq \nu$, the first two terms inside the parenthesis are equal to zero, which implies

$$\Gamma^{\mu}{}_{\nu\nu} = -\frac{1}{2\,g_{\mu\mu}}\,g_{\nu\nu,\mu}. \tag{B.3}$$

At last we consider the Christoffel symbols which have all three indices different, $\lambda \neq \mu$, $\lambda \neq \nu$, and $\mu \neq \nu$. Then the indices of the metric components in all three terms of Eq. (B.1) are different. Thus, all the terms are equal to zero, which leads to

$$\Gamma^{\mu}{}_{\nu\lambda} = 0, \quad \text{if} \quad \lambda \neq \mu, \quad \lambda \neq \nu, \quad \text{and} \quad \nu \neq \mu.$$

Inserting $\mu = \nu$ and $\mu = r$ in Eq. (B.2), we get

$$\Gamma^{r}{}_{rr} = \frac{1}{2\,g_{rr}}\,g_{rr,r}.$$

Inserting $g_{rr} = \mathrm{e}^{\lambda(r)}$ from Eq. (13.15),

$$\Gamma^{r}{}_{rr} = \frac{1}{2\,\mathrm{e}^{\lambda}}(\mathrm{e}^{\lambda})'. \tag{B.4}$$

We now use the chain rule, $[f(\lambda)]' = f'(\lambda)\,\lambda'$, where $f'(\lambda)$ is f differentiated with respect to λ, and (in the present case) λ' is λ differentiated with respect to r. Inserting $f(\lambda) = \mathrm{e}^{\lambda}$, and using Eq. (3.59) gives

$$(\mathrm{e}^{\lambda})' = \mathrm{e}^{\lambda}\,\lambda'.$$

Inserting this into Eq. (B.4),

$$\Gamma^{r}{}_{rr} = \frac{1}{2\,\mathrm{e}^{\lambda}}\,\mathrm{e}^{\lambda}\,\lambda' = \frac{\lambda'}{2}. \tag{B.5}$$

Next we calculate

$$\Gamma^{t}{}_{tr} = \frac{1}{2\,g_{tt}}\,g_{tt,r}.$$

Due to the symmetry of the Christoffel symbols in their lower indices, we have $\Gamma^{t}{}_{rt} = \Gamma^{t}{}_{tr}$. Inserting $g_{tt} = -c^2\,\mathrm{e}^{\nu}$ from Eq. (13.15), we can replace λ by ν in Eq. (B.5),

$$\Gamma^{t}{}_{tr} = \Gamma^{t}{}_{rt} = \frac{\nu'}{2}. \tag{B.6}$$

Furthermore

$$\Gamma^{\theta}{}_{r\theta} = \Gamma^{\theta}{}_{\theta r} = \frac{1}{2\,g_{\theta\theta}}\,g_{\theta\theta,r}.$$

Inserting $g_{\theta\theta} = r^2$,

$$\Gamma^{\theta}{}_{r\theta} = \Gamma^{\theta}{}_{\theta r} = \frac{1}{2\,r^2}\,\left(r^2\right)' = \frac{1}{2\,r^2}\,2r = \frac{1}{r}. \tag{B.7}$$

Next

$$\Gamma^{\varphi}{}_{r\varphi} = \Gamma^{\varphi}{}_{\varphi r} = \frac{1}{2\,g_{\varphi\varphi}}\,g_{\varphi\varphi,r}.$$

(B.16)
Inserting $g_{\varphi\varphi} = r^2 \sin^2\theta$,

$$\Gamma^{\varphi}{}_{r\varphi} = \Gamma^{\varphi}{}_{\varphi r} = \frac{1}{2\,r^2\,\sin^2\theta}\,\left(r^2\,\sin^2\theta\right)_{,r}. \tag{B.8}$$

Since θ is constant under a partial differentiation with respect to r,

$$\left(r^2\,\sin^2\theta\right)_{,r} = \left(r^2\right)_{,r}\,\sin^2\theta = 2r\,\sin^2\theta.$$

Inserting this into Eq. (B.8),

$$\Gamma^{\varphi}{}_{r\varphi} = \Gamma^{\varphi}{}_{\varphi r} = \frac{1}{2\,r^2\,\sin^2\theta}\,2\,r\,\sin^2\theta = \frac{1}{r}. \tag{B.9}$$

Then we calculate

$$\Gamma^{\varphi}{}_{\theta\varphi} = \Gamma^{\varphi}{}_{\varphi\theta} = \frac{1}{2\,g_{\varphi\varphi}}\,g_{\varphi\varphi,\theta} = \frac{1}{2\,r^2\,\sin^2\theta}\,\left(r^2\,\sin^2\theta\right)_{,\theta}.$$

We apply the rule that r is constant during partial differentiation with respect to θ, then the chain rule for differentiation, and obtain

$$\left(r^2\,\sin^2\right)_{,\theta} = r^2\,\left(\sin^2\theta\right)_{,\theta} = r^2\,2\,\sin\theta\,\left(\sin\theta\right)_{,\theta}.$$

According to Eq. (4.24), $(\sin\theta)_{,\theta} = \cos\theta$. This gives

$$\begin{aligned}\Gamma^{\varphi}{}_{\theta\varphi} = \Gamma^{\varphi}{}_{\varphi\theta} &= \frac{1}{2\,r^2\,\sin^2\theta}\,r^2\,2\,\sin\theta\,\cos\theta \\ &= \frac{\cos\theta}{\sin\theta}.\end{aligned} \tag{B.10}$$

We now use Eq. (B.3) to calculate the remaining non-vanishing Christoffel symbols. Inserting $\mu = r$ and $\nu = \theta$,

$$\Gamma^{r}{}_{\theta\theta} = -\frac{1}{2\,g_{rr}}\,g_{\theta\theta,r} = -\frac{1}{2\,e^{\lambda}}\,2\,r = -r\,e^{-\lambda}. \tag{B.11}$$

Inserting $\mu = r$ and $\nu = \varphi$,

$$\Gamma^r{}_{\varphi\varphi} = -\frac{1}{2\,g_{rr}}\,g_{\varphi\varphi,r} = -\frac{1}{2\,\mathrm{e}^{\lambda}}\,2\,r\,\sin^2\theta$$
$$= -r\,\mathrm{e}^{-\lambda}\,\sin^2\theta. \tag{B.12}$$

Inserting $\mu = r$ and $\nu = t$,

$$\Gamma^r{}_{tt} = -\frac{1}{2\,g_{rr}}\,g_{tt,r} = -\frac{1}{2\,\mathrm{e}^{\lambda}}\left(-c^2\mathrm{e}^{\nu}\right)_{,r}$$
$$= -\frac{1}{2\,\mathrm{e}^{\lambda}} - c^2\,\mathrm{e}^{\nu}\,\nu' = \frac{c^2}{2}e^{\nu-\lambda}\,\nu'. \tag{B.13}$$

Finally we insert $\mu = \theta$ and $\nu = \varphi$, which results in

$$\Gamma^\theta{}_{\varphi\varphi} = -\frac{1}{2\,g_{\theta\theta}}\,g_{\varphi\varphi,\theta} = -\frac{1}{2\,r^2}\left(r^2\,\sin^2\theta\right)_{,\theta}$$
$$= -\frac{1}{2\,r^2}\,r^2\,2\,\sin\theta\,\cos\theta = -\sin\theta\,\cos\theta. \tag{B.14}$$

The calculation of all the non-vanishing Christoffel symbols of the spacetime described by the line element (13.14) has now been completed.

From Eqs. (9.29) and (11.15) we get the following expression for the components of the Ricci tensor in terms of the Christoffel symbols and their derivatives

$$R_{\mu\nu} = \Gamma^\beta{}_{\mu\nu}\Gamma^\alpha{}_{\beta\alpha} - \Gamma^\beta{}_{\mu\alpha}\Gamma^\alpha{}_{\beta\nu} + \Gamma^\alpha{}_{\mu\nu,\alpha} - \Gamma^\alpha{}_{\mu\alpha,\nu}. \tag{B.15}$$

Since there are only two unknown metric functions, $\lambda(r)$ and $\nu(r)$, it is sufficient to find two of the field equations (11.35), say $R_{tt} = 0$ and $R_{rr} = 0$. However, it will turn out to be convenient also to calculate $R_{\theta\theta}$.

Inserting $\mu = \nu = t$ in Eq. (B.15),

$$R_{tt} = \Gamma^\beta{}_{tt}\Gamma^\alpha{}_{\beta\alpha} - \Gamma^\beta{}_{t\alpha}\Gamma^\alpha{}_{\beta t} + \Gamma^\alpha{}_{tt,\alpha} - \Gamma^\alpha{}_{t\alpha,t}.$$

Performing the summation over α and β (note, for example, that $\Gamma^\alpha{}_{\beta\alpha} = \Gamma^r{}_{\beta r} + \Gamma^\theta{}_{\beta\theta} + \Gamma^\varphi{}_{\beta\varphi} + \Gamma^t{}_{\beta t}$), and including only the non-vanishing Christoffel symbols, we deduce

$$R_{tt} = \Gamma^r{}_{tt}\left(\Gamma^r{}_{rr} + \Gamma^\theta{}_{r\theta} + \Gamma^\varphi{}_{r\varphi} + \Gamma^t{}_{tr}\right)$$
$$- \Gamma^t{}_{tr}\Gamma^r{}_{tt} - \Gamma^r{}_{tt}\Gamma^t{}_{rt} + \Gamma^r{}_{tt,r}.$$

Inserting the expressions calculated above for the Christoffel symbols, we get

$$R_{tt} = \frac{\nu'}{2} e^{\nu-\lambda} \left(\frac{\lambda'}{2} + \frac{1}{r} + \frac{1}{r} + \frac{\nu'}{2} \right) - 2\frac{\nu'}{2}\frac{\nu'}{2} e^{\nu-\lambda} + \left(\frac{\nu'}{2} e^{\nu-\lambda} \right)' . \tag{B.16}$$

Differentiation of the last term, using the product rule and the chain rule,

$$\left(\frac{\nu'}{2} e^{\nu-\lambda} \right)' = \frac{\nu''}{2} e^{\nu-\lambda} + \frac{\nu'}{2} e^{\nu-\lambda} \left(\nu' - \lambda' \right) .$$

Inserting this in Eq. (B.16), and putting the common factor $e^{\nu-\lambda}$ outside a parenthesis,

$$R_{tt} = \left[\frac{1}{4}\nu'\lambda' + \frac{\nu'}{r} + \frac{1}{4}\left(\nu'\right)^2 - \frac{1}{2}\left(\nu'\right)^2 + \frac{1}{2}\nu'' + \frac{1}{2}\left(\nu'\right)^2 - \frac{1}{2}\nu'\lambda' \right] e^{\nu-\lambda},$$

or

$$R_{tt} = \left[\frac{\nu''}{2} + \frac{1}{4}\left(\nu'\right)^2 + \frac{\nu'}{r} - \frac{1}{4}\lambda'\nu' \right] e^{\nu-\lambda}. \tag{B.17}$$

Inserting $\mu = \nu = r$ in Eq. (B.15) gives

$$R_{rr} = \Gamma^{\beta}{}_{rr}\Gamma^{\alpha}{}_{\beta\alpha} - \Gamma^{\beta}{}_{r\alpha}\Gamma^{\alpha}{}_{\beta r} + \Gamma^{\alpha}{}_{rr,\alpha} - \Gamma^{\alpha}{}_{r\alpha,r} .$$

Performing the summation over α and β, and including only non-vanishing terms, gives

$$\begin{aligned} R_{rr} = {} & \Gamma^{r}{}_{rr}\left(\Gamma^{r}{}_{rr} + \Gamma^{\theta}{}_{r\theta} + \Gamma^{\varphi}{}_{r\varphi} + \Gamma^{t}{}_{rt}\right) \\ & - \Gamma^{r}{}_{rr}\Gamma^{r}{}_{rr} + \Gamma^{t}{}_{rt}\Gamma^{t}{}_{tr} - \Gamma^{\theta}{}_{r\theta}\Gamma^{\theta}{}_{\theta r} - \Gamma^{\varphi}{}_{r\varphi}\Gamma^{\varphi}{}_{\varphi r} \\ & + \Gamma^{r}{}_{rr,r} - \Gamma^{r}{}_{rr,r} - \Gamma^{\theta}{}_{r\theta,\theta} - \Gamma^{\varphi}{}_{r\varphi,r} - \Gamma^{t}{}_{rt,r} . \end{aligned}$$

The first term coming from the first line and the first on the second line cancel each other, and the two first terms on the third line. Also, using the symmetry of the Christoffel symbols in the lower indices, we find

$$\begin{aligned} R_{rr} = {} & \Gamma^{r}{}_{rr}\left(\Gamma^{\theta}{}_{r\theta} + \Gamma^{\varphi}{}_{r\varphi} + \Gamma^{t}{}_{rt}\right) - \left(\Gamma^{t}{}_{rt}\right)^2 - \left(\Gamma^{\theta}{}_{r\theta}\right)^2 \\ & - \left(\Gamma^{\varphi}{}_{r\varphi}\right)^2 - \Gamma^{\theta}{}_{r\theta,r} - \Gamma^{\varphi}{}_{r\varphi,r} - \Gamma^{t}{}_{rt,r} . \end{aligned}$$

Inserting the expressions for the Christoffel symbols leads to

$$R_{rr} = \frac{\lambda'}{2}\left(\frac{2}{r} + \frac{\nu'}{2}\right) - \frac{\nu'^2}{4} - \frac{2}{r^2} + \frac{2}{r^2} - \frac{\nu''}{2},$$

which gives

$$R_{rr} = -\frac{1}{2}\nu'' + \frac{1}{4}\nu'\lambda' - \frac{1}{4}\nu'^2 + \frac{\lambda'}{r}. \tag{B.18}$$

Next we insert $\mu = \nu = \theta$ in Eq. (B.15),

$$R_{\theta\theta} = \Gamma^{\beta}{}_{\theta\theta}\Gamma^{\alpha}{}_{\beta\alpha} - \Gamma^{\theta}{}_{\theta\alpha}\Gamma^{\alpha}{}_{\beta\theta} + \Gamma^{\alpha}{}_{\theta\theta,\alpha} - \Gamma^{\alpha}{}_{\theta\alpha,\theta}.$$

Performing the summation over α and β, and including only non-vanishing terms,

$$\begin{aligned} R_{\theta\theta} &= \Gamma^{r}{}_{\theta\theta}\left(\Gamma^{r}{}_{rr} + \Gamma^{\theta}{}_{r\theta} + \Gamma^{\varphi}{}_{r\varphi} + \Gamma^{t}{}_{rt}\right) - \Gamma^{\theta}{}_{\theta r}\Gamma^{r}{}_{\theta\theta} \\ &\quad - \Gamma^{r}{}_{\theta\theta}\Gamma^{\theta}{}_{r\theta} - \Gamma^{\varphi}{}_{\theta\varphi}\Gamma^{\varphi}{}_{\varphi\theta} + \Gamma^{r}{}_{\theta\theta,r} + \Gamma^{\varphi}{}_{\theta\varphi,\theta}. \end{aligned}$$

Inserting the expressions for the Christoffel symbols leads to

$$\begin{aligned} R_{\theta\theta} &= -r\,\mathrm{e}^{-\lambda}\left(\frac{\lambda'}{2}\frac{1}{r} + \frac{1}{r}\frac{\nu'}{2}\right) - \frac{1}{r}r\,\mathrm{e}^{-\lambda} + r\,\mathrm{e}^{-\lambda}\frac{1}{r} \\ &\quad - \frac{\cos^2\theta}{\sin^2\theta} - \left(\frac{\cos\theta}{\sin\theta}\right)_{,\theta} - \left(r\,\mathrm{e}^{-\lambda}\right)_{,r}. \end{aligned} \tag{B.19}$$

The two first terms are cancelled against the second and third terms in the parenthesis. The the second last term is differentiated by using the rule (2.49) for differentiation of fractions of functions, and the rules (4.24) and (4.25) for differentiating $\sin\theta$ and $\cos\theta$, respectively. Thus

$$\begin{aligned} \left(\frac{\cos\theta}{\sin\theta}\right)_{,\theta} &= \frac{(\cos\theta)'\sin\theta - \cos\theta\,(\sin\theta)'}{\sin^2\theta} \\ &= \frac{-\sin^2\theta - \cos^2\theta}{\sin^2\theta} \\ &= -\frac{\sin^2\theta}{\sin^2\theta} - \frac{\cos^2\theta}{\sin^2\theta} = -1 - \frac{\cos^2\theta}{\sin^2\theta}. \end{aligned} \tag{B.20}$$

Using the product rule (2.24) and the chain rule (2.31), we differentiate the last term

$$\left(r\,\mathrm{e}^{-\lambda}\right)_{,r} = \mathrm{e}^{-\lambda} + r\,\mathrm{e}^{-\lambda}\lambda'. \tag{B.21}$$

Inserting the Eqs. (B.20) and (B.21) into Eq. (B.19) gives

$$R_{\theta\theta} = -\frac{r}{2}\,e^{-\lambda}\lambda' - \frac{r}{2}\,e^{-\lambda}\nu' - \frac{\cos^2\theta}{\sin^2\theta} - e^{-\lambda} + r\,e^{-\lambda}\lambda' + 1 + \frac{\cos^2\theta}{\sin^2\theta},$$

which finally leads to

$$R_{\theta\theta} = 1 - \left(1 + \frac{r}{2}\,\nu' - \frac{r}{2}\,\lambda'\right) e^{-\lambda}. \tag{B.22}$$

Equations (B.17), (B.18) and (B.22) provide the expressions for the components of the Ricci curvature tensor of a static, spherically symmetric spacetime that are needed in Ch. 13 in order to investigate the consequences of the general theory of relativity for this type of spacetime.

Appendix C
The Ricci tensor of the Robertson–Walker metric

The expanding universe models that are still the most prominent relativistic models of the universe, were deduced as solutions of Einstein's field equations by A. Friedmann in 1922. These models are isotropic and homogeneous, and the most suitable line element for such models were found by H. P. Robertson and A. G. Walker about 1930.

Before we can solve the field equations for these models, we must find the form of the field equations for them, that is for the Robertson–Walker line element. Then we need to know the components of the Ricci tensor for this line element. In order to find these components of the Ricci tensor, we shall first calculate the non-vanishing Christoffel symbols for the line element (14.9).

Some of the Christoffel symbols come from the part $r^2\,d\theta^2 + r^2\sin^2\theta\,d\varphi^2$ of the line element (14.9), which is also a part of the line element (13.14). They are given in Eqs. (B.7), (B.9), (B.10), and (B.14). We list them here for easier reference

$$\Gamma^{\theta}{}_{\theta r} = \Gamma^{\theta}{}_{r\theta} = \frac{1}{r}, \tag{C.1}$$

$$\Gamma^{\varphi}{}_{r\varphi} = \Gamma^{\varphi}{}_{\varphi r} = \frac{1}{r}, \tag{C.2}$$

$$\Gamma^{\varphi}{}_{\theta\varphi} = \Gamma^{\varphi}{}_{\varphi\theta} = \frac{\cos\theta}{\sin\theta}, \tag{C.3}$$

$$\Gamma^{\theta}{}_{\varphi\varphi} = -\sin\theta\,\cos\theta. \tag{C.4}$$

Using Eq. (B.2) with $\nu = \theta$ and $\lambda = t$, and then Eq. (14.10), we get

$$\begin{aligned}\Gamma^{\theta}{}_{\theta t} &= \frac{g_{\theta\theta,t}}{2\,g_{\theta\theta}} = \frac{(a^2r^2)_{,t}}{2\,a^2\,r^2} = \frac{r^2\,(a^2)_{,t}}{2\,a^2\,r^2} \\ &= \frac{(a^2)_{,t}}{2\,a^2} = \frac{2\,a\,\dot{a}}{2\,a^2} = \frac{\dot{a}}{a},\end{aligned} \tag{C.5}$$

Ø. Grøn and A. Næss, *Einstein's Theory: A Rigorous Introduction for the Mathematically Untrained*, DOI 10.1007/978-1-4614-0706-5,

where $\dot{a}$ means the derivative of the function a with respect to t. Due to the isotropy of the spacetimes described by the line element (14.9), and the symmetry of the Christoffel symbols in their lower indices, we get

$$\Gamma^{r}{}_{tr} = \Gamma^{r}{}_{rt} = \frac{\dot{a}}{a}, \tag{C.6}$$

$$\Gamma^{\varphi}{}_{t\varphi} = \Gamma^{\varphi}{}_{\varphi t} = \frac{\dot{a}}{a}, \tag{C.7}$$

$$\Gamma^{\theta}{}_{t\theta} = \Gamma^{\theta}{}_{\theta t} = \frac{\dot{a}}{a}. \tag{C.8}$$

Putting $\nu = \lambda = r$ in Eq. (B.2) we get

$$\begin{aligned} \Gamma^{r}{}_{rr} &= \frac{g_{rr,r}}{2\, g_{rr}} = \frac{1-k\,r^2}{2a^2}\left(\frac{a^2}{1-k\,r^2}\right)_{,r} \\ &= \frac{1-k\,r^2}{2\,a^2}\,\frac{a^2\,2\,k\,r}{(1-k\,r^2)^2} \\ &= \frac{k\,r}{1-k\,r^2}. \end{aligned} \tag{C.9}$$

The rest of the Christoffel symbols are calculated using Eq. (B.3), which gives

$$\begin{aligned} \Gamma^{t}{}_{rr} &= -\frac{g_{rr,t}}{2\, g_{tt}} = \frac{1}{2\,c^2}\left(\frac{a^2}{1-k\,r^2}\right)_{,t} \\ &= \frac{2\,a\dot{a}}{2\,c^2\,(1-k\,r^2)} = \frac{a\dot{a}}{c^2\,(1-k\,r^2)}, \end{aligned} \tag{C.10}$$

$$\begin{aligned} \Gamma^{t}{}_{\theta\theta} &= -\frac{g_{\theta\theta,t}}{2\, g_{tt}} = \frac{1}{2\,c^2}\left(a^2\,r^2\right)_{,t} = \frac{1}{2\,c^2} 2\,a\dot{a}\,r^2 \\ &= \frac{a\,\dot{a}\,r^2}{c^2}, \end{aligned} \tag{C.11}$$

$$\begin{aligned} \Gamma^{t}{}_{\varphi\varphi} &= -\frac{g_{\varphi\varphi,t}}{2\, g_{tt}} = \frac{1}{2\,c^2}\left(a^2\,r^2\,\sin^2\theta\right)_{,t} \\ &= \frac{a\,\dot{a}}{c^2}\,r^2\,\sin^2\theta, \end{aligned} \tag{C.12}$$

$$\begin{aligned} \Gamma^{r}{}_{\theta\theta} &= -\frac{g_{\theta\theta,r}}{2\, g_{rr}} = -\frac{1-k\,r^2}{2\,a^2}\left(a^2\,r^2\right)_{,r} \\ &= -\left(1-k\,r^2\right)\frac{a^2\,2\,r}{2\,a^2} = -r\,\left(1-k\,r^2\right), \end{aligned} \tag{C.13}$$

$$\Gamma^r{}_{\varphi\varphi} = -\frac{g_{\varphi\varphi,r}}{2\,g_{rr}} = -\frac{1-k\,r^2}{2\,a^2}\left(a^2\,r^2\,\sin^2\theta\right)_{,r}$$
$$= -r\left(1-k\,r^2\right)\sin^2\theta. \tag{C.14}$$

There are no more non-vanishing Christoffel symbols. Note in particular that $\Gamma^\beta{}_{tt} = 0$, which is of physical significance, as shown in Ch. 14.

We now proceed by calculating the components of the Ricci tensor which we need in order to find the form of the field equations for the present application. Inserting $\mu = \nu = t$ in Eq. (B.15), and including only the non-vanishing Christoffel symbols in the summation over α and β, we get

$$R_{tt} = \Gamma^\beta{}_{tt}\,\Gamma^\alpha{}_{\beta\alpha} - \Gamma^\beta{}_{t\alpha}\,\Gamma^\alpha{}_{\beta t} + \Gamma^\alpha{}_{tt,\alpha} - \Gamma^\alpha{}_{t\alpha,t}$$
$$= -\left(\Gamma^r{}_{tr}\,\Gamma^r{}_{rt} + \Gamma^\theta{}_{t\theta}\,\Gamma^\theta{}_{\theta t} + \Gamma^\varphi{}_{t\varphi}\,\Gamma^\varphi{}_{\varphi t}\right.$$
$$\left.+\Gamma^r{}_{tr,t} + \Gamma^\theta{}_{t\theta,t} + \Gamma^\varphi{}_{t\varphi,t}\right).$$

Substituting the expressions (C.6)–(C.8) for the Christoffel symbols leads to

$$R_{tt} = -\left[3\frac{\dot a^2}{a^2} + 3\left(\frac{\dot a}{a}\right)_{,t}\right] = -\left(3\,\frac{\dot a^2}{a^2} + 3\,\frac{\ddot a\,a - \dot a^2}{a^2}\right)$$
$$= -\left(3\,\frac{\dot a^2}{a^2} + 3\,\frac{\ddot a}{a} - 3\,\frac{\dot a^2}{a^2}\right),$$

which gives

$$R_{tt} = -3\,\frac{\ddot a}{a}. \tag{C.15}$$

Inserting $\mu = \nu = \theta$ in Eq. (B.15), we get

$$R_{\theta\theta} = \Gamma^\beta{}_{\theta\theta}\,\Gamma^\alpha{}_{\beta\alpha} - \Gamma^\beta{}_{\theta\alpha}\,\Gamma^\alpha{}_{\beta\theta} + \Gamma^\alpha{}_{\theta\theta,\alpha} - \Gamma^\alpha{}_{\theta\alpha,\theta}.$$

Performing the summation over α and β we find

$$R_{\theta\theta} = \Gamma^r{}_{\theta\theta}\left(\Gamma^r{}_{rr} + \Gamma^\theta{}_{r\theta} + \Gamma^\varphi{}_{r\varphi}\right) + \Gamma^t{}_{\theta\theta}\left(\Gamma^r{}_{tr} + \Gamma^\theta{}_{t\theta} + \Gamma^\varphi{}_{t\varphi}\right)$$
$$- \Gamma^\theta{}_{\theta t}\,\Gamma^t{}_{\theta\theta} - \Gamma^t{}_{\theta\theta}\,\Gamma^\theta{}_{t\theta} - \Gamma^r{}_{\theta\theta}\,\Gamma^\theta{}_{r\theta} - \Gamma^\theta{}_{\theta r}\,\Gamma^r{}_{\theta\theta}$$
$$- \Gamma^\varphi{}_{\theta\varphi}\,\Gamma^\varphi{}_{\varphi\theta} + \Gamma^t{}_{\theta\theta,t} + \Gamma^r{}_{\theta\theta,r} - \Gamma^\varphi{}_{\theta\varphi,\theta}.$$

Substituting the expressions (C.9)-(C.14) for the Christoffel symbols leads to

$$R_{\theta\theta} = -r\left(1-k\,r^2\right)\left(\frac{k\,r}{1-k\,r^2} + \frac{2}{r}\right)$$
$$+\frac{a\,\dot a\,r^2}{c^2}\,3\,\frac{\dot a}{a} - 2\,\frac{\dot a}{a}\,\frac{a\,\dot a\,r^2}{c^2}$$

$$+ 2r\left(1 - k\,r^2\right)\frac{1}{r} - \frac{\cos^2\theta}{\sin^2\theta} + \left(\frac{a\,\dot{a}\,r^2}{c^2}\right)_{,t}$$
$$- \left[r\left(1 - k\,r^2\right)\right]_{,r} - \left(\frac{\cos^2\theta}{\sin^2\theta}\right)_{,\theta}. \tag{C.16}$$

From Eq. (B.20) we have

$$\left(\frac{\cos\theta}{\sin\theta}\right)_{,\theta} = -1 - \frac{\cos^2\theta}{\sin^2\theta}. \tag{C.17}$$

Furthermore

$$\left(a\,\dot{a}\,r^2\right)_{,t} = \left(a\,\dot{a}\right)_{,t}\,r^2 = \dot{a}^2\,r^2 + a\,\ddot{a}\,r^2, \tag{C.18}$$

and

$$\left[r\left(1 - k\,r^2\right)\right]_{,r} = \left(r - k\,r^3\right)_{,r} = 1 - 3\,k\,r^2. \tag{C.19}$$

Multiplying out the first terms in (C.16) and using the expressions (C.17)–(C.19) in the last three terms, we get

$$R_{\theta\theta} = -k\,r^2 - 2\left(1 - k\,r^2\right) + \frac{3\,\dot{a}^2\,r^2}{c^2} - \frac{2\,\dot{a}^2\,r^2}{c^2}$$
$$+ 2\left(1 - k\,r^2\right) - \frac{\cos^2\theta}{\sin^2\theta} + \frac{a\,\ddot{a}\,r^2}{c^2} + \frac{\dot{a}^2\,r^2}{c^2}$$
$$- 1 + 3\,k\,r^2 + 1 + \frac{\cos^2\theta}{\sin^2\theta}.$$

Collecting terms we get

$$R_{\theta\theta} = \frac{a\,\ddot{a}\,r^2}{c^2} + \frac{2\,\dot{a}^2\,r^2}{c^2} + 2\,k\,r^2.$$

Putting r^2/c^2 outside a parenthesis we finally have

$$R_{\theta\theta} = \left(a\,\ddot{a} + 2\,\dot{a}^2 + 2\,k\,c^2\right)\,r^2/c^2. \tag{C.20}$$

Due to the isotropy of the models, we only need the two components of the Ricci tensor that we have now calculated. Inserting the expressions (C.15) and (C.20) for R_{tt} and $R_{\theta\theta}$, respectively, into Einstein's field equations, we arrive at a set of two differential equations. The solutions of these equations show how the expanding motion of the universe models varies with time, and how the mass density evolves in the models.

References

1. Ernst Cassirer. *Zur Einsteinschen Relativitätstheorie: erkenntnistheoretische Betrachtungen*. Bruno Cassirer, Berlin, 1921.
2. F. de Felice and C. J. S. Clarke. *Relativity on Curved Manifolds*. Cambridge University Press, Cambridge, 1992.
3. R. D'Inverno. *Introducing Einstein's Relativity*. Clarendon Press, London, 1992.
4. Albert Einstein. *The World as I see it*. Covici Friede, New York, 1934.
5. J. Foster and J. D. Nightingale. *A short course in General Relativity*. Longman, London and New York, 1979.
6. I. R. Kenyon. *General Relativity*. Oxford University Press, 1990.
7. M. Ludvigsen. *General Relativity, A Geometric Approach*. Cambridge University Press, Cambridge, 1999.
8. C. W. Misner, K. S. Thorne, and J. A. Wheeler. *Gravitation*. Freeman and Company, San Francisco, 1973.
9. B. F. Schutz. *A First Course in General Relativity*. Cambridge University Press, Cambridge, 1985.

Ø. Grøn and A. Næss, *Einstein's Theory: A Rigorous Introduction for the Mathematically Untrained*, DOI 10.1007/978-1-4614-0706-5,

Index

Ø. Grøn and A. Næss, *Einstein's Theory: A Rigorous Introduction for the Mathematically Untrained*, DOI 10.1007/978-1-4614-0706-5,